Th. Gaiser M. Krol H. Frischkorn J.C. de Araújo

Global Change and Regional Impacts

Springer
Berlin
Heidelberg
New York
Hong Kong
London
Milan
Paris
Tokyo

Thomas Gaiser M. Krol
H. Frischkorn J.C. de Araújo

Editors

Global Change and Regional Impacts

Water Availability and Vulnerability of Ecosystems and Society in the Semiarid Northeast of Brazil

with 163 Figures, 46 in Colour, and 72 Tables

Springer

DR. THOMAS GAISER
University of Hohenheim
Institute of Soil Science
and Land Evaluation
70593 Stuttgart
Germany

PROFESSOR HORST FRISCHKORN
Universidade Federal do Ceará
Departamento de Engenharia
Hidráulica e Ambiental
60451-970 Fortaleza-Ceará
Brazil

DR. MAARTEN KROL
Potsdam Institute for
Climate Impact Research
Telegrafenberg A31
14473 Potsdam
Germany

PROFESSOR JOSE C. DE ARAÚJO
Universidade Federal do Ceará
Departamento de Engenharia
Hidráulica e Ambiental
60451-970 Fortaleza-Ceará
Brazil

ISBN 3-540-43824-6 Springer-Verlag Berlin Heidelberg New York

Library of Congress Cataloging-in-Publication Data Applied for

Die Deutsche Bibliothek – CIP-Einheitsaufnahme
Global change and regional impacts : water availability and vulnerability of ecosystems and society in the semiarid Northeast of Brazil / Thomas Gaiser...(ed.). Berlin ; Heidelberg ; New York ; Hong Kong; London ; Milan ; Paris ; Tokyo : Springer, 2003
ISBN 3-540-43824-6

Springer-Verlag Berlin Heidelberg New York
a member of BertelsmannSpringer Science+Business Media GmbH

http://www.springer.de

Camera ready by editors assisted by F. Graef and C. Mai-Peter
Cover design: E. Kirchner, Heidelberg
Printed on acid-free paper SPIN 10882909 32/3130/as 5 4 3 2 1 0

Forewords

The steady growth of the global population and the increasing utilization of natural resources urgently require international efforts towards the conservation of living conditions for mankind in an environment with functioning natural cycles. Due to the global significance of the problem, the German government, through its program „Research for the Environment", pays particular attention to bilateral research cooperation and the involvement of German scientists in international programs. Within the framework of the German-Brazilian scientific and technological cooperation, the WAVES program started in 1994. Research regarding the effects of climate variability on water availability and its impacts on ecosystems and social structures in the northeast of Brazil has provided important contributions to the understanding of the complex relationships between ecological, social and economical questions. This program, funded by the German Federal Ministry for Education and Research, is expected to assist considerably in the sustainable development and utilization of this extremely vulnerable region with high drought risk.

Dr. Norbert Binder
German Federal Ministry of Education and Research (BMBF)

The WAVES program is of high relevance for the state government of Ceará, because it is concerned with the living conditions and the sustainable existence in a semiarid environment, which is crucial for the development of the state. We need objective, scientifically based information for the implementation of our development projects in order to enable the population, that is living in these semiarid areas under adverse conditions, to develop adequately. Studying the climatic conditions with the objective of providing the government of Ceará with tools for the precise definition of policies for the semiarid areas is therefore an extremely important activity for us. We are glad to have been offered the privilege of joining the partnership between the research institutions and the German and Brazilian governments, because we have common goals with respect to the improvement in the quality of life for the population, a socially equitable development.

Dra. Mônica Clark Nunes Cavalcante
Minister for Planning and Coordination, State Government of Ceará

Acknowledgements

This book is based on the results of the bilateral research program WAVES. Many organizations, scientists and experts have contributed and their participation is gratefully acknowledged. In the first place, we thank the funding agencies Conselho Nacional de Desenvolvimento Científico e Tecnológico (CNPq) and the German Ministry for Education and Reasearch (BMBF) for financial support of the program. We are grateful to the Associação Brasileira de Recursos Hídricos (ABRH), the University of Hohenheim and the Potsdam Institute for Climate Impact Research for sponsoring this publication. Our compliments go also to the program coordinators Prof. Dr. Luiz Gonzaga Rebouças Ferreira and Prof. Dr. Karl Stahr, admirably assisted in all organizational matters by their staff members Mrs. Heidemarie Friedemann, Mrs. Margrit Fritz, Mrs. Karen Rachel Dantas Santos, Mr. Luiz Camboim de França, Mr. Berthyer Peixoto Lima, Mr. Luciano Marcelo Fallé Saboya, Mrs. Célia Marilac de Oliveira and Mr. Jonas Baima Araújo. We thank Dr. Frieder Graeff and Mrs. Claudia Mai-Peter, who contributed greatly to the edition of this publication. We are grateful to the board of external reviewers that revised the manuscripts of this book critically. Thanks go also to the initiators of the WAVES program Prof. Dr. Horst Frischkorn, Prof. Deocleciano Guedes Ferreira and Prof. Valdemar Rodrigues for their help in so many practical problems. Among the program pioneers, we gratefully remember our colleague Dr. Marcus Forster who unexpectedly died in 1999. His effort and dedication inspired us and will continue to do so. We are particularly indebted to Dr. Vicente Fialho who acted as a special promoter of the research activities and, together with Mr. Bergson Fernandes, as a key person in the transfer of the results to state authorities. Finally, we extend our deepest appreciation to the people of Piauí and Ceará, especially to the inhabitants of the municipalities of Tauá and Picos, who always welcomed us with overwhelming Brazilian hospitality. We will always remember with great respect their effort and struggle to stay in their homeland, the fascinating „Sertão“, in spite of the extremely adverse living conditions. We hope they will be the main beneficiaries of our work.

Preface

As a consequence of the worldwide increase in the utilization of fossil energy sources, changes in the global climate have occurred due to rising concentrations of greenhouse gases in the atmosphere. The global climatic change will most likely affect natural resources and human living conditions in semiarid regions. To assess these impacts, it is necessary to study the interactions between climate change, the hydrological cycle, agricultural production and, above all, the implications for the quality of life.

It was with this concept of a multidisciplinary and integrated approach that the WAVES program was conceived in 1994. Scientists with interest in interactions between water resources and socio-economics in the northeast of Brazil, supported by the funding agencies CNPq and BMBF, decided to elaborate a program with a holistic and imperatively integrated approach to study interactions between climate variability and change, water availability, as well as land use systems and society. Integrated models were designed as an alternative tool to develop sustainable strategies to make semiarid regions less vulnerable to climatic changes. The northeast of Brazil was selected as a reference region, because developing semiarid regions are among the regions most vulnerable to global climatic change. In addition, climate models indicate that a drastic decrease in precipitation within the next 50 years should be considered plausible. Following the preparatory phase, in October 1995, the funding agencies approved the bilateral research concept, which has been regarded as a guide to define all subsequent actions related to the program. After an intensive phase of research from 1996 to 2000, the results were presented and discussed with representatives of public institutions during a series of workshops organized by the State Ministry for Planning in Ceará under the auspices of the state minister Dra. Mônica Clark Nunes Cavalcante.

This book is a summary of the main research results. It shows large expertise and knowledge (ranging from detailed water analyses at the spot to transregional migration) that is necessary to understand the complex relationship between natural and socioeconomic systems and demonstrates methods to integrate this knowledge and make it available for the strategic planning of sustainable development.

T. Gaiser, H. Frischkorn, J.C. de Araújo, M. Krol

Table of contents

Introduction

General View of the WAVES Program

Thomas Gaiser, Luiz.G. R. Ferreira and Karl Stahr

Institute of Soil Science and Land Evaluation, University of Hohenheim, Emil-Wolff-Str.27, 70593 Stuttgart, Germany, tgaiser@uni-hohenheim.de

Introduction

The globalization of the world's economy at the dawn of the new millennium is most likely to continue and will be accompanied by social and environmental changes that will affect the entire earth system. The causes of global environmental problems are not fully understood: emission of greenhouse gases and pollutants by the industrialized world, overuse of soil and biotic resources, destruction of forests and natural vegetation, and the extinction of ecosystems with their genetic resources (WBGU 1993; IPCC 1997). In addition, the consumption of natural resources and environmental pollution are even more pronounced through the world's rapidly growing population. The totality of these processes, their interactions and consequences is called global change. The observed symptoms are interdependent and interrelated in various ways. Global change is therefore a complex phenomenon, which cannot be completely understood through the analysis and evaluation of single phenomena. The understanding of global change and its impacts necessitates the analysis of the relevant system components and their interactions. When using scenario assumptions, the global effects can be described with simulation runs of complex models and can be visualized as global average values of the principal components. However, the results are very general and mask the strong regional differences (Walker and Steffens 1999). At the regional level, the effects of global change may vary considerably due to geographical location, climatic variability, natural resources as well as socio-economic conditions.

Global climatic change will affect in particular semiarid regions, especially those regions where, at present, the overall amount of precipitation is low and its distribution is irregular. These areas cover one-third of the earth's surface, and they host about 20% of its population. The rise in mean temperature will most likely cause degradation of natural resources and human living conditions through prolonged droughts and decrease in water availability. In some of these regions the natural resource capacity is already at its limit and the climatic conditions regularly cause social stress situations. A very probable global climate change will have detrimental effects on the living conditions of the population.

The northeast of Brazil is typical for the semiarid problem regions. Population growth, land use changes and increasing demands on life quality endanger the natural resources (Magalhães et al. 1988). Moreover, the natural climate variability deteriorates the social and economic well being of the majority of the population. The variability of precipitation frequently provokes drought periods, which, in their extreme expression, lead to disastrous effects on water supply and agricultural production with the occurrence of famines as experienced in the years 1983 and 1998 in Ceará. The consequences of droughts can be mitigated through emergency measures. The federal government contributes actively to the alleviation of the problem by investment programs into water supply facilities, subsidies and other economic policies. However, in the context of the ongoing global changes, there is a need to adapt the planning tools and to develop new strategies to meet the future challenges in sustainably securing and improving the living conditions in Northeast Brazil.

Identification of problems and definition of research objectives

The climate variability in northeast Brazil is extremely high, which means that some parts of the northeast are classified as arid regions from an agrometeorological point of view (Hargraeves 1974; Leprun et al. 1995). In years with relatively high amounts of annual rainfall, 90% of the total precipitation can be concentrated to a few storm events with very high rainfall rates, whereas, in dry years, rainfall can fail almost completely (Hall 1976). Recurrent drought periods, which can partly be correlated with the ENSO (El Niño-Southern Oscillation) phenomenon, are the major problem and are threatening the lives of the people. The affected area, the so-called drought polygon, belongs to various Brazilian federal states (Bahia, Alagoas, Paraíba, Rio Grande do Norte, Pernambuco, Ceará and Piauí) and is estimated to cover 950,000 km^2 (Fig. 1).

The probability of drought stress depends on the local conditions and ranges between 20 and 90%. In the states of Piauí and Ceará it is between 40 and 80%. The region is considered to be one of the most problematic areas within the Brazilian territory.

The majority of the population lives under harsh and difficult conditions (Briesemeister 1994). The mean monthly income amounted to 215 US$ in 1996, being 48% below the national average. The agricultural sector is the main source of income for 41% of the working population, whereas the national average is situated around 25%. The same situation holds for the educational sector. About 53% of the working population in northeast Brazil has spent less than 3 years at school (IBGE 1996). About 80% of the population in rural areas is at least periodically jobless and therefore affected by poverty (Magalhães et al. 1988). A large part of the labor force in the agricultural sector belongs to the group of farmers engaged in small-scale subsistence farming. They mainly produce for the needs of their families, selling only the excess of their produce. Or they are working in a tenure system with the objective of subsistence farming. They dispose of small portions of land (rarely more than 10 ha). Farm types of this size account for 53% of the agricultural production units, but cultivate only 3% of the agriculturally used land. However, they are important for the regional supply of food. Therefore, drought events have two negative consequences: physical yields are decreasing, while the proportion of the yield to secure subsistence on the farms is increasing. This causes great variations in market prices, complicating the purchase of food for the poorer part of the population. Income opportunities outside the agricultural sector are often limited to trading activities.

At present, Piauí is excluded from this development because it has a very narrow coastal line and intensive colonization only started in the middle of the nineteenth century. Migration has been and is still a response of the population to the difficult living conditions in the semiarid northeast. In the last three decades, urbanization in Brazil has increased from 44 to 78%, the increase being more pronounced in the industrialized southeast: the demographic census in 1996 states that about 11 million people in the urban centers of southeast Brazil originated from the northeast (IBGE 1997).

Facing the complex problems that exist in northeast Brazil, the objective of the WAVES program was to design tools for the elaboration of sustainable development strategies for the rural areas, taking into account relevant aspects of global change, in particular future climate conditions. The tools should support regional planning authorities in evaluating the potential effects of their decisions on water and land resources, agricultural production and the living conditions in rural areas. In view of this aim, two major tasks were carried out:

- Analysis of the natural and social systems, focusing on the interactions between water availability, agricultural production and life quality
- Integration of the research results in order to simulate and describe quantitatively the potential development of natural and social systems in relation to changing water availability (either due to climate changes or governmental policies).

Due to the spatial extent of the semiarid northeast of Brazil, the investigations were concentrated onto two states, Piauí and Ceará, covering about 400,000 km^2 (Fig. 2). Within each state a focus region had been previously defined according to ecological and agro-economic characteristics. Since it was obvious from the start that a detailed analysis of the system dynamics is not possible at the level of both states, each discipline carried out more detailed investigations in the focus regions (municipality of Picos with adjacent municipalities in Piauí and the municipality of Tauá in Ceará).

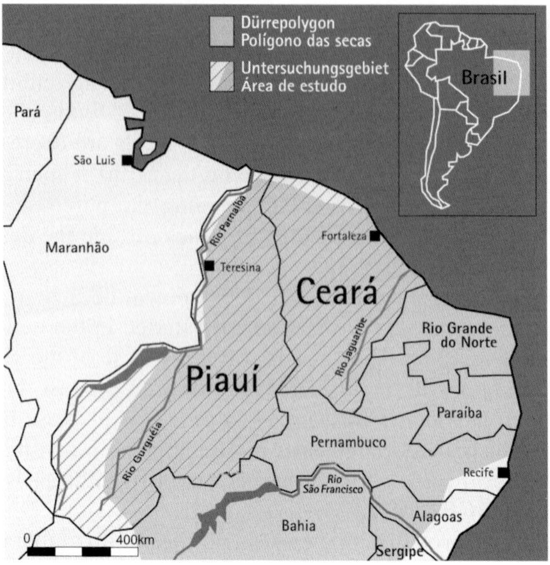

Fig. 1. Spatial extent of the drought polygon (yellow shade) in the northeast of Brazil

Fig. 2. The study area of the WAVES program: The Brazilian federal states Piauí and Ceará with a total of 332 municipalities. The focus regions Picos (PI) and Tauá (CE) are coloured in black

Designing pathways for sustainable development by integration of knowledge

Interdisciplinary and Bilateral System Analysis

Regional impact analysis of changing global framework conditions and governmental interventions requires the identification of basic system components and processes. Internal, interacting processes and their dynamics have to be separated from external forces that are not influenced by the regional system, but they rather create the boundary conditions for regional development (Fig. 3).

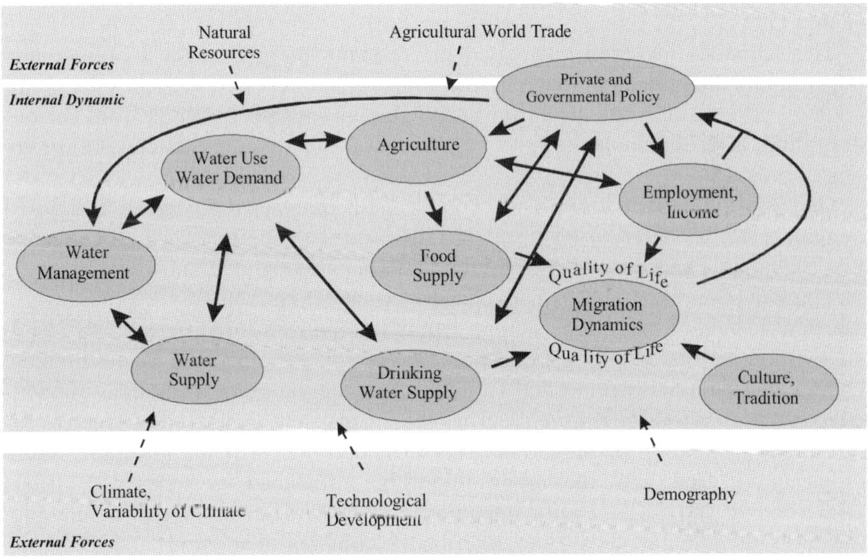

Fig. 3. External forces and internal system dynamics relevant for the description of the causal relation and interactions between water availability and life quality in the northeast of Brazil

Describing the causal relationships between water availability (natural and technical), life quality in the rural areas and migration requires that all system processes that affect life quality must be covered, starting from the limitations and vulnerabilities of water supply and agricultural production (spatial and temporal distribution of precipitation, variability of production in relation to soils, crops and crop management, technical aspects of water storage and distribution) up to the reasons that push (or pull) people to migrate (lack of water supply and food, unsatisfactory job and income situation, land ownership, infrastructure and medical assistance). This necessitates an interdisciplinary approach involving various disciplines (Fig. 4).

The disciplines in the WAVES program were covered by teams of scientists from Brazilian and German research institutions that had interest in interactions between water resources and socio-economic conditions in the northeast of Brazil (Table 1).

The scientists involved elaborated the research program with a holistic and imperatively integrated approach. In the course of the research activities the methodology how to integrate scientific knowledge and research results from various disciplines at different spatial levels turned out to be a major challenge in the research program.

One way to integrate disciplinary knowledge is ***integrated modeling***. Integrated modeling is considered to be a suitable instrument to summarize the knowledge that is available and focus it onto a specific topic, although the knowledge may be incomplete (Daschkeit 2001). Thus, integrated modeling is mainly a tool to assess the significance of individual factors within a complex system (sensitivity analysis), which may be a relevant question for the scientific as well as the political community. Another method that facilitates the integration of disciplinary knowledge is the ***scenario technique***. Scenarios are consistent images of alternative futures. The generation of scenarios helps to understand how the uncertain future might unfold and how the future might be influenced by today's decisions (Nakicenovic and Swart 2000). Integrated models and scenarios can be complementary. Integrated models try to describe the complex interactions and feed-back mechanisms within a system, whereas scenarios may define the external forces and boundary conditions (Döll et al., this Vol.). On the other hand, the results of integrated models support the scenario analysis, in particular with respect to the consistency of the scenarios.

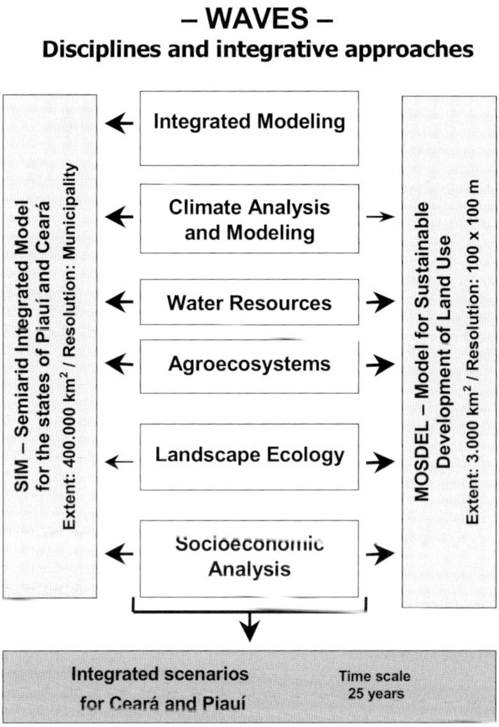

Fig. 4. Scientific disciplines involved in the systems analysis and development of integrative approaches

Adopting these two integrative methods, the results of investigations, surveys and literature reviews were exchanged and analyzed within each discipline in order to obtain a comprehensive image of the situation for the respective spatial scales at the disciplinary level. Then, working groups that were closely linked to each other through common spatial or disciplinary interfaces came together for data exchange, spatial integration and model development. This provided synergistic effects at the lower integrational level and led to the transformation of expert knowledge and research results into disciplinary databases and models. At the highest integrational level, i.e., the integrated modeling and scenario development, all working groups gathered to combine their expertise, data and models (Fig. 4, Table 2). The highest integrational level produced three products for two different spatial scales:

- Integrated scenarios covering the period 2000–2025 at the level of both states
- The Semiarid Integrated Model (SIM) at the level of both states
- The Model for Sustainable Development of Land Use (MOSDEL) at the level of the focus regions

Table 1. WAVES working groups and cooperating research institutions, covering different disciplines

Working group	Brazilian Institutions	German Institutions
Climate	Fundação Cearaense de Meteorologia e Recursos Hídricos	Potsdam Institute for Climate Impact Research
Water resources	Universidade Federal do Ceará	University of Kassel
	Secretaria de Agricultura, Departamento de Hídrometorologia	HYDROISOTOP GmbH
		Potsdam Institute for Climate Impact Research
		University of Potsdam
Agro-ecosystems	Universidade Federal do Ceará	University of Hohenheim
Socioeconomic analysis	Universidade Federal do Piauí	Technical College of Cologne
	Universidade Federal do Ceará	University of Kassel
		University of Hohenheim
Landscape ecology	Universidade Federal do Piauí	Technical University of Munich
	Universidade Federal do Ceará	
Integrated modeling	Universidade Federal do Ceará	Potsdam Institute for Climate Impact Research

Definition of Interfaces and Spatial Scales

The identification of interdisciplinary spatial and temporal interfaces is a prerequisite for integration of disciplinary knowledge through integrated modeling (Wessmann 1992). This holds in particular when global change issues are to be described that deal with a large number of different processes at contrasting spatial and temporal scales. Definition of common spatial interfaces is even more significant when the investigations, aimed at the construction of an Integrated Regional Model, extend over an area of 400,000 km^2 with numerous administrative and ecological units. Figure 5 shows that the temporal and spatial scales of the scientific disciplines involved differ considerably. In order to link the disciplinary scales, common spatial scales were defined for each of the two integrated regional models, and all data, models and model results had to consider or had to be transformed to these scales (Table 3). In order to facilitate communication within the project, three spatial (study and modeling) levels were defined, according to the extent of the integrated models and the spatial scales of the disciplinary investigations:

- „Micro-scale": Observations and measurements at the point or field scale, surveys at the farm or family level
- „Meso-scale": Surveys in small to medium size catchments or in municipalities
- „Macro-scale": Large river basins, federal states

Table 2. Disciplinary models and databases, covering major system components and processes, as basic modules for the integrated models SIM and MOSDEL

System components	SIM	MOSDEL
Climate	CLIMSCENMOD	
Soils and terrain	SPICE	SPICE
Physical water availability	WASA	WARIG
Water demand	NoWum	WARIG
Erosion/sedimentation	HidroSed	(EPIC)
Water costs	VALOR	
Agricultural production	EPIC/CROPWAT/ITC	EPIC/ITC
Regional agricultural economy and income	RASMO	MOSDEL
Environmental degradation		MOSDEL
Quality of life	MIGFLOW	MOSDEL
Demographic growth/migration	MIGFLOW	

In general, two techniques are available to link different spatial or temporal scales:

- Upscaling, i.e., generalization of detailed (spatial and temporal) information to generate knowledge for larger areas (or periods).
- Downscaling, i.e., linking information about large spatial units (or periods) with knowledge about process dynamics at a smaller (or shorter) scale to produce subscale information for selected areas (or periods).

Both approaches were applied depending on the respective discipline and (integrated) model. The disciplines that relied heavily on observations in the field, experiments or interviews with individuals had to transform their results to the higher spatial (and temporal) level, whereas those working with global climate models or secondary statistical data had to apply the downscaling approach to match the common spatial scales. This type of mixed approach is also termed „strategic cyclical scaling“ (Root and Schneider 1995).

Table 3. Common spatial extent and scales within the two integrated regional models SIM and MOSDEL

	Spatial scale		Temporal scale	
	Extent	Resolution	Extent	Resolution
SIM	„Macro-scale“ (States of Piauí and Ceará)	Municipality	25 years (2000–2025)	Variable (1 day to 1 year)
MOSDEL	„Meso-scale“ (Guaribas water shed and municipality of Tauá)	1 ha	5 reference years	Variable (1 day to 1 year)

Besides the definition of common scales, there was also a need to define and adjust the interfaces between the individual disciplines for data exchange and coupling of the disciplinary models. Table 4 gives a simplified overview of the interfaces and data requirements of each working group. The data requirements are composed of numerous parameters or variables, each specified according to its spatial and temporal extent and resolution.

Table 4. Interfaces between the working groups

Working group	Data demand interfaces	Data requirements
Climate (1)	(2) and (3)	Definition of boundary conditions
Water Resources (2)	(1), (3), and (4)	Climatic conditions, soil water balance, water demand
Agro-ecosystems (3)	(1) and (2)	Climatic conditions, water availability
Socio-economic analysis (4)	(2) and (3)	Water availability, agricultural production
Landscape ecology (5)	All working groups	Initial parameters, disciplinary models
Integrated modeling (6)	All working groups	Initial parameters, disciplinary models

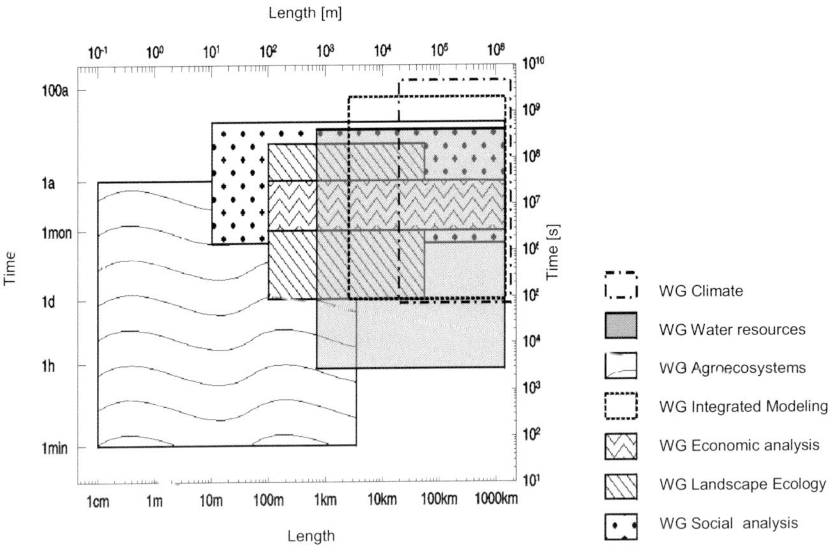

Fig. 5. WAVES working groups and their specific temporal and spatial scales

Achievements and potential application

Major achievements of the WAVES program are in the development of innovative approaches to integrate disciplinary knowledge with respect to the description of complex system processes and their interactions. This was achieved through the elaboration of scenarios of possible future development strategies and two integrated regional models (see Part 1, this Vol.; Döll et al.; Krol et al.; Printz and Lang). The integrated scenarios are the result of a series of workshops held in 2000 and 2001 together with representatives of state authorities of Ceará headed by the Ministry of Planning and Coordination (SEPLAN; Table 5). These workshops served on the one hand as a test as to whether the system processes and output variables covered by the integrated models are relevant for potential users. On the other hand, they revealed an extremely high interest and cooperation by the planning authorities to use scenario analysis in connection with the integrated regional models for the future regional management of water and land resources. The scenario methodology assisted the state government in coordinating and integrating various ministries with respect to a common goal, and to promote „interdisciplinary" communication between the representatives of the ministries. The workshops indicated an urgent need for the state ministries to have a support system for medium-term scenario analysis that is able to locate municipalities that will suffer yield losses during the growing period or such municipalities that may suffer from water scarcity during the subsequent dry season. Furthermore, the joint use and updating of databases has been proposed.

Table 5. Major results of the individual working groups and potential applications

	Achievements	Potential application
	Integrated scenarios for the states of Ceará and Piauí in cooperation with state authorities	
Macro-scale	Semiarid Integrated Model (SIM)	Long-term strategic planning of sustainable development of land use and water management
	Climate scenarios (CLIMSCENMOD)	Long-term planning for the period 2000–2050
	Soil and Land Resource Information System (SPICE)	Database of soils and land resources
		Large-scale agro-ecological zoning
		Identification of areas with high production potential or high erosion risk in relation to climate, soil and management
	Model of Water Availability in Semiarid Environments (WASA)	Simulation of natural water availability in surface waters and dams
		Simulation of the effects of investments in the water storage or water transposition on availability of water in the individual municipalities and river catchments
	Northeast Water Use Model (NoWUM)	Estimation of water demand per municipality as a function of population growth, infrastructure, water pricing and irrigated areas
	HidroSed	Calculation of sedimentation in large water reservoirs
	Data base of water costs (VALOR)	Estimating the costs for the supply of water as a function of water extraction technologies
	Regional Agricultural Sector Model of Ceará and Piauí (RASMO)	Spatial distribution of water demand in the agricultural sector, in relation to farm size and farm technology level
		Income structure according to the regional availability of resources
	Migration Flow (MigFlow)	Estimation of changes in life quality per municipality
		Estimation of net migration rates

Table 5. (Continued)

	Achievements	Potential application
Meso-scale	Model for sustainable development of land use (MOSDEL)	Medium-term strategic planning of sustainable development of land use and water management
	Soil and Land Resource Information System (SPICE)	Soil database Estimation of yield losses
	Modeling of Water Availability and Quality in Rio Guaribas Basin (WARIG)	Estimation of water availability in surface and ground waters
	Survey of water quality and water infrastructure in the municipalites of Picos and Tauá	Recommendations for improving drinking water supply and suitability of waters for irrigation
Micro–scale	Erosion Productivity Impact Calculator (EPIC)	Identification of optimal crop management (fertilization, irrigation) in relation to soil, weather conditions and crop
	Testing of crop management options in maize and cowpea cropping systems	Management recommendations for agricultural extension services
	Assessment of soil water balance as a function of weather conditions, soil and vegetation	Calculation of irrigation requirements
	Botanical survey in Caatinga vegetation	Identification of genetic resources for potential economic use

Particularly in the state of Ceará, where the obligation for sustainable development is fixed in its constitution, state committees have been formed at the governmental level (e.g., Comitê de Desenvolvimento do Ceará; COMDECE, headed by the Minister for Planning and Coordination) and carry out pilot programs to identify and test measures towards sustainable development in selected municipalities. The municipality of Tauá has been included in these pilot programs due to the activities of the WAVES project and its location in the driest part of Ceará with low potential water resources. The databases (GIS layers, satellite images, data about socio-economic conditions and natural resources, SPICE, etc.) established by the WAVES program as well as the integrated model MOSDEL for the municipality of Tauá will be of great help for this development program (Table 5).

In future, the demand for tools that assist in regional planning of water management will increase in Brazil due to the new federal water legislation that was enacted in 1997. The federal law prescribes the establishment of „river basin committees“ for the participatory management of river basins. For medium- and long-term planning, these „river basin committees“ will need scientifically based decision support instruments. In this respect, the state of Ceará plays a pioneer role within the Brazilian federation, being in the process of establishing river basin committees, after having defined the administrative units of all basins within the state territory. Therefore, models such as WASA, VALOR, NoWUM or WARIG (Table 5) for large-scale as well as medium-scale catchments are most welcomed by the state ministries such as the Ministry for Water Resources in Ceará and its executing agencies FUNCEME (Fundação Cearaense de Meteorologia e Recursos Hídricos) and COGERH (Companhia de Gerenciamento de Recursos Hídricos). The mesoscale model WARIG in connection with the integrated model MOSDEL had been designed especially to be applied in the „Guaribas river“ basin which has been designated as a „pilot river basin“ by the Ministry for Agriculture in Piauí.

However, although the data density increases from the macroscale to the mesoscale, none of the implemented databases can claim to be sufficiently complete. In addition to the continuous need of database updating, several data gaps have been identified (e.g., water demand monitoring, spatial resolution of climate data, georeferenced land use data, quantification of storage capacities of small dams, etc.). One powerful instrument to satisfy some of the data requirements could be the application of remote sensing data and analysis. Furthermore, some of the developed models still lack comprehensive validation at different spatial and temporal scales. This is mainly due to the shortage of appropriate input and control data, especially for models that deal with human perception and behavior as well as for the integrated models.

This has been confirmed by Daschkeit (2001) who identified the main weakness of integrated modeling in the context of global change research in the lack of modeling techniques for human dimensions and the difficulties in validating them. The problem of validation is the reason why it is extremely difficult to quantify uncertainties of the simulation results. Nevertheless, as demonstrated by the WAVES program, integrated regional modeling in connection with integrated scenarios is one of the most promising approaches to integrate knowledge of different disciplines at varying spatial scales and to describe complex system interactions. Provided that data are available at the respective temporal and spatial resolution, they could be adequately validated, which would increase their predictive power and their reliability when applied in regional planning.

Conclusions

Six years of intensive bilateral and interdisciplinary research in the semiarid northeast of Brazil have produced remarkable results with respect to the development of integrative methods for sustainable water and land use management. Integrated regional scenarios for both states have been elaborated in close cooperation with state institutions. The scenario technique has promoted not only interdisciplinary communication between scientists, but also between institutions that deal with regional planning. A series of databases and partly validated sectorial models have been developed, which constitute a framework for future efforts of the state government and local planning authorities to efficiently store and retrieve data from surveys and monitoring programs and to apply them in strategic water and land use planning. The two integrated regional models are probably the most important innovations for tropical semiarid regions. However, they still lack comprehensive testing and validation, which depends on the adequate resolution and density of input and control data. Hence, there is a future need to increase the quality of the databases and to include important system processes, in particular at the state level, into the integrated regional models. The application of integrated regional models is certainly not a modern way of prophesizing, but they can assist in pinpointing alternative and robust long-term strategies for the extremely vulnerable semiarid regions and could therefore contribute to a sustainable improvement in the living conditions of the people. The availability of water with adequate quality as well as its ecologically and socially sound use will be the key issue for a sustainable development of the northeast of Brazil and other semiarid regions on this earth.

References

Briesemeister D (1994) Brasilien heute. Politik, Wirtschaft, Kultur. Vervuert, Frankfurt am Main

Daschkeit A (2001) Symposium „Integrative Modellierung zum Globalen Wandel". Tagungsbericht. TA-Datenbank-Nachrichten 10:109–115

Hall AL (1976) Drought and irrigation in Northeast Brazil. Cambridge Latin American Studies. Cambridge University Press, Cambridge

Hargraeves GH (1974) Climatic zoning for agricultural production in Northeast Brazil. Utah State University, Utah, 6 pp

IBGE (1996) Minimum National Social Data Set. URL: http://www.ibge.gov.br/english/

IBGE (1997) Contagem da População. Rio de Janeiro

IPCC (1997) IPCC Special Report – the regional impacts of climate change: an assessment of vulnerability. Summary for policymakers. IPCC Working Group II, Geneva, Switzerland. URL: http://www.ipcc.ch/pub/reports.htm

Leprun JC, Molinier M, Cadier E, Fotius G, Galindo O, Ramos FSR (1995) Les sécheresses de la région Nordeste du Brésil et leurs conséquences. Sécheresse 1(6):23–33

Magalhães AR, Filho FC, Garagorry FL, Gasques JG, Molion LCB, da Neto SA, Nobre CA, Port ER, Reboucas OE (1988) The effects of climatic variations on agriculture in Northeast Brazil, part III. In: Parry ML, Carter TR, Konijn NT (eds) The impact of climatic variations on agriculture, vol 2. Assessments in semiarid regions. Kluwer Academic Publishers, London, pp 273–380

Nakicenovic N, Swart R (2000) Emission scenarios. IPCC special report on emission scenarios. Cambridge University Press, Cambridge

Root TL, Schneider SH (1995) Ecology and climate: research strategies and implications. Science 269:334–341

Walker B, Steffen WL (1999) The nature of global change. In: Walker B, Steffen WL, Canadell J, Ingram J (eds) The terrestrial biosphere and global change – implications for natural and managed ecosystems. Cambridge University Press, Cambridge, pp 1–19

WBGU (1993) Welt im Wandel: Grundstruktur globaler Mensch-Umwelt-Beziehungen. Wissenschaftlicher Beirat der Bundesregierung Globale Umweltveränderungen. Jahresgutachten. Economica, Bonn

Wessmann CA (1992) Spatial scales and global change: bridging the gap from plots to GCM grid cells. Annu Rev Ecol Syst 23:175–200

Part 1: Integrative Methods and Results

Integrated Scenarios of Regional Development in Ceará and Piauí

Petra Döll, Maarten Krol, Dagmar Fuhr, Thomas Gaiser, Joachim Herfort, Sabine Höynck, Annekathrin Jaeger, Christoph Külls, E. Mario Mendiondo, Andreas Printz and Susanne Voerkelius

Center for Environmental Systems Research, University of Kassel, 34109 Kassel, Germany; doell@usf.uni-kassel.de

Introduction

One aim of the WAVES program (http://www.usf.uni-kassel.de/waves) was to identify sustainable development paths for Ceará and Piauí, two states in the semi-arid northeast of Brazil. In these states, regional development is negatively influenced by the high seasonality of rainfall and El-Niño-related drought years. The research focused on the interrelation between climate variability, water availability, agriculture and quality of life, and it was geared towards supporting regional planning in the study area. For sustainability-oriented research and regional planning, scenarios are important tools. Many of today's activities and decisions will have a significant impact in the quite distant future, and it is therefore necessary to get an idea about their impact if sustainability is to be achieved. The generation of scenarios helps to see how the uncertain future might unfold (Nakicenovic and Swart 2000) and how the future might be influenced by today's decisions. In our understanding, scenarios are consistent and plausible images of alternative futures.

Figure 1 illustrates how regional planning for sustainable development is related to integrated systems analysis (which was also a WAVES research approach) and to scenario generation. On the one hand, the goal of sustainability requires regional planning in which policy measures and projects are not only assessed with respect to financial and technological feasibility, but also with respect to – ideally – all technological, ecological, economic and social aspects. Thus, an integrated analysis of the system of interest is needed. On the other hand, regional planning that takes into account inter-generational equity, a core idea of sustainability, needs to consider the (long-term) future and is thus supported by the generation of scenarios. Scenarios show different possibilities of how the future might look like. A scenario neither predicts the future nor should it be qualified by a probability. Scenarios should be just images of possible futures that are rich enough to help decision-making. Usually, they do not represent a situation at a certain point in time but the dynamic development of the system. Regional scenarios also show the consequences of larger-scale (global, continental, national) developments for the region.

The robustness of a certain policy can be tested by assessing its impact in different possible future situations (Fig. 1). Thus, scenarios can help to demonstrate what impact regional planning might have. Scenarios are always integrated to some extent, as it is not possible to derive images or stories of the future that do not include demographic, economic and technological aspects. Obviously, the planning (or research) aim and scope will determine which parts of the image, i.e., which aspects and processes, are refined and which are only considered in a broad manner. In general, scenario building is based on an integrated systems analysis. Both integrated systems analysis and scenario building require, and thus encourage, interdisciplinary communication (Minx et al. 1993).

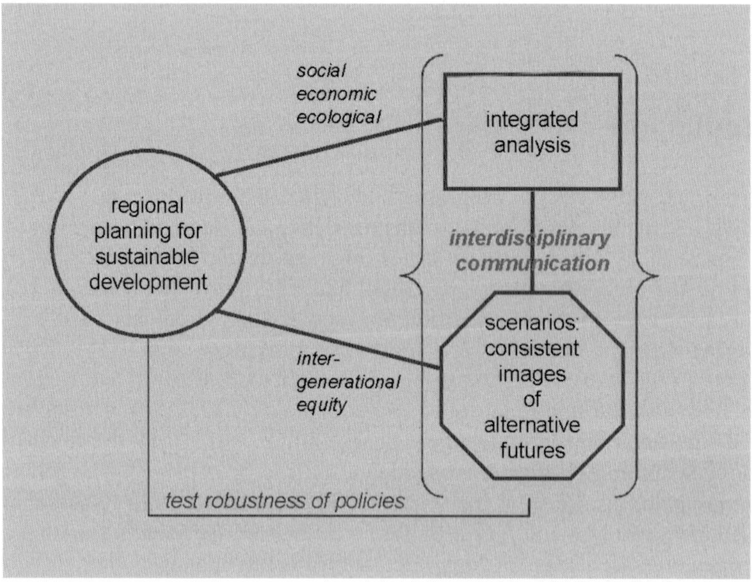

Fig. 1. Relationship between regional planning for sustainable development, integrated analysis and scenarios

In the WAVES program, integrated analysis was mainly achieved by two integrated models (at two different spatial scales) that combined various newly developed disciplinary models. Scenarios for Ceará and Piauí were generated by a multidisciplinary Scenario Group, which experienced that scenario development is a means for integration and interdisciplinary communication. In order to improve the relevance of the developed scenarios for regional planning, three scenario workshops with the regional planning authorities of Ceará were organized during the last year of the WAVES program. In these workshops, policy options for reducing vulnerability to drought were discussed, as well as the plausibility of the scenario assumptions (cf. Sect. 3). In addition, interesting policy measures (interventions) were defined; the impacts of which were then assessed by applying simulation models (cf. Sect. 4).

This paper is organized as follows. In the next section, the methodology to derive qualitative-quantitative scenarios is described. In Section 3, the two reference scenarios that were derived within the WAVES program are presented, including the storylines, the quantification of the driving forces and the modeled system indicators. The impact of selected policy measures is assessed in Section 4. In Section 5, conclusions are drawn with respect to scenario development in general and to the situation in Ceará and Piauí until 2025. Finally, recommendations for an improved integration of scenarios into regional planning are given.

Methodology To Derive Qualitative-Quantitative Scenarios

Scenarios can be qualitative (pure narratives of alternative futures), quantitative (only numerical output) or a combination of both. In two recent global scenario-building efforts, the global greenhouse gas emission scenarios of the Intergovernmental Panel on Climate Change (Nakicenovic and Swart 2000) and the World Water Vision scenarios (Rijsberman 2000), qualitative-quantitative approaches were taken. In WAVES, we adapted these approaches and derived qualitative-quantitative scenarios in which qualitative knowledge and ideas are combined with quantitative mathematical modeling.

The methodology to derive qualitative-quantitative scenarios encompasses the following steps:

1. Identification of the problem field
2. System definition including driving forces as well as temporal and spatial resolution and extent (base year, time horizon and time step, scenario regions)
3. Definition of indicators of the system state (in practice related to the mathematical models that are available to compute the indicators)
4. Generation of qualitative reference scenarios (storylines)
5. Generation of quantitative reference scenarios
   a) Quantification of the driving forces
   b) Computation of the indicators using mathematical models
6. Generation of intervention scenarios
   a) Modification of selected driving forces or parameters of reference scenarios
   b) Computation of the indicators using mathematical models

Evaluation of the Scenarios

Ideally, scenarios are derived by an interdisciplinary group of scientists, policy makers and stakeholders. Döll et al. (2001) describe the scenario-building methodology in more detail and show in an exemplary fashion how the methodology was applied in the WAVES program to derive reference and intervention scenarios for Ceará and Piauí. In this paper, we concentrate on steps 4 to 6.

A storyline is the narrative description of a scenario where an image of a future is created by considering the behavior of the main driving forces of the system (step 4). Reference scenarios describe futures without any specific policy measures. (Implicitly, however, they necessarily reflect certain policies that are part of the respective futures.) They serve as the baselines to assess the impact of policy measures on the future state of the system. It is recommended to produce two or four reference scenarios with quite distinct storylines. With three reference scenarios, there is the tendency to adopt the „intermediate" scenario as the „the most probable", while the others are not given full consideration. Intervention scenarios are defined in order to assess the impact of a certain intervention on the future situation. „Baseline scenario" and „policy scenario" are synonyms for „reference scenario" and „intervention scenario", respectively.

Once the storylines have been written, the driving forces of the reference scenarios must be quantificd (step 5). Here, we define driving forces as those scenario variables that cannot be computed by the applied models but are needed as model input. Both the quantified driving forces and the computed model variables (indicators) are part of the quantitative scenario. In order to make assumptions about the future development of certain driving forces, their historical development is first analyzed. Then, numerical values of the driving forces that reflect the respective qualitative scenario are defined for future times. Care must be taken to quantify driving forces that are known to be correlated in a consistent manner. In addition, it must be taken into account that different driving forces act at different spatial scales. For example, driving forces like greenhouse gas concentrations or some food prices are global-scale driving forces which affect smaller spatial units, like federal states, but which can barely be influenced by any actions at that smaller (e.g. state) scale. Now, if smaller-scale driving forces are defined, they should be consistent with the development of the global-scale driving forces (scale consistency). When all input necessary for the various models is quantified, the models can be used to compute the system indicators for the various reference scenarios.

Intervention scenarios are defined by modifying one or more driving forces or parameters of the reference scenarios (step 6). In general, to test the robustness of the intervention, the impact of an intervention is assessed against the background of all reference scenarios.

It is important to note that the study area is generally subdivided into homogeneous spatial subunits, the scenario regions, for which the storylines and the quantitative assumptions of the development of the driving forces are developed (step 2). The scenario regions are usually larger than the spatial subunits of the mathematical models used to compute the systems indicators.

Scenario building is an iterative process, and it is desirable to discuss preliminary scenarios available after step 5 has been performed for the first time, preferably among scientists, decision makers and stakeholders, and then to revise storylines and quantifications. In a regional planning process, steps 3 to 7 are likely to be repeated. After a first computation and evaluation of the system indicators, in particular the indicator definition can be refined, and new interventions might become interesting.

Reference Scenarios of Regional Development

In the WAVES program, qualitative-quantitative scenarios of the situation in Ceará and Piauí up to the year 2025 were generated (base year 1996/98). Two reference scenarios and a larger number of intervention scenarios were devised. Reference scenario A (RS A) is called „Coastal Boom and Cash Crops", reference scenario B (RS B) „Decentralization – Integrated Rural Development". For scenario building, the study area was subdivided into eight scenario regions (Fig. 2), which are assumed to differ with respect to the future development of the driving forces. Criteria for the configuration of the scenario regions were: similar agro-economic conditions, similar natural conditions (precipitation, position within river basin, sedimentary vs. crystalline subsurface) and administrative boundaries. The driving forces were applied to each of the 332 municipalities of the states of Ceará and Piauí, which are the spatial units to compute system indicators.

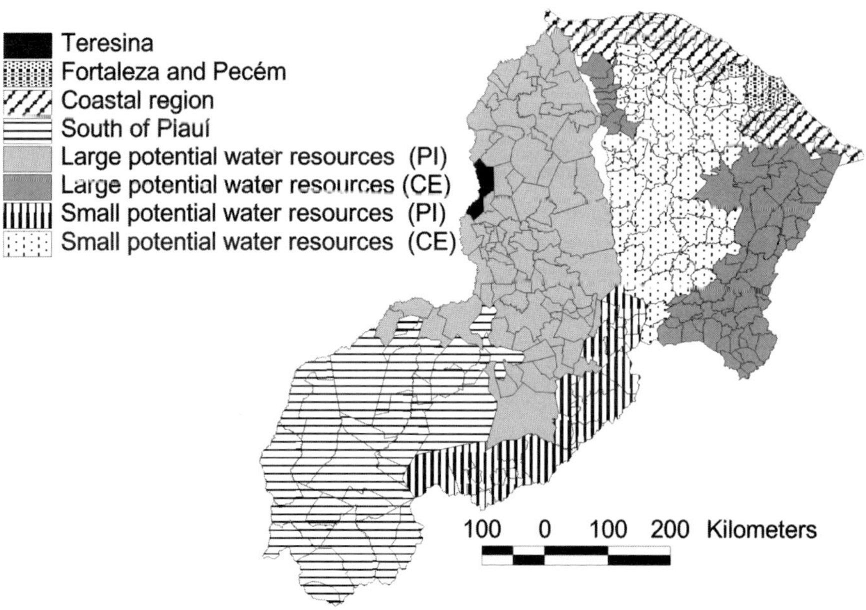

Fig. 2. The eight scenario regions, the spatial units for developing storylines and quantitative assumptions about the development of driving forces. The white areas between Piauí and Ceará are disputed between both states and are almost uninhabited. The outlines of the municipalities, for which the system indicators are computed, are shown too

Storylines

For each of the two reference scenarios, a storyline was written that covers aspects that are important with respect to rural development and to the water scarcity problems in the study area. The detailed storylines for each of the scenario regions can be found at http://www.usf.uni-kassel.de/waves/szenarien/reference_scenarios .html. Table 1 provides a concise characterization of RS A and RS B. Each reference scenario continues certain existing trends. RS A („Coastal Boom and Cash Crops") carries on the current trend of increased cash crop production for the Brazilian and external market, the efforts to promote tourism mainly along the coast, and the fast economic development in the growing metropolitan area of Fortaleza, the capital of Ceará. RS B („Decentralization – Integrated Rural Development") takes up the strengthening of regional centers, e.g., by the establishment of universities, which has recently begun in the study area.

As regional development is strongly influenced by global development, the two reference scenarios were designed such that they can be embedded into global scenarios. For the Third Assessment Report of the Intergovernmental Panel on Climate Change (IPCC), four global scenario families were developed (Nakicenovic and Swart 2000). A comparison of the storylines of the global scenarios with those of our regional scenarios shows that RS A can be embedded in global IPCC scenario family A1 and RS B in scenario family B2 (Fig. 3).

Table 1. Characteristics of the two reference scenarios

Reference scenario A (RS A) „Coastal Boom and Cash Crops"	Reference scenario B (RS B) „Decentralization – Integrated Rural Development"
– Strong economic development (commerce, industry, tourism) in the coastal regions of Piauí* and Ceará – Fortaleza grows very fast – Where water is available for irrigation, the production of cash crops by large companies dominates over subsistence farming – Brazilian and global markets for agricultural products dominate – Centralized governance prevails	– Regional centers prosper (attractive medium-sized towns with improved infrastructure) – Regional centers have become the markets for local and regional agricultural products – Small-scale agro-industry has extended – Local initiatives prevail – Piauí and Ceará show a high degree of autonomy in relation to the Brazilian south – International agencies support sustainable agriculture in crisis-prone regions

* The small coastal region in Piauí consists of only two municipalities.

	Oriented mainly towards economic growth	Oriented mainly towards the environment and social innovation
Globalized world	Global A1/regional RS A	Global B1/none
Regionalized world	Global A2/none	Global B2/regional RS B

Fig. 3. Correspondence of regional reference scenarios for Piauí and Ceará with the global IPCC scenario families (Nakicenovic and Swart 2000)

The A1 storyline describes a future world of very fast economic growth and the rapid introduction of new and more efficient technologies. A strong globalization with intensive economic and cultural exchanges leads to a convergence of wealth and lifestyle among the world regions. In such a world, strong economic development in the coastal region and cash crop production in the hinterland with enough water for irrigation that is assumed in the regional reference scenario RS A are plausible. The globally strong economic development results in buyers for industrial and agricultural products, increased investments of foreign companies and an increasing number of tourists from countries that do not yet participate in global tourism. The B2 storyline describes a world in which the emphasis is on local solutions to economic, social, and environmental sustainability. The B2 world of 2025 is less globalized than the A1 world but still more globalized than today. It is a world with intermediate levels of economic development, less rapid and more diverse technological change than in the A1 storyline, and with strong social innovation. In such a world, we can imagine Piauí and Ceará as described in the integrated rural development scenario RS B. Local and regional markets become more important, while global (and Brazilian) markets are rather weak. Without an economic boom and intensive cash crop production, it might be easier to achieve environmental sustainability. Finally, the diversification of production in agriculture and (small-scale) agro-industry requires social and economic innovations.

Quantification of Driving Forces

The mathematical models applied for the quantification of scenarios require many input variables, i.e., driving forces which need to be quantified beforehand for each of the eight scenario regions. In general, it is assumed that all municipalities within a scenario region are subject to the same change of driving force. The Scenario Group quantified only those driving forces that were of overall importance (the main driving forces population, gross domestic product GDP, irrigated area, etc.). For the latest documentation of the quantified main driving forces, see: http://www.usf.uni-kassel.de/waves/szenarien/tabelas_cenario.pdf, in Portuguese, while the scenario assumptions are discussed in more detail in Döll et al. (2000). The disciplinary working groups then quantified the remaining driving forces (and parameters) of their models in accordance with the main driving forces and the storylines. Here we only present the quantification of population, per-capita GDP, irrigated areas, areas that are potentially available for agricultural production, reservoir construction and climate change. Other driving forces are described in Krol et al., Döll et al., and Höynck (all this Vol.).

Population. Population scenarios are necessary for simulating migration, water use and the agro-economy (agricultural production and income). For RS A and RS B, two different population scenarios were derived. They differ with respect to the distribution of population among scenario regions (Table 2), but total population growth in the study area is assumed to be the same in both scenarios. In RS A, the fraction of the population living along the coast and in the two capitals increases strongly, while in RS B, the current trend of migration from the hinterland to the coast weakens. The historical population development of the study area is strongly influenced by migration towards the economic centers in the south of Brazil. Fertility has been above the Brazilian average, but with a decreasing trend. Using a simple cohort population model which was calibrated for the period 1991–1996, the total population scenario was derived by analyzing fertility, mortality and migration rates of Piauí and Ceará, by comparing them with the respective values for the whole of Brazil, and by taking into account IPCC population scenarios (Nakicenovic and Swart 2000) and a population scenario for Brazil (IBGE 1997). It is assumed that fertility and mortality will continue to decrease in the future, and that the more recent trend of decreasing net migration continues. By 2025, both the fertility rate and the net migration will have dropped to approximately 65% of the 1996 value and the mortality rate to approximately 85%. The resulting fertility and mortality rates of 2025 are still higher than the values for the Brazilian average as predicted by the Brazilian population scenario for 2020. According to the population model, the number of inhabitants of the study area will increase from 9.5 million inhabitants in 1996 to 11.9 million in 2025.

Gross domestic product. Like population scenarios, GDP scenarios determine migration, water use and the agro-economy. Per-capita GDP in the study area is approximately one-half of the Brazilian average, and the average growth rate since 1970 has been somewhat higher than the Brazilian one. The global IPCC scenario A1 (to which RS A fits) shows a higher growth rate of per-capita GDP than the global scenario B2 (RS B; Nakicenovic and Swart 2000). In the case of both IPCC scenarios, the growth rates for the global scenario region to which Brazil belongs are much higher than the average historical growth rates from 1950 to 1990. We think that this type of GDP growth assumption, with an average of 4.7%/year until 2025, is not plausible for the regional scenarios for Ceará and Piauí; per-capita GDP growth is likely to be smaller, and more similar in RS A and RS B. Therefore, we assume a value of 2.7%/year from 1996 to 2025 in RS A (except for Teresina and the regions with small potential water resources, with 2.5%/year), and 2.2%/year in RS B (except for regions with large potential water resources, with 2.4%/year). The sectorial distribution of the GDP is assumed to shift differently in the two reference scenarios. In Ceará, the agricultural sector remains unchanged at 7%, while the service sector increases from 68% of total GDP today to 74% in RS A and 71% in RS B. In Piaui, it is assumed that the agricultural sector will become more important than it is today (14% in RS A and 18% in RS B as compared with 13% today), while the fraction of the industrial sector will remain virtually unchanged at 21%.

Table 2. Spatial distribution of population: historical data and scenarios

	Fraction of total population of Ceará and Piauí in each scenario region [%]			
Scenario region	1991	1996	2025 RS A	2025 RS B
Teresina	6.7	6.9	8.1	7.3
Metropolitan area of Fortaleza and Pecem	26.1	27.6	35.4	30.6
Coastal region	12.7	12.8	17.4	13.3
Southern part of Piauí	3.1	2.9	2.4	2.9
Regions with large potential water resources (PI)	14.5	13.9	11.4	13.5
Regions with large potential water resources (CE)	19.4	19.0	15.5	18.4
Regions with small potential water resources (PI)	2.9	2.7	1.6	2.3
Regions with small potential water resources (CE)	14.7	14.1	8.2	11.7

Irrigated areas. Irrigated areas have a very strong impact on water use and thus water scarcity as well as on the agro-economy of a region. In Brazil, irrigated areas were extended at a rate of 7%/year in the 1970s, by 5% in the 1980s but by only 1.2%/year between 1990 and 1998 (FAO 2000). Here, and in the following, the term „irrigated areas" refers to the areas that are equipped for irrigation. This decreasing trend is observed in most countries and is due to restricted water and land resources. In Ceará and Piauí, however, rural development is still considered to be strongly linked to irrigated agriculture. Therefore, large increases of irrigated area are foreseen in both reference scenarios, with an average over the study area of 3.8%/year in RS A and 2.5%/year in RS B. For RS A, all public irrigation projects planned in 1998 according to Lopes Neto (1998) are assumed to be implemented, while private irrigation, which accounts for approximately half of the current irrigated areas, increases by 2%/year from 1996/98 to 2025. The 2% increase represents the average for the study area Piauí and Ceará and differs among the scenario regions. These assumptions lead to an increase of irrigated areas in the study area from 56,000 ha in 1996/98 to 161,000 ha in 2025, with 84,000 ha in new public projects and 21,000 as new private irrigation. For RS B, with its decentralized development based on small-scale local initiatives, it is assumed that only a quarter of the area in each of the planned public irrigation projects is implemented up to 2025, but that private irrigated areas increase by 2.9%/year. Irrigated area in 2025, for RS B, is 112,000 ha (new projects: 21,000 ha, new private irrigation: 35,000 ha). In both scenarios, the additional private irrigated areas of a scenario region are distributed among the municipalities according to the following rule: 50% of the additional areas are distributed homogeneously over all municipalities, and 50% in proportion to the 1996/98 irrigated area.

Potential agricultural areas. The potential agricultural area is a constraint in the regional economic optimization model of the agricultural sector in Ceará and Piauí RASMO (Höynck, this vol.). In the case of both reference scenarios, it is assumed that it will increase in all scenario regions. In Teresina, Fortaleza and the coastal region, the increase in RS A is somewhat larger than in RS B, while in the other scenario regions, the opposite is true. In the whole study area, the potential agricultural area increases from 5.65 mio ha 1996 to 7.25 mio ha in 2025 in the case of RS A, and to 7.45 mio ha in the case of RS B. In RS A, the fraction of large farms increases, in RS B, the fraction of medium-size farms, and in RS A, the fraction of small farms remains higher than in RS B. The reference scenarios differ appreciably with respect to the increases in the coastal region and the metropolitan area of Fortaleza (higher increase in RSA due to the more favorable marketing conditions) and the regions with low potential water resources (higher increase in RSB due to governmental policies). The areas with high potential water resources are expected to be favored for agricultural production in both scenarios.

Reservoir construction. In particular in Ceará, where groundwater resources are very low, reservoirs are the main source of water supply. Over the last 100 years, the construction of reservoirs has been one of the main strategies to extend the availability of surface water into the dry season or into coming drier years and to perennize rivers. The largest number and total capacity (not considering the Represa Sobradinho in Bahia) of reservoirs in northeastern Brazil is found in Ceará. There, the large storage capacity installed already limits the possibilities to further increase water availability by this strategy with a high efficiency. While in the case of Piauí we do not take into account the construction of new reservoirs, in the case of Ceará we assume, for both reference scenarios, that the Castanhão Dam in the Jaguaribe Basin will be finalized in 2001 (storage capacity $4450 \cdot \times 10^6$ m^3). The investments for additional reservoirs will continue at the present rate (8 mio R$/yr) until 2025, which results in a total of 51 reservoirs with $3150 \cdot \times 10^6$ m^3 of additional storage capacity in Ceará. The sequence of reservoirs follows a prioritization list of the State of Ceará (PROGERIRH), but in RS A, investments are more concentrated in the coastal zone.

Climate change. Climate change scenarios for the years 2000–2050 (daily values) were derived by a statistical downscaling method, taking into account precipitation changes in northeastern Brazil as computed by global climate models for a 1% yearly increase of greenhouse gas concentrations, and historic station data for 1921–1980 (Gerstengarbe and Werner, this Vol.). The applied climate models were the ECHAM4 climate model of the Max-Plank-Institute, Hamburg, Germany, and the HADCM2 model of the Hadley Centre for Climate Prediction and Research, Bracknell, Great Britain. Both models result in quite different precipitation developments in the study area. However, the major changes with respect to today, as well as pronounced differences between the two scenarios, only appear after 2025 (Krol et al., this Vol.). The average precipitation over the study area of the period 2011–2040 is 864 mm/year according to ECHAM4 and 1053 mm/year according to HADCM2, compared with 875 mm/year for 1939–1968 and 905 mm/year for 1969–1998. Climate variability dominates climate change.

Model-Based Quantification of Indicators

Within WAVES, a number of disciplinary models (climate, water availability, water use, agricultural productivity, agro-economy/land use and migration) were developed for both the state level (spatial resolution: municipalities; spatial extent: Ceará and Piauí) and the focus region level, and were coupled to form two integrated models. All these models can be used to quantify the indicators of the scenarios. In the following, we focus on a few integrating indicators that are derived by the large-scale integrated model SIM (Krol et al., this Vol.): water demand, water supply, a water supply sufficiency index for irrigation, agricultural productivity of maize, tomato production, farm income and migration. Only the results based on the ECHAM4 climate model are shown.

Water scarcity. In principle, water scarcity is a function of water supply and water demand. Water scarcity is the most direct impact of drought conditions or overuse of the natural resources. In both reference scenarios, water demand shows strong increases until 2025 (Fig. 4), reaching approximately 110% in the case of RS A, and 60% in the case of RS B. Almost 90% of these increases are due to the assumed extension of irrigated agriculture, while increased water demands in the domestic, industrial, tourism and livestock sectors may be regionally important but overall are less relevant (Döll et al., this Vol.[1]). Water demand for irrigation shows a strong interannual variability as it is negatively correlated with precipitation volumes. The increase in water extraction (volume of water that can actually be withdrawn to fulfill the demand) is smaller than that of water demand, as water supply does not meet demand, especially for increased irrigation (see below).

Even though the total renewable water resources (annual surface runoff and groundwater recharge, modeled as described in Güntner and Bronstert, this Vol.) will not change appreciably until 2025, water supply does, due to the construction of new reservoirs, adding to the surface storage capacity. Potential water supply can be partially represented by the total volume of water that is stored in large reservoirs at the beginning of the dry season. This value is mainly determined by the water storage infrastructure and the climatic conditions, with small differences between the two scenarios (Fig. 5). The water volume stored is generally very large, but large parts of it do not cover large water demands. For example, the Boa Esperança Reservoir (Parnaíba Basin) at the western border of Piauí, with a ca-

[1] Please note that the irrigation water demand given in Döll et al. (2001) is smaller than the values computed by the integrated model for this publication. The latter are based on different estimates of precipitation and potential evapotranspiration and take into account multi-cropping.

pacity of 5000· × 10^6 m^3, is used for hydropower generation. The increase in the year 2001 is due to the newly constructed Castanhão Dam.

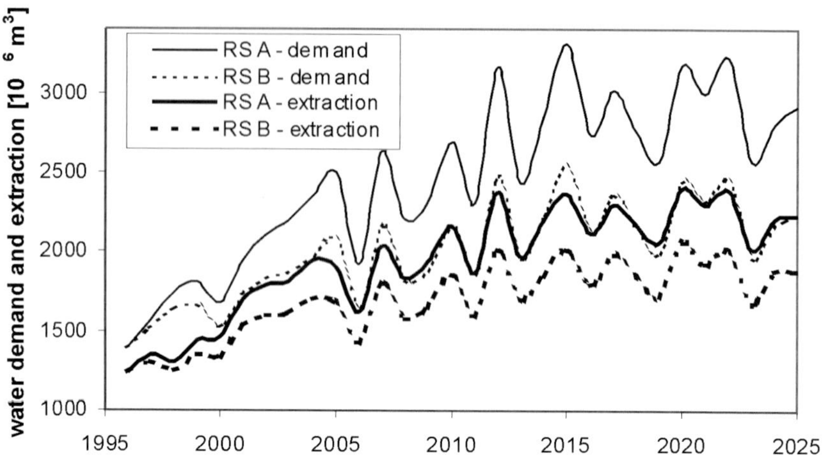

Fig. 4. Total annual water demand and water extraction in Ceará and Piauí

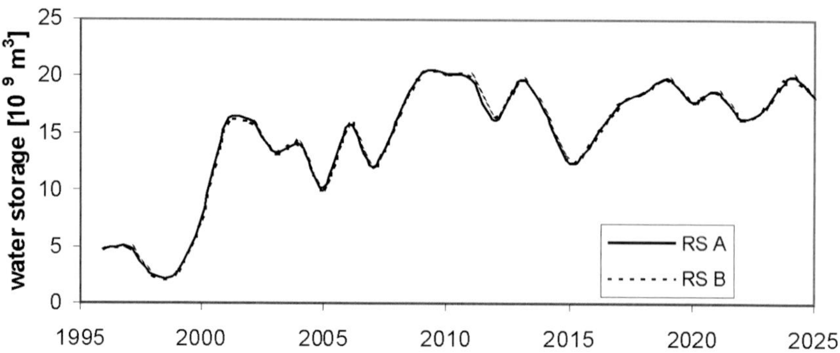

Fig. 5. Water storage in large reservoirs at the beginning of the dry season (1st July)

The net effect of the increasing trends of both water demand and supply on water scarcity depends critically on the spatial distribution of water availability and water demand. To quantify scarcity, a water sufficiency index was calculated that evaluates which part of the demand can actually be covered by the availability. This indicator is computed for each of the water use sectors: irrigation, livestock, households, industry and tourism, which differ with respect to their sources of water (groundwater, surface water). In the case of water scarcity, the irrigation sector will only be provided by the water volume that is not needed by the other four sectors, which is according to the legal requirements in the study region. Figure 6 shows the temporal development of water supply sufficiency for the irrigation sector, where a value of 1 indicates a complete coverage of demand (no scarcity impact). The irrigation sector is selected here because it is the dominant use sector and because it mainly depends on surface water. There is a gradual downward trend in water sufficiency for RS A from 88% down to 80%, whereas for RS B the value remains at 88%. The decreasing water sufficiency in the case of RS A is fully attributable to negative tendencies in the regions with large potential water resources and the coastal zone. This indicates that in these regions water scarcity is increasing with time, and the additional water demands from expansion of the irrigated area cannot be met. In the case of RS B, the expansion in irrigated areas in all regions is better adapted to the available and increased water supply.

In the scenario assumptions of RS A and RS B, precipitation reduces by 7% until 2025, but the effects of climate change until 2025 are small compared with the effects of increasing demands and the expansion of supply systems.

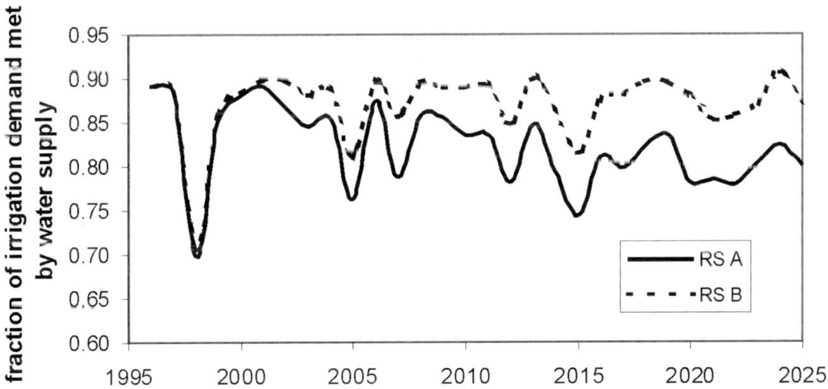

Fig. 6. Water supply sufficiency for irrigation: fraction of the water demand for irrigation that can be covered by water supply

Agricultural production. Agricultural production is not only strongly affected by climate, but also by the possibilities for irrigation. Both rainfed and irrigated production depend on precipitation. Rainfed production is directly affected by negative anomalies in precipitation, especially when the rainy season is significantly shortened or when longer dry spells within the rainy season occur, depleting soil moisture. Irrigated production is affected by unfavorable climatic conditions in two ways. Firstly, water supply for irrigation is less in drier years, especially after a sequence of drier years. Secondly, irrigation water demand is higher.

Rainfed yields strongly depend on the climate, and the differences between RS A and RS B play a negligible role (only by the way in which they impact on the spatial distribution of cropping areas). Until 2025, climate change shows little influence on the yield of, for example, maize (Fig. 7): both mean yield and variability do not exhibit a significant trend.

For the irrigated production of, e.g., tomato, clear trends and differences between the two scenarios appear, which are due to a higher increase of irrigated tomato production in RS A (Fig. 8). In RS A, production increases by 340%, in RS B by 220%, even though in RS A, a larger part of the area equipped for irrigation cannot be irrigated due to lack of water (Fig. 6).

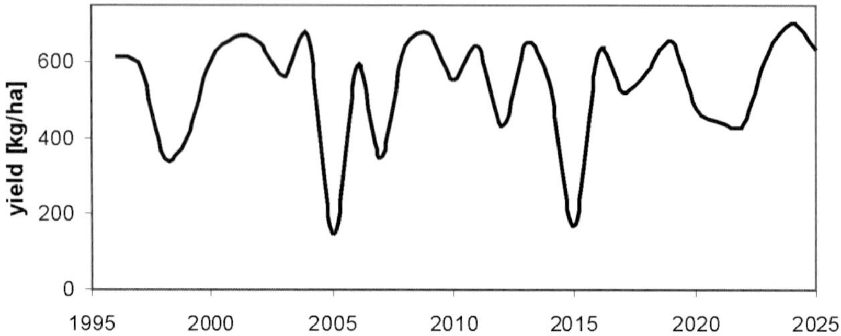

Fig. 7. Temporal development of mean rainfed maize yield in both reference scenarios

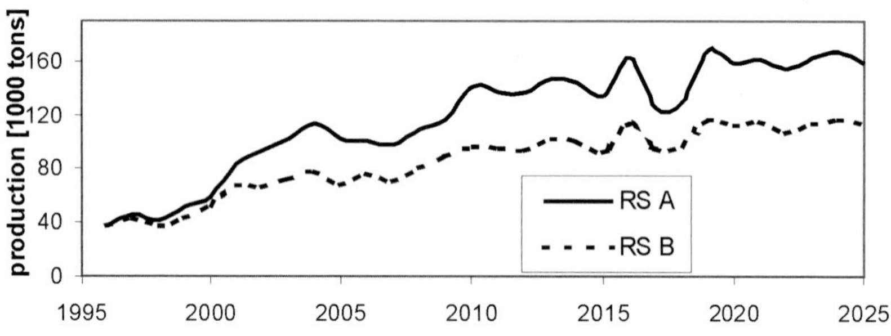

Fig. 8. Tomato production in Ceará and Piauí in RS A and RS B

For almost all crops, production increases are found in both reference scenarios, except for cassava, mostly cultivated on smaller farms, whose areas are decreasing. For crops that are mostly grown under rainfed conditions, production increases are very similar for the RS A and RS B scenarios; for crops with a relevant share of irrigated production, production increases in RS A are stronger due to the faster expansion of irrigated areas.

As a result, total farm income grows faster in RS A than in RS B. The interannual variability of agricultural GDP remains very large, with negative net values for the three driest years until 2025. Due to the large variability, a statistically significant growth trend cannot be identified. The difference between RS A and RS B, however, shows an increasing trend, which is due to the different proportions of high-value irrigated crops. The difference reaches approximately half of the mean agricultural GDP of the 1990s, but results are very uncertain as the assumed costs and prices are uncertain; in particular, the profitability of smaller farms is uncertain (Höynck, this Vol.).

Migration. Scenarios of migration are to be interpreted with caution. Migration is simulated as being mainly driven by gradients in income, mimicking large-scale migration within the states of Ceará and Piauí and outside of these states over longer periods (Fuhr et al., this vol.). This makes GDP (total and agricultural) and its growth rates the main drivers of migration, but the income drop in drought years, enhancing the rural-metropolitan income gradient, is the direct incentive for migration. Model simulations successfully reproduce the assumed tendencies in population development and its regional distribution. Considering net migration from the Sertão (scenario regions with large and small potential water resources), the dynamical feature of high migration in drought years remains due to the continued variability of agricultural GDP (Fig. 9), with persistently lower numbers for RS B (total migration is 25% higher in RS A). Unfortunately, there are no historic data of migration from the Sertão nor annual values of migration from the whole study area that could be used to check the plausibility of the computed migration; the annual net migration from the study area, averaged over the period 1991–1996, was 60,000 (Döll et al. 2000). In the migration simulation, the effect of emergency measures, which support the rural population in drought years, is not taken into account explicitly, such that migration peaks are likely to be highly overestimated. In summary, the present (incomplete) quantitative understanding of migration indicates that there will be no consistent decrease in migration from the Sertão, even under the assumption of economic growth in the Sertão.

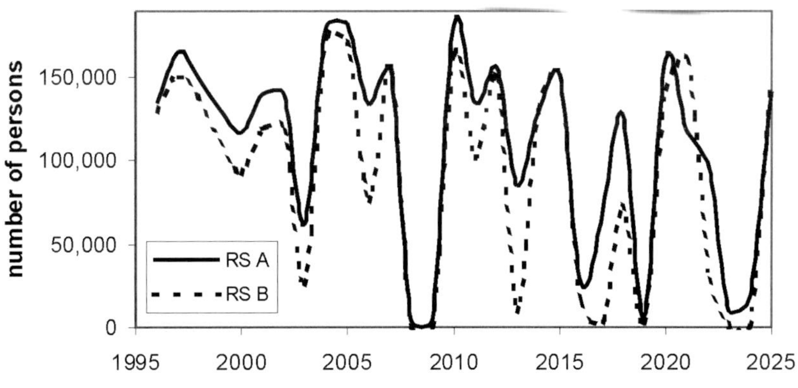

Fig. 9. Annual net migration from the Sertão

Intervention scenarios

In an exemplary manner, the impact of three different policies or management measures is analysed against the background of the two reference scenarios: water pricing, reduced reservoir construction and introduction of high-yield cashew varieties.

Water Pricing

Appropriate pricing of water helps (1) to achieve the economically most efficient use of water, (2) to avoid its wasteful use and thus to protect the environment, and (3) to generate resources to improve water supply and water (demand) management. In the regional water use model NoWUM, water use in households connected to the public water supply and water use by industry are simulated as being influenced by the water price due to the price elasticity of water use (Döll et al., this Vol.). With a price elasticity of –0.5, for example, water use decreases by 5% if water price increases by 10%. Price elasticity in the domestic sector is lower than in the industrial sector. While in the domestic sector public-supplied and self-supplied households are distinguished (the latter not being affected by water pricing), this distinction cannot be made in the industrial sector due to lack of data; this is likely to lead to an overestimation of the responsiveness of industrial water use to water pricing by the public water suppliers. Unfortunately, there is a lack of reliable data on price elasticities in Brazil, which leads to significant uncertainties in the computed impact of water price on water use. Besides, it is at present not possible to apply the price elasticity concept for irrigation water, as there is no knowledge about price elasticities in this water use sector.

In general, the current water prices in Piauí and Ceará cover only operation and maintenance costs, and not investment costs, e.g., for an extension of water supply, or environmental costs. However, it is widely recognized that sustainable development requires a full-cost pricing for water services (World Water Commission for the 21st Century 2000). Currently, approximately 1% of the average per-capita GDP in the study area is spent on (public) water supply, which is comparable to the situation in Germany, with a tenfold per-capita GDP and the same per-capita water use. The increase assumed for both reference scenarios (6%/year) from 1996/98 to 2025 (cf. Sect. 3.2) is smaller than the historic price increases in Ceará during the last decade (which was 8–11%/year, based on 1989–1998 data from CAGECE, the main water provider in Ceará). A price increase of only 2.5%/year results in an approximately constant fraction of the total income that is spent on water, and it is unlikely that with such a price development investment costs could be covered. A price increase of 11%/year (the price increase of the fixed fee component of the water tariff in Ceará during the last decade) leads, on average, to a water price that, with the same per-capita water use, amounts to about 10% of the per-capita GDP. Please note that prices per volume of freshwater will effectively double when households are connected to the sewage system (in

1996/98, only about 10% of urban withdrawals are discharged into a sewage system).

With an 11%/year increase in water price, total domestic water use (including self-supply) decreases by approximately 45% as compared with the reference scenarios (where 6%/year was assumed), while with 2.5%/year, it increases by about 40%. The impact of different water pricing is approximately the same in both reference scenarios. By 2025, industrial water use intensity (withdrawal water use divided by industrial GDP) will have decreased to 16, 37 and 66% of the present value in the cases of 11, 6 and 2.5% water price increase per year, respectively. Consequently, industrial water use in the study area decreases by 58% in the case of 11%/year, and increases by 80% in the case of 2.5%/year as compared with the reference scenarios.

Reduced Reservoir Construction

To assess the beneficial effects of the reservoirs that it is assumed will be constructed before 2025 in the reference scenarios, variants of the scenarios were defined in which less dams are constructed. While in both RS A and RS B it is assumed that annual investments in new reservoirs remain constant at year 2000 level, the investment volume is assumed to be only half this size in the intervention variants. As in the reference scenarios, dam construction has priority in the scenario region with high potential water resources in scenario B, and in both these regions and in the coastal region in scenario A. In Fig. 10, the intervention scenarios are denoted by IS A-res and IS B-res, IS standing for Intervention Scenario and –res for reduced dam construction.

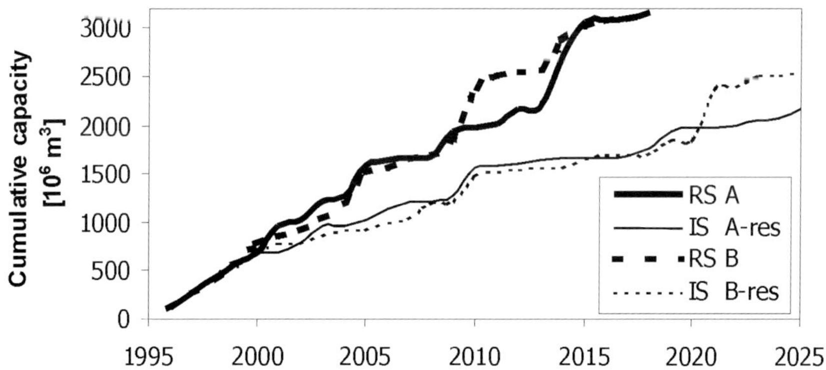

Fig. 10. Cumulative capacity of reservoirs constructed from 1996 to 2025 for reference and intervention scenarios

In the model simulations of IS A-res and B-res, stored water at the beginning of the dry season is less than in RS A and RS B (Fig. 11) for most years, illustrating the positive effect of newly installed water storage capacity on the availability of water. However, in the dry years after 2010, the reduced reservoir construction in the intervention scenarios does not lead to a reduction in the volume of stored water. The relative filling of the new reservoirs at the start of the dry season is higher in the intervention scenarios (Fig. 12). This holds for both the scenarios A and B and indicates that the total installed storage capacity in the case of the interventions scenarios is nearing a level where the efficiency of new dams is highly reduced.

Reduced dam construction has only a very small impact on water scarcity. The smaller number of new reservoirs in IS A-res as compared to RS A lowers the mean sufficiency of water supply for irrigation between 2016 and 2025 from 80.5 to 80.1%; for the B scenarios, the sufficiency declines from 88.0 to 87.8%. The impact of climate change on the efficiency of water storage infrastructure and water sufficiency is not apparent until 2025, but is a potentially very significant process afterwards (Krol et al., this Vol.).

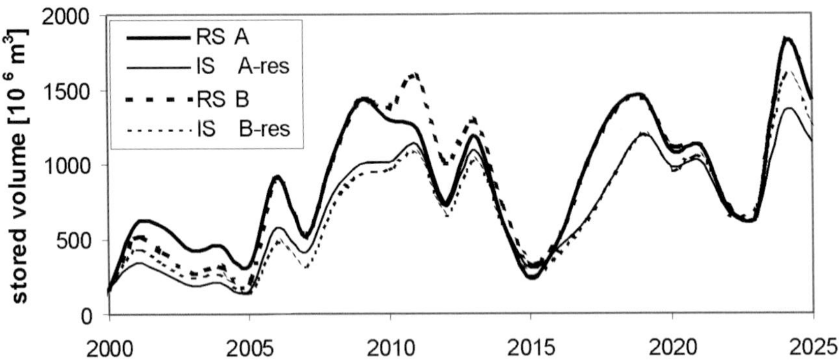

Fig. 11. Increase in stored water volume at the beginning of the dry season by reservoirs constructed after 1996 for reference and intervention scenarios

Introduction of High-Yield Cashew Varieties

Improved crop varieties, especially of cash crops, are being developed to increase yields and to support market orientation of the agricultural sector. In the intervention scenario „Cashew“ it is assumed that the traditional cashew varieties are replaced by improved varieties with a fourfold yield potential from 2010 on. Prices for cashew nuts are assumed to remain stable. The potential benefit through the marketing of cashew fruits is not considered. The impact of this intervention is only assessed against the background of RS A.

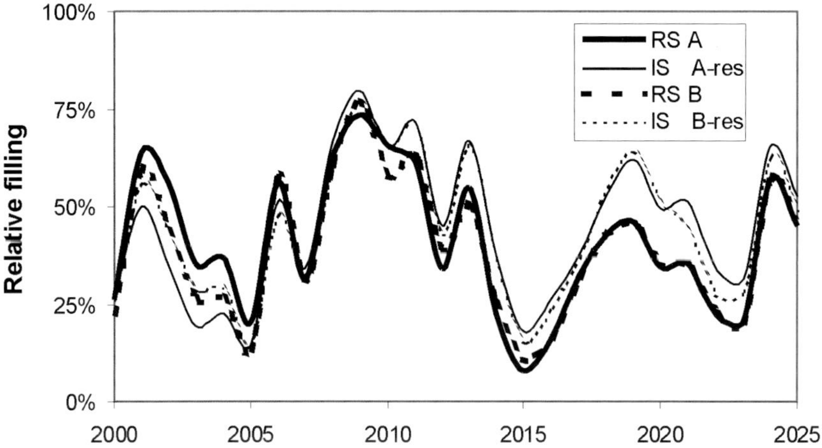

Fig. 12. Relative filling at the beginning of the dry season by reservoirs constructed after 1996 for reference and intervention scenarios

RASMO (Höynck, this Vol.) computes the changes in agricultural land use resulting from the introduction of the high-yield cashew varieties. The cashew-growing area (both rainfed and irrigated) increases by 44% (from 363 to 522·× 10^3 ha), especially in northern Piauí, the coastal municipalities and the lower part of the Jaguaribe catchment (Fig. 13). Cashew production increases by 440% (Fig. 14). The net effect on agricultural GDP amounts to about +10% in the first decade after 2010, dropping to about 8% to the end of the simulation period.

The projected net effect on migration is very small (0.5%), partly because most migration-prone areas are simulated to benefit the least from the intervention due to restrictions by soil (clayey and stony soils in the Sertão), climate and irrigation possibilities. The intervention would thus implicitly result in supporting regional development in the regions with competitive advantages (coastal zone and regions with large potential water resources).

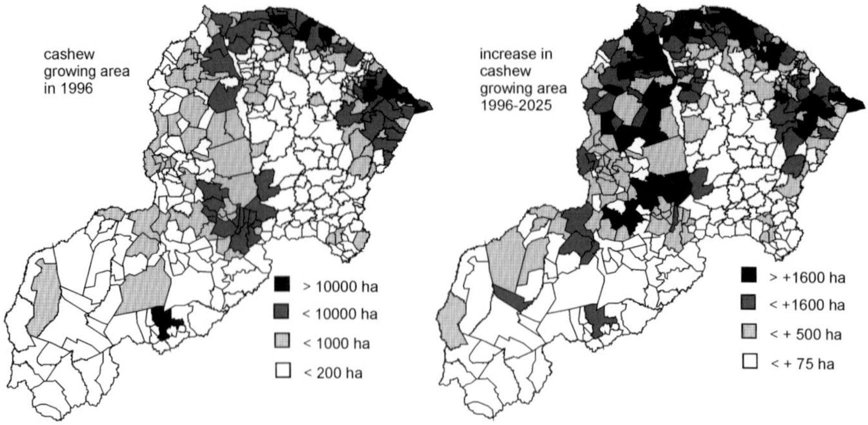

Fig. 13. Distribution of cashew cultivation in 1996 and change in cashew cultivating area as induced by the intervention, which results in a very significant productivity increase

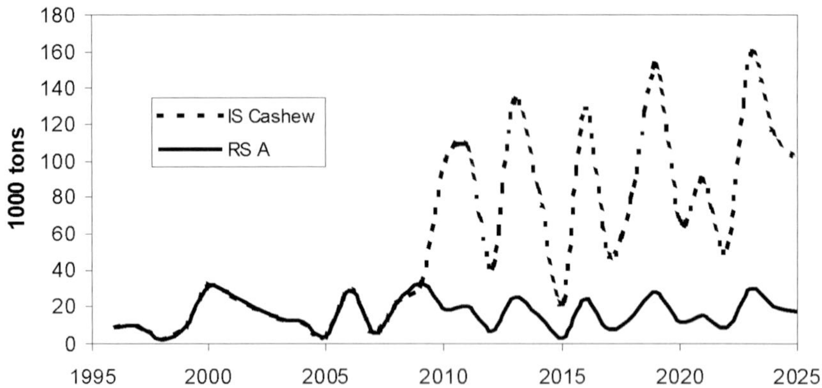

Fig. 14. Total production of cashew nuts in Ceará and Piauí in scenario A: reference scenario and intervention scenario „Cashew“

Conclusions and Recommendations

Integrated scenario development has the potential to support sustainability-oriented regional planning in Ceará and Piauí. Scenario development requires, and thus encourages, interdisciplinary communication as well as the exchange with stakeholders and policy makers, which is a prerequisite for the applicability of research results. Scenarios combine a large amount of qualitative and quantitative knowledge into consistent descriptions of alternative future development paths. They can be presented in the form of stories and therefore transmit the results of the integrated analysis in a transparent and understandable way, facilitating communication and diffusion of results. The consistency of the derived scenarios is strongly supported by the application of models, and of an integrated model in particular. (In the presented scenarios, for example, the extension of irrigated crop production is restricted by water availability, while both variables are affected by climate change.) The quantification of appropriate system indicators (e.g., for water scarcity) is decisive for a successful integration of scenarios into regional planning, but it is limited by the capabilities of the applied mathematical models, which again are often restricted by data availability.

The model-based quantification of the system indicators related to water stress, agricultural production and migration results in the following conclusions with respect to the development of Ceará and Piauí until 2025:

1. Water stress increases significantly only in case of RS A („Coastal Boom and Cash Crops"), which is mainly due to the extension of irrigation (implementation of all currently planned public irrigation projects) and not caused by climate change.
2. If, in Ceará, investments for new reservoirs do not continue until 2025 at the current rate of 8 million R$/year but drop to only half that value, the volume of water which is stored in the new reservoirs at the beginning of the dry season will not be reduced much after 2010, and not at all in drought years. More importantly, water scarcity, expressed as the sufficiency of water supply to fulfill water demands, will not be significantly affected by reduced dam construction.
3. Therefore, it is likely that the implementation of all currently planned irrigation projects would lead to low sufficiency of water supply for irrigation even if more reservoirs were constructed than assumed in the reference scenarios (investment rate 8 million R$/year).
4. Due to the high price elasticity of domestic and industrial water demand, domestic and industrial water use can be efficiently managed by water price, in particular if all the industries were charged. (However, there is a lack of reliable data on price elasticities.) If the annual (inflation-corrected) water price increase is less than 6%/year (assumed in the reference scenarios), water stress will increase as domestic and industrial water demand will increase significantly. However, the impact of water price on the dominant water use sector, irrigation, could not be assessed.
5. In RS A, the agricultural GDP increases more strongly than in RS B („Decentralization – Integrated Rural Development") which is due to higher production of irrigated crops. However, in both scenarios, interannual variability remains very high.
6. Assuming constant prices, introduction of high-productivity cashew varieties increases agricultural GDP by 8–10%.
7. Migration from the Sertão is less in the case of RS B than in the case of RS A, but will not decrease significantly until 2025.

Please note that the uncertainty of the computed system indicators and thus of the above conclusions is high. The lack of knowledge about present-day irrigation, in particular about irrigated areas and multicropping, leads to significant uncertainty of the modeled water demand. Modeling of agricultural production and GDP suffers from a lack of reliable data on present-day cropping as well as coarse representation of price formation processes, credit availability and informal marketing. As to migration, only few of the reasons for migration that are well known qualitatively can be quantified, such that the impact of important drivers could not be included in the scenario generation.

A very important source of uncertainty are climate data, both historic data and climate change scenarios, which have a strong impact on computed water availability, irrigation water demand and crop productivity. For example, when the precipitation over Ceará is estimated based on about 200 measurement stations as compared with about 30 in the standard climate data set, 1969–98 long-term average annual precipitation is estimated to be 844 mm instead of 924 mm, which results in a 50% increase of computed irrigation water demand in Ceará. However, the conclusion concerning the modest effect of reduced investments for reservoir construction with respect to alleviating water scarcity (conclusion 2) is quite robust as to different climate change scenarios. It holds for the climate change scenario with declining precipitation volumes (ECHAM4, applied in this paper), but similarly for a constant climate, for which the installed capacity is high already, and also for the scenario with increasing precipitation (HADCM2), for which water scarcity will mainly decrease due to a reduced water demand and a better filling of the already available storage capacity.

Ideally, scenario development should start at the beginning of projects that are concerned with sustainability-oriented research. To make it relevant for regional planning, it should be performed in cooperation with policy makers and stakeholders. An extended scenario group consisting of researchers, policy makers and stakeholders would first define whether explorative or normative scenarios should be developed. Explorative scenarios are the type that have been developed in the WAVES program, while normative (or sustainability) scenarios are based on the definition of „desirable futures“ and help to identify the conditions and interventions that are necessary to reach the desirable future system state. Explorative scenarios should be formally evaluated by the extended scenario group, e.g., by performing a multi-criteria analysis. The evaluation of explorative scenarios and the development of normative scenarios require the identification of suitable sustainability indicators and the definition of threshold values.

Acknowledgements

We thank SEPLAN and SRH for their interest in our work and for their cooperation in the policy workshops, CAGECE, AGESPISA, COGERH, CPRM and FUNCEME for supplying data, and BMBF and CNPq for financial support.

References

Döll P, Fuhr D, Herfort J, Jaeger A, Printz A, Voerkelius S (2000) Szenarien der zukünftigen Entwicklung in Piauí und Ceará. WAVES Report, Working Group Scenarios, Center for Environmental Systems Research, University of Kassel (can be downloaded from http://www.usf.uni-kassel.de/waves/vorl_endbericht/3_szenarien.pdf)

Döll P, Hauschild M, Fuhr D (2001) Scenario development as a tool for integrated analysis and regional planning. Proceedings of the German-Brazilian Workshop on Neotropical Ecosystems (CD-ROM). GKSS, Hamburg, Germany (can be downloaded from http://www.usf.uni-kassel.de/waves/english/vorl_ergebnisse/prel_results.html)

FAO (2000) Irrigation in Latin America and the Caribbean in figures. Water reports 20, Rome, Italy

Lopes Neto A (1998) Possibilidades de Modernização Rural do Ceará através da Agricultura Irrigada e da Fruticultura. CNPq, SECITECE, Fortaleza, Brazil

Minx E, Neuhaus C, Waschke T (1993) Vom Brückenbauen oder: Wie machbar ist Interdisziplinarität? Organisationsentwicklung 93(1):52–64

Nakicenovic N, Swart R (eds) (2000) Emission scenarios. IPCC Special Report on Emission Scenarios. Cambridge University Press, Cambridge

Rijsberman FR (ed) (2000) World water scenarios: analyzing global water resources and use. Earthscan, London

World Water Commission for the 21st Century (2000) A water secure world – vision for water, life and the environment. Report to World Water Council, World Water Vision (can be downloaded from http://www.worldwatervision.org/reports.htm)

Integrated Modeling of Climate Change Impacts in Northeastern Brazil

Maarten S. Krol, Ann-Kathrin Jaeger and Axel Bronstert

Potsdam Institute for Climate Impact Research (PIK), P.O. Box 601203, 14412 Potsdam, Germany, Maarten.Krol@pik-potsdam.de

Introduction

Societies in semiarid areas in developing regions are among those most vulnerable to climate variability and potentially most vulnerable to climate change. The vulnerability to climate variability is caused by the strong restrictions that limited water availability poses on the use of natural resources, by the generally low reliability of water availability and, on the other hand, an often appreciable density of population strongly depending on these resources with little short-term options to reduce the dependency. Reasonable conditions in wetter years support the persistence of population in the area; marginal or poor conditions in drier years and arrears in development hamper significant improvements in the quality of life. Regional tradition and adaptation to local conditions serve as experiences to cope with droughts. This may help to survive emergency situations, but does not really reduce vulnerability, as it does not reduce development deficits. Present-day climate conditions are already marginal to sustain societal demands; therefore, climate change poses a serious potential threat.

Clearly, the study of climate change impacts in developing semiarid regions calls for an integrated approach. Climate impacts are not merely an effect of changes in water availability, but emerge from the confrontation of availability and societal demands and the role these demands play in society. Therefore, the study should include not only the physical understanding of climate impacts on the water balance and on crop yields, but also include the analysis of water use, agricultural economy and societal impacts. In northeastern Brazil, one of the most marked societal impacts of droughts is the emigration of population from rural areas in the urban centers and to destinations outside the region (Magalhães et al. 1988).

Latest assessments of available climate observations conclude that global climate has been warming over the last century and that human influences play a major role in climate change (IPCC 1999). For the twenty-first century, an enhanced warming is projected to take place under a broad variety of assumptions on global developments. Global precipitation is projected to increase, but at low latitudes, where most of the semiarid regions in developing countries are situated;

both regional increases and decreases are projected. Where precipitation increases, an increased inter-annual variability is assessed to be very likely.

Scientific consensus on climate change projections for specific semiarid regions is not available at present. This may partly be due to the small scale of these regions, compared with the global dimension represented in the climate models supporting the present assessments. Both the spatial resolution of global models and the model skill at the regional scale complicate the interpretation of results. Downscaling techniques are thus required (Sect. 3).

The research area of integrated modeling in the context of climate change studies is developing rapidly. Integrated models initially originated from natural science models (energy balance, carbon cycle, crop yield), where models for greenhouse gas emissions from energy use and representations of climate change impacts were included, like IMAGE (Rotmans 1990) and IMAGE 2 (Alcamo et al. 1994). In an alternative approach, macro-economic models were extended with cost estimates for impacts of climate change (like GCAM, Edmonds et al. 1994). Later, integrated models emerged with a more centrally defined integration approach (like ICLIPS, Toth et al. 1997 or TARGETS, Rotmans and de Vries 1997).

Regional climate impact assessments also tend to include ever more integrative approaches, extending from agricultural impacts (MINK study, Rosenberg et al. 1993) to representation of interests of various socio-economic sectors (MBIS study, Cohen 1997). Earlier regional studies on North-eastern Brazil (Magalhães et al. 1988) were broad and rich in information but did not arrive at an integrative concept as adopted in the present (Bronstert et al. 2000).

The implementation of this concept in a fully coupled integrated model forms a very challenging task (Sect. 2). The resulting model dynamically describes the relationships between climate forcing, water availability, agriculture and selected societal processes, allowing the analysis of possible climate change impacts (Sect. 4) or simulation of scenarios of regional development (Döll et al., this Vol.).

The Semiarid Integrated Model (SIM)

Although a generally accepted approach to integrated modeling does not exist, some characteristics are commonly found. Basic characteristics of an integrated model are its focus on a problem rather than on individual processes, and its focus on dynamic behavior rather than on exact replication of the present.

Integrated modeling starts with a systems analysis of the problem, the assessment of relationships between water availability and quality of life and migration in rural semiarid northeastern Brazil at the meso-/macro-scale in the context of global change, especially climate change. A top-down analysis of the problem identifies which are the basic variables, processes, and external forcing that should be accounted for in the integrated model (Fig. 1).

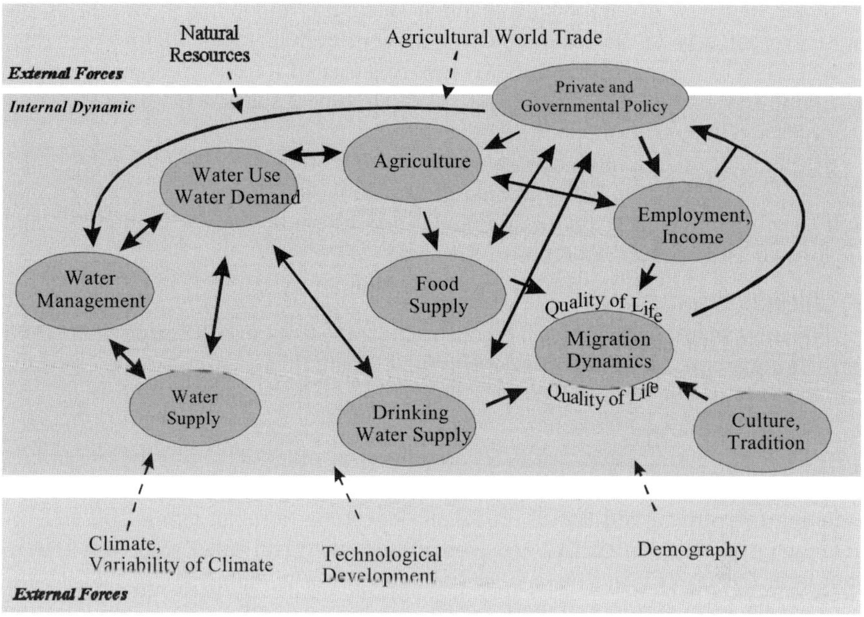

Fig. 1. Conceptual model of dynamic relationships between water availability, agriculture and migration in northeastern Brazil as influenced by global change

In this analysis, the focus is explicitly on internal features of the most relevant dynamic behavior rather than on what is best understood, as, for instance, the influence of water scarcity on water consumption. The specific challenge is to find a good compromise between the clarity in the representation and the comprehensiveness regarding the main dynamical processes. Special attention should be paid to cross linkages and feedback processes, which already are important in the present functioning of the system (Bronstert et al. 2000), but can especially influence long-term dynamics, as, for instance, the effects of regional development on the agricultural sector and on water extraction.

A specifically interesting feature of the dynamics is the ability of the system to dampen or enhance the variability resulting from the variable climate forcing. Here, e.g., water infrastructure and water policy can cause a reduced variability in availability of water resources, but increased water demands may amplify variability of impacts.

The context of global change forces the analysis to consider long-term behavior of the system. A reasonable dynamic understanding is preferred over an excellent static understanding, and processes influencing long-term changes are more in focus than processes just explaining heterogeneity at the micro-scale. Clearly, only (aggregated) process-based models can simulate both the effects of individual influences of global change and policy interventions. In this sense, integrated models generally show a mix of deductive and inductive approaches.

In the analysis, geographically explicit integrated modeling is chosen as the most explicit way to approach consistency considerations and regional validity. The Semiarid Integrated Model (SIM) was constructed for this regional integrated modeling at the meso-/macro-scale (Krol et al. 2001), and is extending the commonly found integrated modeling further into human dimensions.

SIM is built up in a modular way, where each module consists of a disciplinary contribution. Integrative tasks include guaranteeing the consistency between modules, defining interfaces between modules and filling up gaps with plausible and clear basic assumptions. Modules of SIM are:

- Climate
  Climate scenarios or a historic reconstruction are input to SIM, and results from long-term daily observations combined with climate trends from general circulation models (Gerstengarbe and Werner, this Vol.).
- Water availability
  A large-scale water balance model describes run-off, storage in water reservoirs and soil moisture based on a hydrotope-approach (typical size 10–50 km^2), accounting for vertical and lateral processes depending on topography, soil and vegetation cover, with an explicit representation of the main water reservoirs (Bronstert et al. 2000). A water use model simulates water withdrawal, considering the various water use sectors and their specific intensity and efficiency (Hauschild and Döll 2000), directly coupled to the agricultural and demographic models.
- Agriculture
  Crop yields are simulated using FAO schemes for yield response to water stress, as used in CROPWAT (FAO 1979), considering the 14 regionally most important crops (including maize, bean, rice, cashew) and for impacts of soil quality (FAO 1983). Agro-economy is represented as optimizing farm income by varying cropping and husbandry activities, under restrictions of available land, technical and financial opportunities, and feed and food requirements, and accounting for production costs and prices (Höynck, this Vol.).
- Socio-economy
  Migration amongst municipalities and to external destinies is preliminarily modeled to occur when gradients in the quality of life exceed migration costs (fixed and distance-dependent; Fuhr et al., this Vol.). Here quality of life is a composite indicator, where mean municipal income per head is the dominant influence. Migration is accounted for in a demographic model, which resolves population development for age and gender.

Regional Interpretation of Climate Change Projections

Complex physically based climate models (as general circulation models, GCMs) show an increasing ability to simulate present-day climate as well as historic trends over the last centuries at the global to continental scale (IPCC 2000). They project significant global climate warming (1.4–5.8°, 1990–2100) and precipitation increase to take place in the current century, under the assumption of a continuous increase in atmospheric greenhouse gas concentrations, as would be caused by a continued intensive use of fossil fuels.

Still, the skill of these models in representing climate at the scale of northeastern Brazil (NEB) is modest. Of the seven climate GCMs whose climate change experiments were made available for climate impact assessments by the IPCC Data Distribution Centre (IPCC-DDC 1999), only three are able to represent in their simulations the semiaridity and strong seasonal cycle that are characteristic for this region, see Fig. 2. One of these three models has a serious flaw in representing global precipitation, hampering serious interpretation of its results on changes in precipitation. This lack in skill may be caused by the relatively coarse resolution of GCMs, 300 to 900 km, leaving 2 to 12 grid cells only to cover all of northeastern Brazil. An alternative explanation may be the imperfect representation of regionally important physical processes. Either way, the lack in skill seriously affects the applicability of model results for impact assessments.

The recommended approach to critically review regional performance in selecting model results to be used in assessments (IPCC-TGCIA 1999) is often ignored, for instance, in a specific assessment of plausible climate change in Brazil, including a focus on NEB (Hulme and Sheard 1999). This can easily lead to inconsistent interpretations; for example, in one GCM, NEB turns from very arid into arid between 2000 and 2100, which, by only using climate change output, would be interpreted as a transformation from semiarid into sub humid.

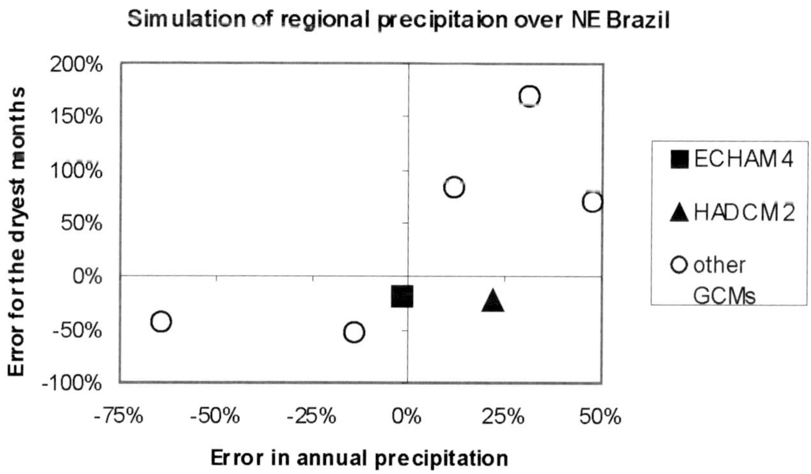

Fig. 2. Simulation of regional precipitation over semiarid northeastern Brazil in the period 1961 to 1990 by seven GCMs (IPCC-DDC 1999) compared with the observational database (New et al. 1999), illustrating the skill to represent regional climate characteristics. For one GCM, NEB is hyper-arid, for three GCMs climate does not show a marked seasonality (the driest months are too wet), one GCM gives a reasonable representation of regional climate but overestimates global precipitation by 40%

The two models involved that still reasonably allow a regional interpretation of their results for northeastern Brazil are ECHAM4 (Roeckner et al. 1996) and HADCM2 (Johns et al. 1997). Following the recommendations of IPCC-TGCIA (1999), we selected results from these models for our analyses. For an annual increase of greenhouse gases by 1% per year as of 1990, projections of precipitation changes over NEB (2070–2099 compared to 1961–1990) are –50% for ECHAM and +21% for HADCM.

Given the very small number of models meeting the minimal criteria adopted above for a direct regional interpretation of its climate change results, conclusions on likely precipitation changes in NEB, on median changes or probable ranges of precipitation change cannot be drawn. Still, in climate change studies for semiarid NEB, both the possibilities of a strong decrease in precipitation (–50%) and an appreciable increase in precipitation (+21%) should be considered as plausible to take place in the current century.

Assessment of climate change impacts on, e.g., surface hydrology and agricultural production for the states of Ceará and Piauí requires, at the coarsest, a resolution of climate data at the scale of subregions in Ceará and Piauí with marked differences in hydrometeorological or agrometeorological conditions, i.e., the scale of 10–100 km. This seriously hampers the direct (grid-cell based) regional interpretation of GCM-simulated climate change, whose resolution is much coarser. Indirect methods, using local area models (LAMs) of climate or statistical downscaling of large-scale features to derive regional climate may overcome this problem. The latter method was applied in the WAVES project for the generation of regional climate scenarios (Gerstengarbe and Werner, this Vol.), as no LAMs are presently operational with sufficient skill for climate studies in the study region (Böhm et al., this Vol.). Future development of regional LAMs or GCMs with increased resolution could yield improved regional simulations of the climate of semiarid NEB, as is hinted at by the fact that the two models with reasonable reproduction of regional and global climate exhibit the highest resolutions in the model set considered.

The downscaling method adopted combines observed daily historic climate data at the level of climate stations with long-term climate trends from GCM projections. Here the tendency in annual precipitation at the large scale was taken as the regionally most relevant trend. Simultaneously observed daily data on precipitation and temperature were used to interpret these tendencies into projections of these variables at the station level. Other meteorological variables like relative humidity and short-wave radiation were added using regression relations derived from the few available daily time series of a more complete set of meteorological variables. This resulted in climate scenarios at the level of the climate stations. Interpolation routines were used to transform these scenarios into a climate scenario defined at the level of municipalities. This additional step was necessary, as the municipality was taken as the common spatial unit used in the integrated simulations of hydrological, agricultural and socio-economic processes.

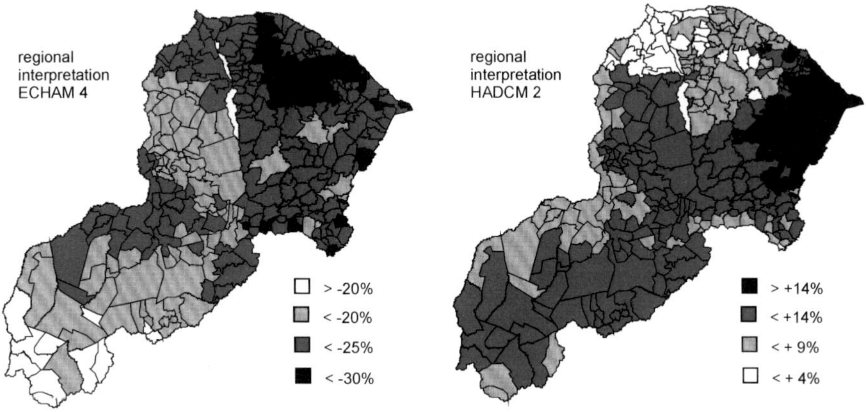

Fig. 3. Regional interpretations of GCM results for precipitation trends over the period 2001 to 2050, assuming continued growth of atmospheric greenhouse gas concentrations

Results for the two selected GCMs (Fig. 3) show well-defined spatial patterns of precipitation trends, arising from station-specific correlations between local and large-scale precipitation amounts. The difference between the spatial patterns indicates that this correlation is different for anomalously dry years than for anomalously wet years. For a description of the methodology, see Gerstengarbe and Werner (this Vol.).

Climate Change Impacts for Northeastern Brazil

Assessing the effects of climate change on NEB, simulations for one fixed reference scenario of regional and global developments with the three different climate scenarios were compared. These combined scenarios are afterwards referred to as the ECHAM scenario, the HADCM scenario and the Constant scenario. The reference scenario „Globalisation and Cash Crops“ (Döll et al., this Vol.) was selected, being one of the central reference scenarios developed in WAVES. This scenario assumes a continuation of historic trends in demographic and economic development, an increasing of international markets with development focusing on the coastal region and the interior, where water resources are potentially available (especially the downstream river valleys and mountainous areas). Especially in these areas water supply is enhanced through additional dam construction. Agriculturally used area expands gradually; the irrigated area more than doubles.

The first impact of climate change in the causal chain of processes (Fig. 1) is the impact on the availability of water resources. Here precipitation changes have a direct effect on the water balance, affecting run-off generated river flow and surface water storage.

River runoff shows a strong reaction on the precipitation changes (Fig. 4), with a strong decrease in runoff using the ECHAM scenario, with smaller, statistically insignificant increases for the HADCM scenario. Until 2025, results are statistically similar to historic simulations.

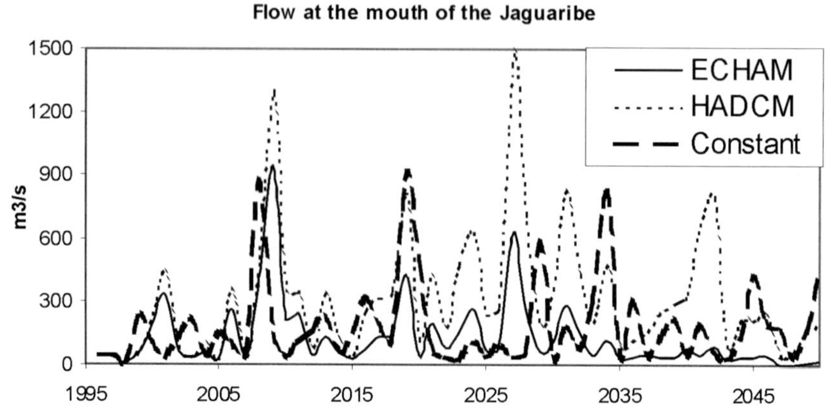

Fig. 4. Outflow of the River Jaguaribe, showing a marked reduction in runoff for the ECHAM scenario after 2025, and statistically insignificant tendencies for HADCM and the scenario with constant climate

Similar tendencies appear for the water storage in reservoirs. Dam construction is one of the main regional strategies to reduce water shortages in the dry period (July to November) and to carry water availability from wetter years to following drier years. In SIM, the larger reservoirs, with capacity over 50 Mm^3, are simulated explicitly. The total water volume stored in these reservoirs at the beginning of the dry season (July 1st) shows a strong increase between 1995 and 2015, the period where a marked increase in storage capacity occurs in the scenario (by 7000 Mm^3). Total storage capacity in Ceará and Piauí then reaches almost 22000 Mm^3, of which 16400 Mm^3 is installed in Ceará. Afterwards, in the HADCM scenario and the scenario with constant climate, the reservoirs show a variable degree of stored water, without a significant trend; for the ECHAM scenario, stored volume in Ceará shows a marked decline (Fig. 5).

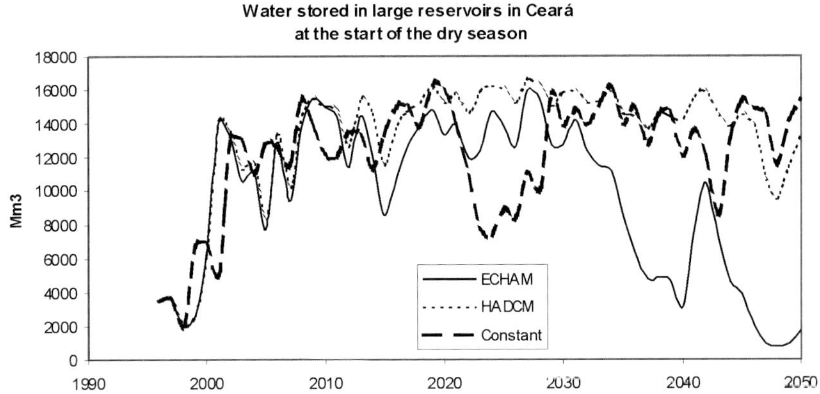

Fig. 5. Water stored in large reservoirs in Ceará (>50 Mm^3) at the beginning of the dry period for scenarios using climate interpretations based on ECHAM, HADCM simulations and on a constant climate

The general decline in water availability for the ECHAM scenario after 2025 should be seen in relation to the trends in water use. Due to population growth, increased connection to public water supply systems, intensification of industry and, above all, a strong increase in irrigated area, water demand grows strongly in the reference scenario until 2025; afterwards irrigation area is assumed constant while price increases of water drive increases in water efficiency, combining with increased demands into a stable or declining net extraction (Fig. 6a). The decline in water extraction for the ECHAM scenario is caused by unfulfilled demands due to insufficient water supply. The increasing demand cannot always be met in all three assumed climate scenarios. An index of water sufficiency measures which part of total irrigation water demand can be covered by water availability during the dry season (Fig. 6b). Here it shows that the increases in water storage capacity in the reference scenarios allow a 90% sufficiency until 2025, but afterwards sufficiency is only preserved in the climatically favorable scenarios.

Important impacts of climate trends on agricultural production are found both in rain-fed and irrigated production (Fig. 7). For the ECHAM scenario, declining rainfall amounts reduce rain-fed yields, especially after 2025, when the climate trend becomes apparent. After 2025 until the end of the simulation, mean maize yield in Ceará falls by 35%. For the HADCM and Constant scenario, no clear trends are found. For the irrigated production of, e.g., tomato, the increase in irrigated areas dominates all other effects until 2020. Afterwards, a declining tendency is simulated, due to the decreasing sufficiency of water supply to support the irrigation. For the ECHAM scenario, this tendency is strongest, with mean production losses of 50% after 2025.

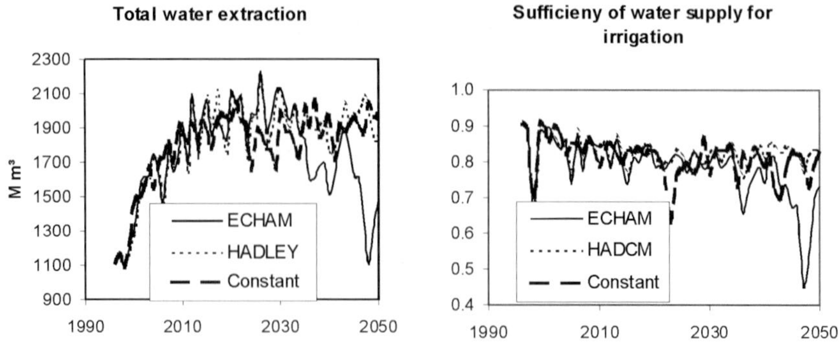

Fig. 6. a Total water extraction in Ceará for the three climate scenarios and **b** the relative sufficiency of water supply for irrigation water demand

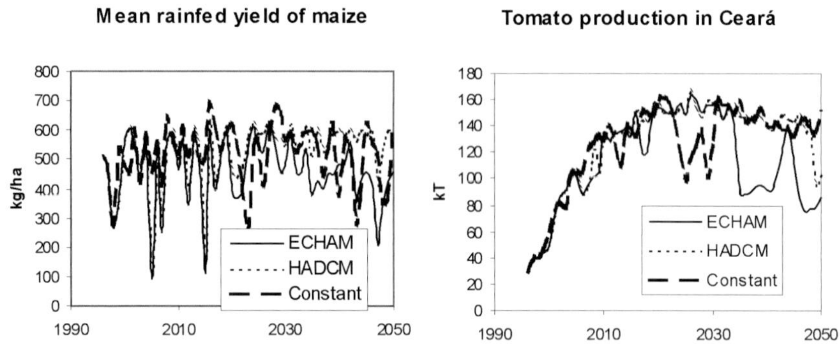

Fig. 7. a Mean rain-fed maize yield in Ceará and **b** total irrigated production of tomato for the three climate scenarios

Decreases in agricultural production lead to decreasing contributions of the agricultural sector to the economy. An interesting indicator for production losses is the number of municipalities where the area-weighted loss in yield, relative to local potentials, exceeds 60%. Under this condition, the municipality is considered to be severely affected by drought. For the ECHAM scenario, this number of municipalities shows an increasing tendency after 2025; no clear trends for HADCM and the Constant scenario (Fig. 8).

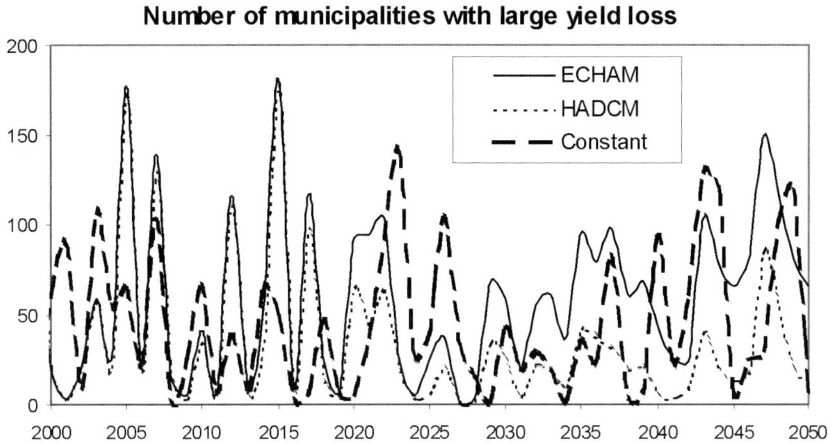

Fig. 8. Number of municipalities in Ceará with large yield losses (of 184 municipalities) for the three climate scenarios

Production losses lead to significant losses in municipal income in the rural areas, where the importance of the agricultural sector is assumed to remain large. This implies an increased gradient in income between the rural areas and the metropolitan areas in drought years, thus driving simulated migration. In this way, enhanced migration in drought years keeps on appearing, even when population development levels off.

Discussion and Conclusions

Global climate change will have an effect on the climate of northeastern Brazil. The direction of precipitation changes, however, cannot be determined with certainty. Both very significant precipitation losses and moderate precipitation increases should be considered plausible.

Under current climatic conditions, surface water availability and crop production exhibit a major variability, showing the regional vulnerability. The impacts of precipitation losses, as projected by one of the climate models with best regional performance (ECHAM scenario), would be of significant importance for the region, even enhancing the vulnerability.

For the states of Ceará and Piauí, large-scale reductions in the availability of flowing and stored surface water leads to an increasing imbalance between water demand and water supply after 2025, under the assumptions of the reference scenario, where future water demands grow until 2025, but then stabilize.

Agricultural production would also show negative tendencies after 2025, both by reductions in rain-fed yields and by reductions in irrigated production due to insufficiency of water supply to meet irrigation water demands.

For the climate scenarios with a constant or moderately increasing precipitation, no apparent tendencies in impacts are found.

In the climate scenarios, trends in precipitation are entangled with the natural variability, leaving long periods for tendencies to become statistically significant. This applies to tendencies in impacts as well. Even for the climate scenario with the most marked trends, significance of impacts is found after 2025 only. Before this, impact levels do not exceed the levels of impacts emerging from natural variability.

This should not discourage consideration of possible climate change when preparing policies to increase drought resilience. Measures to increase resilience will largely rely on long-term policies. In discussions with responsible regional agencies, focus was on measures in water infrastructure and its management, water use efficiency improvements, and structural changes in the agricultural sector. All these items refer to long-term changes, where possible climate change could have a significant influence. The present analysis suggests that the efficiency of various measures under different future climatic conditions (which can be considered as the robustness of the measure) might be a more relevant criterion in selecting measures than optimizing the measure for present climatic conditions.

Integrated modeling proved an important instrument in evaluating climate impacts. Feedbacks between trends in agriculture, water use, and insufficiency of surface water supply have a relevant influence on model results, especially for the scenario with diminishing precipitation volumes. Such feedbacks would not be addressed by single thematic studies or direct sequential couplings of models.

Many uncertainties remain, not only in the possible future climate developments, but also in the regional responses to water shortage and trends in water use; descriptions of specific water use, water management, tendencies in (irrigated) agriculture and societal processes are based on scarce data, and other relevant themes are still lacking representation in the scenarios and models, e.g., planned adaptation strategies as the connection of large catchments to reduce impacts of sub-regional droughts. These uncertainties were partly studied for the isolated contributing models, see various contributions in Araújo et al. (this Vol.), but an ample study of uncertainties in the integrated model is lacking. Nevertheless, model results are well interpretable, and the strong impacts of climate change in the ECHAM scenario are considered to be a stable result, with large regional relevance.

Acknowledgements

The research presented here would have been impossible without the contributions of data, models and discussions from partners in the WAVES project, as amply reported upon in Araújo et al. (this Vol.). This research was supported by the

German Bundesministerium für Bildung und Forschung (BMBF) under grant 01LK9713/4, as part of the WAVES program, funded by BMBF and the Brazilian Conselho Nacional de Desenvolvimento Científico e Tecnológico (CNPq). The analyses were partly made possible by the generous supply of data by many regional institutions, especially COGERH, FUNCEME, and CPRM.

References

Alcamo J, Kreileman GJJ, Krol MS, Zuidema G (1994) Modeling the global society-biosphere-climate system: Part 1: Model description and testing. Water Air and Soil Pollution 76:1-35

Bronstert A, Jaeger A, Güntner A, Hauschild M, Döll P, Krol M (2000) Integrated modelling of water availability and water use in the semi-arid northeast of Brazil. Phys Chem Earth B Hydr Oc Atm 3(25):227—232

Cohen SJ (1997) MacKenzie Basin Impact Study (MBIS), Final Report, Atmospheric Environment Service, Environment Canada, Toronto.

Edmonds JA, Wise M, MacCracken C (1994) Advanced energy technologies and climate change: an analysis using the global change assessment model (GCAM). PNL-9798, UC-402. Richland, Washington: Pacific Northwest Laboratory.

FAO (1979) Yield response to water. FAO irrigation and drainage paper 33, Rome

FAO (1983) Guidelines: land evaluation for rainfed agriculture. FAO soils bulletin 52, Rome

Hauschild M, Döll P (2000) Water use in semi-arid Northeastern Brazil – modelling and scenario analysis. Kassel world water series 2, Univ Kassel

Hulme M, Sheard N (1999) Cenários de alteraçoes climáticas para o Brasil. Climate Research Unit, Norwich, UK

IPCC (1999) GCM climate change scenarios and observed climate datasets, http://ipcc-ddc.cru.uea.ac.uk

IPCC-TGCIA (1999) Guidelines on the use of scenario data for climate impact and adaptation assessment, Version 1. Prepared by TR Carter, M Hulme and M Lal: Intergovernmental panel on climate change. Task Group on Scenarios for Climate Impact Assessment, 69 pp

Johns TC, Carnell RE, Crossley JF, Gregory JM, Mitchell JFB, Senior CA, Tett SFB, Wood RA (1997) The Second Hadley Centre coupled ocean-atmosphere GCM: model description, spinup and validitation. Climate Dynam 13:103–134

Krol MS, Jaeger A, Bronstert A, Krywkow J (2001) The Semi-arid Integrated Model (SIM), a regional integrated model assessing water availability, vulnerability of ecosystems and society in NE Brazil. Phys Chem Earth (B) 26(7-8):529–533

Magalhães AR, Filho HC, Garagorry FL, Gasques JG, Molion LCB, da Neto MSA, Nobre CA, Porto ER, Rebouças OE (1988) The effect of climatic variations on agriculture in Northeast Brazil. In: Parry ML, Carter TR, Konijn NT (eds) The impact of climatic variations on agriculture, vol 2. Assessments in Semi-Arid Regions. Kluwer Academic Publ, Dortrecht, pp 273–380

New MG, Hulme M, Jones PD (1999) Representing 20th century space-time climate variability. I. Development of a 1961–1990 mean monthly terrestrial climatology. J Climate 12:829–856

Roeckner E, Arpe K, Bengtsson L, Christoph M, Claussen M, Dümenil L, Esch M, Gioretta M, Schlese U, Schulzweida U (1996) The atmospheric general circulation model ECHAM-4: model description and simulation of present-day climate. Max-Planck Institute for Meteorology, report no 218, Hamburg, Germany, 90 pp

Rosenberg NJ, Crosson PR, Frederick KD, Easterling WE, McKenny MS, Bowes MD, Sedjo RA, Dramstadter J, Katz LA, Lemon KM (1993) The MINK methodology: background and baseline. In: Towards an integrated impact assessment of climate change: the MINK study. Reprinted from Climatic Change 24:1-2, Rosenberg NJ (ed), Dordrecht, The Netherlands, Kluwer Academic Publishers.

Rotmans J (1990) IMAGE: an Integrated Model to Assess the Greenhouse Effect. Kluwer Academic Publishers, Dordrecht, The Netherlands.

Rotmans J, de Vries HJM (1997) Perspectives on Global Change: The TARGETS approach. Cambridge University Press, Cambridge, UK.

Toth FL, Bruckner T, Füssel H-M, Leimbach M, Petschel-Held, G, Schellnhuber HJ (1997) The tolerable windows approach to integrated assessments. Proceedings of the IPCC Asia-Pacific Workshop on Integrated Assessment Models, Tokyo, Japan, 10-12th March 1997.

The GIS-Based Model for Sustainable Development of Land Use (MOSDEL)

Andreas Printz and Ruth Lang

Munich Polytechnic University (TUM), Institute for Landscape Ecology, Am Hochanger 6, 85350 Freising, Germany, printz@wzw.tum.de

Introduction

The WAVES working group „Landscape Ecology" developed the MOdel for Sustainable DEvelopment of Land use (MOSDEL) as a Geographic Information System (GIS) tool to facilitate the integration of the conclusions from the WAVES research into regional land use planning. Scaling techniques were used in the model to extrapolate the largely point-based results of the different WAVES working groups to the entire region. Thus, the model shows the causal link between site potential, agricultural productivity, economic benefits and ecological impacts. Assessments concerning the degree of local self-sufficiency and natural resources management can be more fairly represented. Three classes – economic, hydrological and ecological sustainability and/or risk – can be queried and allocated as results of a model run. The following article gives a short overview of the model's functioning and potential applications.

Methods and Area Descriptions

Importance of Spatial Integration and Regional Scale

MOSDEL, a GIS-based simulation model, was developed as one of the three WAVES integration levels (Fig. 1, spatial level marked with oval). At a regional scale (1:100,000), the creation of a valid database in spite of low data availability was possible only through the combined efforts of the various WAVES working groups (Landscape Ecology, Agroecosystems, Socio-economic Analysis, etc.) at the local level. The regional scale offers the option for up and down extrapolation and enables a better understanding of important man-environment system features that can be represented on maps. Therefore, adapted simulation tools of this nature can meet the needs of decision makers interested in integrated and sustainable development.

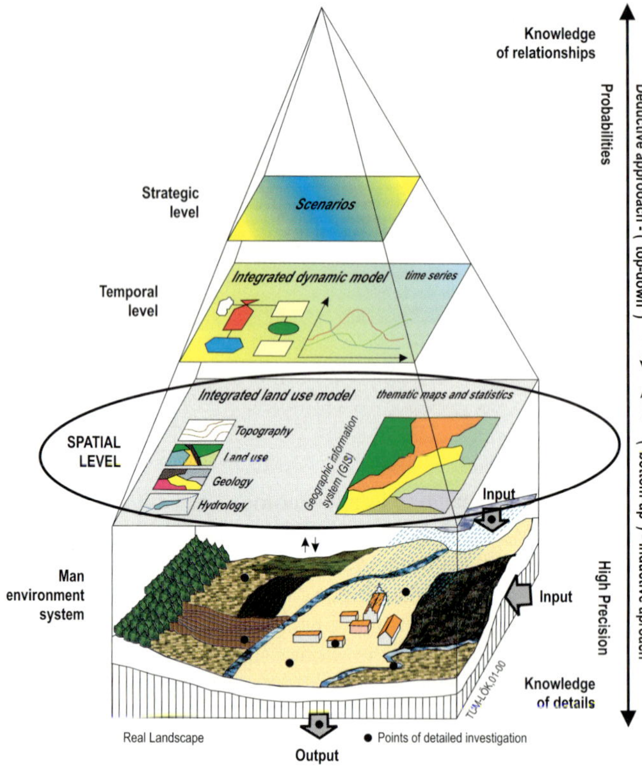

Fig. 1. Integration levels of the WAVES program (according to Haber 1993, modified)

WAVES Reference Areas

MOSDEL was first successfully tested as a prototype version during a 3-year period in the Picos region (approx. 3000 km^2) in the State of Piauí. This region has proper conditions for high agricultural production potential (medium precipitation, large alluvial areas, good water availability, market accessibility). In order to address the goals of other WAVES working groups, the focus of the prototype model was on productivity of food and cash crops.In the last research year, the simulation model faced several challenges:

- Transfer of the model to other regions and circumstances;
- Implementation of animal husbandry;
- Evaluation and simulation of degradation caused by grazing;
- Implementation and application of the model.

The model was applied also to the second reference region, the municipality of Tauá – Ceará (4000 km^2), and was supported extensively by the Brazilian Landscape Ecology partner, which was already involved in an overall ecological survey of the region (Oliveira et al, this Vol.). The following discussion and results are based on the final model version for Tauá.

Input Data

For the integration of single-point results into the model, a homogenous spatial database was created and a grid cell size of 100×100 m (1 ha) was chosen as a common spatial reference. This size was small enough to extrapolate the point results, and large enough to correspond with data from the best topographic map (1:100,000 scale with contour lines at 40 m or 50 m spacing) or remotely sensed data (original 30×30m pixel size). At this scale, acre-sized plots and landscape mosaic patches are represented with sufficient resolution. The highest temporal resolution of the model is 1 day (correlating with precipitation data). The time base for individual simulation runs is 1 year. The number of consecutive runs was technically unlimited.

Data Processing

Input parameters embraced two central themes: natural site conditions and land use management practices (Fig. 3). The data for each item are stored in the GIS database in different classes (e.g., physical, administrative and systemic data) as thematic layers. According to the selection menu, various algorithms were programmed to query and link these layers through defined rules. The data are calculated for each grid cell (Fig. 2). As mentioned above, the model was designed to create three sustainability or risk qualities. These results are obtained from separate modules that will be described in the following.

Modeling Economic Sustainability/Risks

Due to limited project resources, the study had to be concentrated exclusively on the dominant economic activity in rural areas: agriculture. The economic success of different farm types was assessed relative to specific site conditions.

For the simulation of economic and ecological consequences of different land use management practices, it was necessary to classify farms according to their environment and their major agricultural activities. The correlation of a model cell with a particular farm type was a major step towards the integration of socioeconomic and ecological conditions at site level. Exclusively dependent on local resources, the agricultural production of the region is characterized by a high degree of local, traditional practices, adapted to the particular conditions (e.g., crop selection, use of irrigation, etc.) of each site. Addressing the most influential geographic features in Tauá – topography and soils – the municipality was subdivided into four landscape units: mountain region, alluvial areas, Pediplain east and Pediplain west (Fig. 4 in this chapter and Fig. 5 in Oliveira et al., this Vol.). Each landscape unit was correlated with two specific farm types (subsistence farming and a more market-oriented type). Knowledge from local experts and the WAVES working groups „Socio-economic Analysis“ was used to create a framework of eight different farm characteristics (farm size, technology, crop selection, number of economically dependent persons, etc.). For example, only farms in the alluvial areas use irrigation technologies; the farms in the mountain region cultivate cotton, cassava and castor beans; in the eastern Pediplain, a strong focus is on animal husbandry; while in the western part more intensive agriculture is practiced due to better soil conditions, etc.

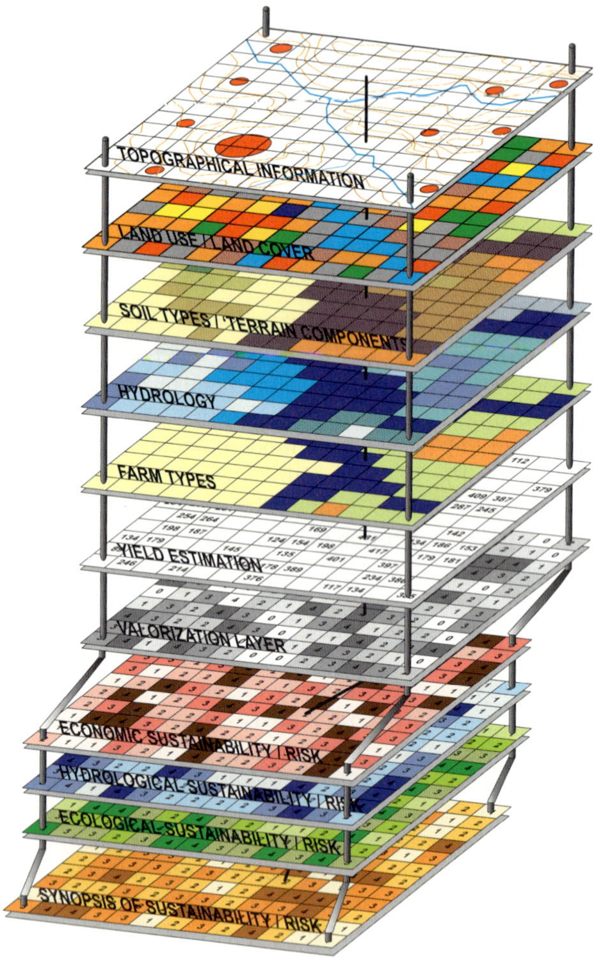

Fig. 2. Overlay of thematic layers

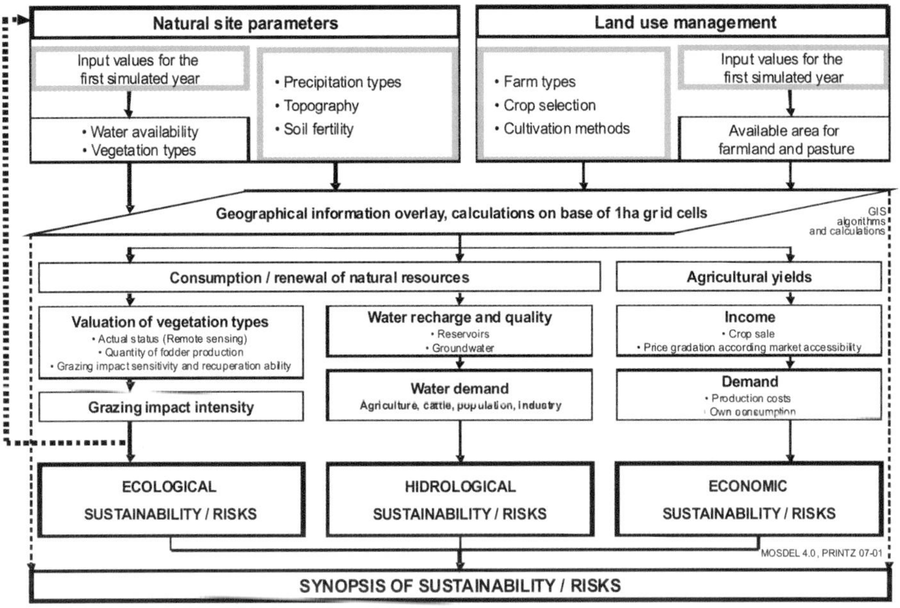

Fig. 3. MOSDEL - scheme of model's run

The identification of available arable area for agricultural production was based on the interpretation of satellite imagery (Landsat TM 5, 10.07.99 and 22.08.00; see text and Fig. 10 in Oliveira et al., this Vol.). Within the WAVES program, an extensive database (EPIC and SPICE, see Gaiser et al., this Vol.) was created, which provided, based on soil information („Terrain Components"), the yields for the major cultivated crops under different precipitation regimes and management systems. The distribution of crops within the cultivated areas was randomized. If the same crop was located in the same cell in consecutive model runs, their yields were calculated along a declining gradient in order to represent the rapid impoverishment process of tropical soils. However, if land was left fallow, soil fertility recuperation was taken into account.

Calculated yields in tons were translated into monetary values based on Tauá city market prices of the reference year 1996, provided by the working groups „Socio-economic Analysis". It was thus possible to compare specific yields of the different farm types.

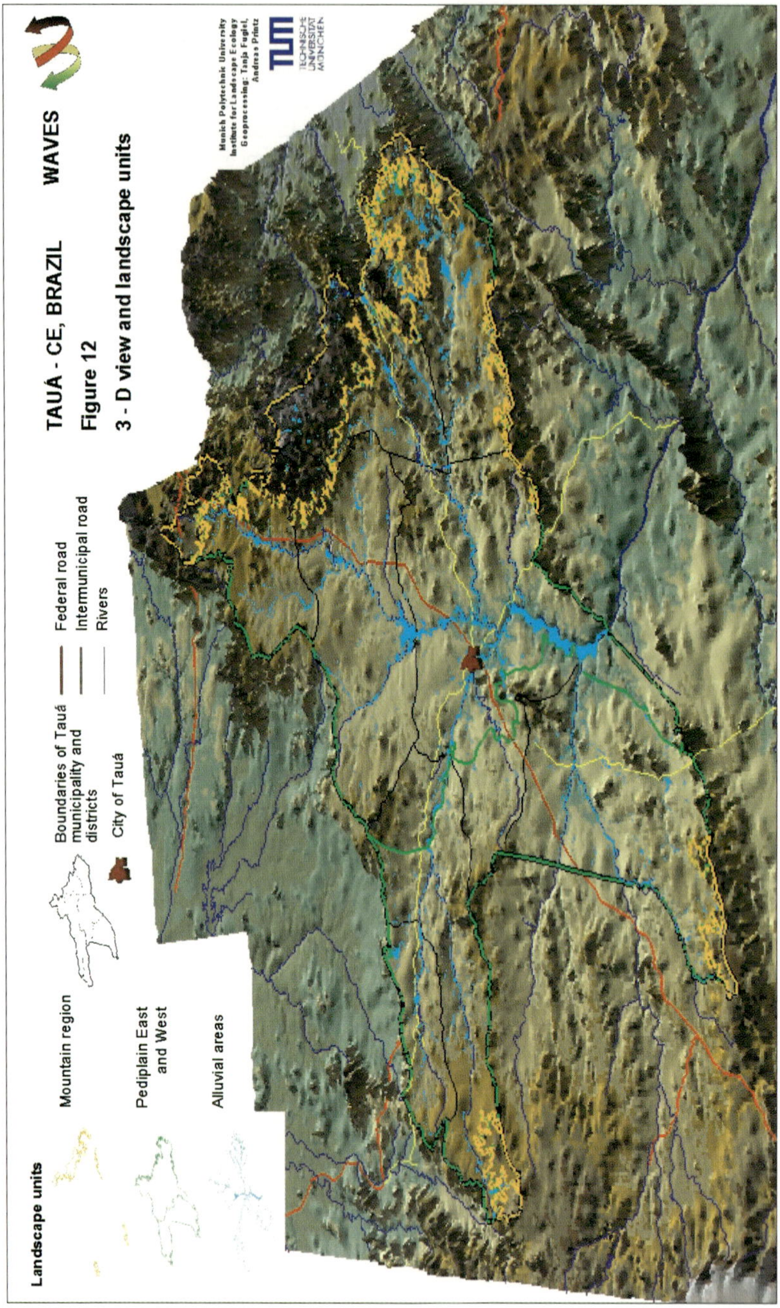

Fig. 4. 3-D view and landscape units

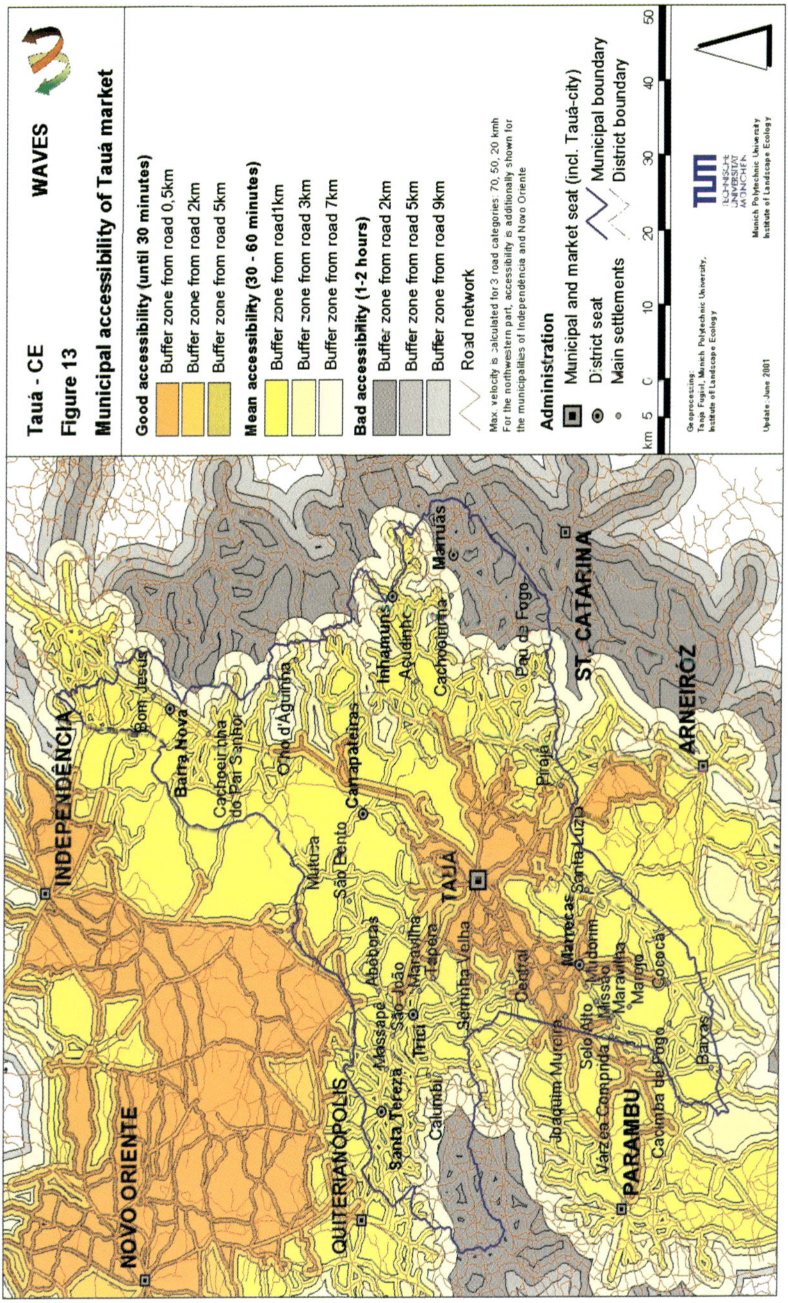

Fig. 5. Municipal accessibility of Tauá market

However, within the municipality of Tauá, a strong price gradient for agricultural goods exists, correlated with the distance or accessibility to the market (city of Tauá). In order to consider in the model the losses which farmers in remote rural areas incur paying for transportation if they wish to sell their products in town or to street traders, a road network analysis was performed using ARC-INFO. Three different road qualities were defined with respect to travel velocity and buffer areas. Depending on the travel time to the market center, three accessibility zones were delineated: <20 min, 20–60 min and >1 h. Losses of 0, 10 and 20% were considered for the respective zones (Fig. 5). As practical economic models to integrate these parameters into the procedure (especially for the subsistence farm types) were not available, production costs and auto-consumption were considered as a specific monetary value percentage for subsistence and market-oriented farms.

Finally, the results were divided by the number of economically dependent persons of each farm type. The model allows for the running of various precipitation types within the same socioeconomic configuration setting. Economic risk areas were defined as those clusters with the highest deviation of income per capita during a dry year in relation to the results of a medium year.

Modeling Hydrological Sustainability/Risks

Hydrological sustainability and risks are based on a calculated water balance for the research area. Three information layers are implemented in the water module: surface runoff, soil water dynamics and groundwater. Climate variations are represented by five precipitation types which constitute the initial input data used for calculations.

Surface runoff calculation is based on the SCS (US Soil Conservation Service) approach (USDA 1972). The „Terrain Components" were grouped into four hydrological soil classes with respect to their infiltration rates. Thirteen land cover classes (see remote sensing; Oliveira et al, this Vol.) were estimated for each of the hydrological soil classes (CN values). Empirical data for similar site conditions were adapted for completion (Smith and Williams 1980; Williams et al. 1984). Due to time constraints, it was not possible to take into consideration infiltration and seepage effects from surface flow through simulation cells. However, both inputs gave a good idea of the location of the major runoff areas within the research area. The part of precipitation not lost through runoff, interception or evapotranspiration was considered seepage. Based on daily precipitation data, plant available water in the soil was calculated by the SIMPEL model (Hörmann 1999a,b); the spatial distribution of the hydrological characteristics was derived from the „Terrain Components" and was provided by the SPICE database (Gaiser et al., this Vol.). It was thus possible to identify areas threatened by „drought", where soil water content is below necessary levels for plant growth. In addition, groundwater recharge estimation data were gathered from field measurements. These values were adapted according to the respective hydrogeological site conditions by the WAVES working group „Water Resources and Management", and areas with homogenous recharge rates were delineated (Fig. 6). To define available water, quantity and quality data from a well cadastre were used. Available water quantity was estimated calculating groundwater recharge rates for wells with a defined watershed (25 ha).

Due to high groundwater salinity in Tauá, it was important to separate three water quality classes and assign sustainable use to each of them (following standard criteria, unless reality in the „Sertão" demanded different priorities). To avoid risk of soil salinization, only fresh water (water with low mineralization) was considered suitable for irrigation. It should be mentioned that, in the Tauá region, only four wells produce fresh water (<500 μS cm^{-1}, Fig. 6). Thus, most of the fresh water in the model originates from reservoirs. Although it is known that several reservoirs have serious salinization problems, it was assumed that all reservoir water was fresh, as there were no reliable field data available. Water with medium mineralization (500–1500 μS cm^{-1}) was considered to still be tolerable for human consumption; salinized water (>1500 μS cm^{-1}), however, can be used only for cattle (>1500 μS cm^{-1}).

The supply/demand balance was calculated for each landscape unit and administrative district, and the areas with the clusters showing highest deficits were assigned as hydrological risk areas.

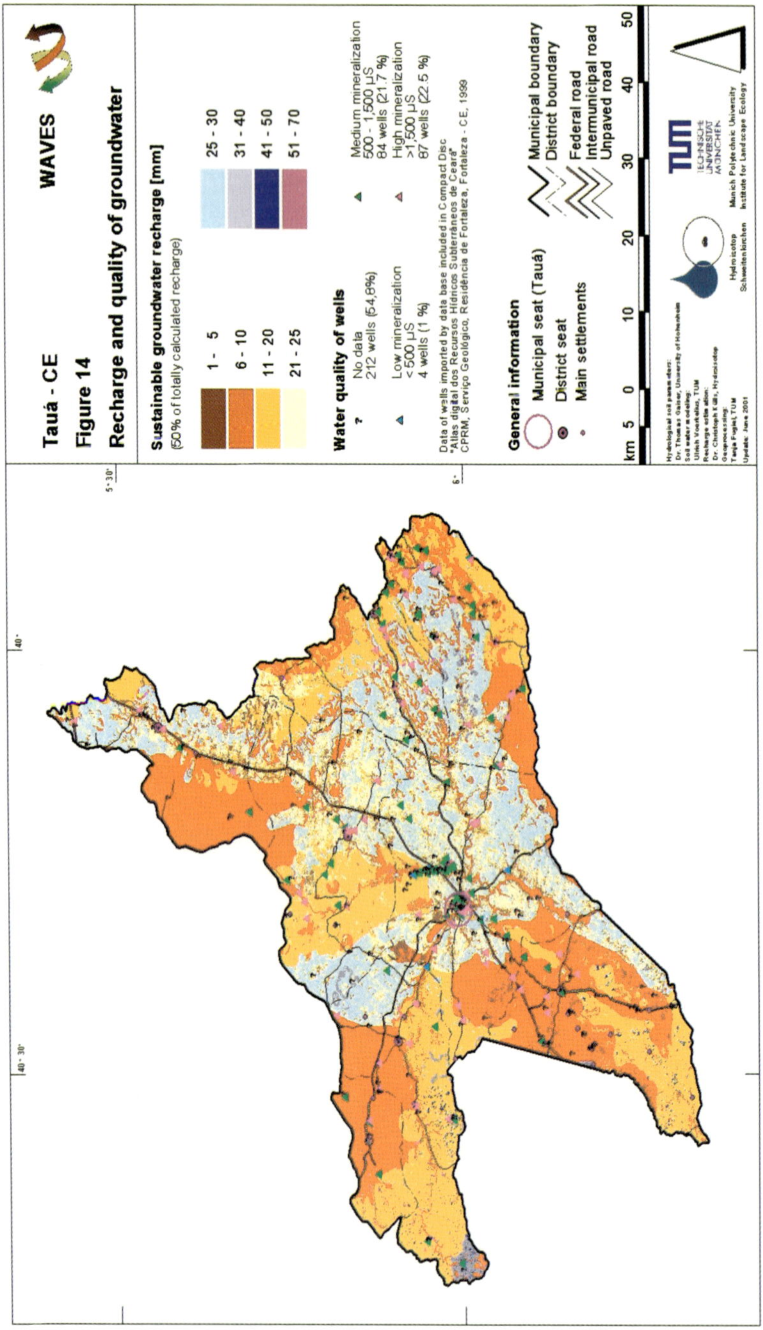

Fig. 6. Recharge and quality of groundwater

Modeling Ecological Sustainability/Risks

Due to low data availability, an assessment of ecological sustainability had to focus on northeastern Brazil's main ecological problem: land degradation by overgrazing. An erosion susceptibility assessment was performed (Oliveira et al, this Vol.), but was not considered in the model. The primary research focus here was on the modeling of the relationship between land degradation and grazing in the medium and long run. Up to the present time, the major fodder source for cattle in northeastern Brazil has been woodland grazing. Overgrazing effects, often occurring in non-regulated public lands, can be readily observed over a large portion of the semiarid Sertão. In Tauá, the situation is particularly accentuated (Oliveira et al., this Vol.). We found that many farmers, and especially the „vaqueiros" (cowboys), had surprisingly keen instinct for the causes and effects of land degradation. However, rarely is such intuitive knowledge translated into changes in land use practice. Even if a natural resources consciousness exists, it does not play a major role. Questions of ownership, short-term economic gains or simple subsistence are prioritized. Due to this fact, communication and valid conclusions for the model parameters concerning sustainable development were extremely difficult. Consequently, the modeling of the ecological risk had to be performed with only a few scientifically approved values that were found in the literature (de Araújo Filho and de Carvalho 1997) and through expert consultations.

The scale of the model did not allow for the consideration of nutritional values of natural vegetation. The focus was rather on the carrying capacity and regeneration of natural vegetation under different grazing scenarios. Firstly, the quantitative impact of animals (cattle, sheep and goats quantified as „animal units" [AU]) and their spatial distribution was defined. An approximate distribution of animals for each district of Tauá was given by the staff of the local state agriculture agency. To consider the fodder supply by crop residues, the modeling of agricultural output of the farm types included crop residues for the most important market crops (SPICE database). These values were then translated into nutritional values per animal unit. Another source of fodder are areas of planted pasture. Due to a lack of financial means of farmers to purchase fodder in the „Sertão" hinterland, these two fodder sources (crop residues and planted pasture) represent the sum of all available (endogenous) fodder. Calculated as „animal units", a certain cattle number per district could be satisfied from these fodder sources. The remaining unsatisfied „animal units" were considered as grazing impact in areas of the natural vegetation.

Through remote sensing techniques, thirteen (13) land cover classes and vegetation types were identified (reduced and aggregated to eight (8) for valorization). As woodland grazing in natural vegetation is common within the entire region, all areas were included in a randomized distribution of animals (differentiated according to their distribution throughout the districts). Model cells previously defined as agriculture, reservoir, alluvium and settlement areas were excluded. Eight classes of natural vegetation were ranked (values range from „0", for bare soil, up to

„100“, for vigorous dense Caatinga vegetation; Table 1). Furthermore, a recuperation or succession time for each vegetation type was defined (used in the model as parameter for the vegetation value, either for overgrazing or for recuperation, if more than 100% of sustainable carrying capacity was exceeded). Two zones were distinguished and simulated separately, as the vegetation „climax“ is different for mountains compared with lowland plains (Table 1). The mountain zone includes, besides the same vegetative types as the plain zone, an additional type, well adapted to the more humid conditions of the mountains.

Specific sustainable carrying capacity values for different vegetation types of the Caatinga with respect to grazing impacts do not exist; a general coefficient of 10 ha/AU is included in the discussion instead. With this in mind, a Gaussian equation was implemented, adapting each factor according to local information. Figure 7 shows the Gaussian equation curve and vegetation values that were used for the calculation of ecological sustainability. The carrying capacity peak for grazing in Caatinga vegetation is found for the most part in the vegetation units with a vegetation value of 55–75, where the canopy cover is dispersed enough to allow for greater diversity, and hence for a wider nutritional range at the ground level. In comparison, a mature, dense wood with a large amount of standing biomass presents greater difficulty of access for animals. As the parameters of the curve were not assured by empirical studies, the approach had to be considered as hypothetical for predicting future vegetation development.

Finally, using the carrying capacity values of the curve and balancing the demand of animal units in randomized vegetation model cells, the vegetation value of cells with deficits decreased and the value of cells with surplus increased according to Table 1.

Table 1. Vegetation values of different vegetation types in relation to their location in mountains (SUC_VALUE1) or lowland plains (SUC_VALUE2)

		Valorization	
Vegetation types	Class no.	SUC_ VALUE1	SUC_ VALUE2
Dense arborous Caatinga with „Mata Seca“	1	90-100	-
Dense arborous Caatinga	2	80-90	90-100
Arborous-shrub Caatinga	3	65-80	65-90
Dense shrub Caatinga	4	55-65	55-65
Open shrub Caatinga	5	32-55	30-55
Succession with dominant woody plants	9	30-32	28-30
Bare soils with single cactus, shrubs and herbs	10	25-30	22-28
Bare soils with rocks	11	0-25	0-22
Total recuperation years (from class no. 11 to 1 or 2)		61	51

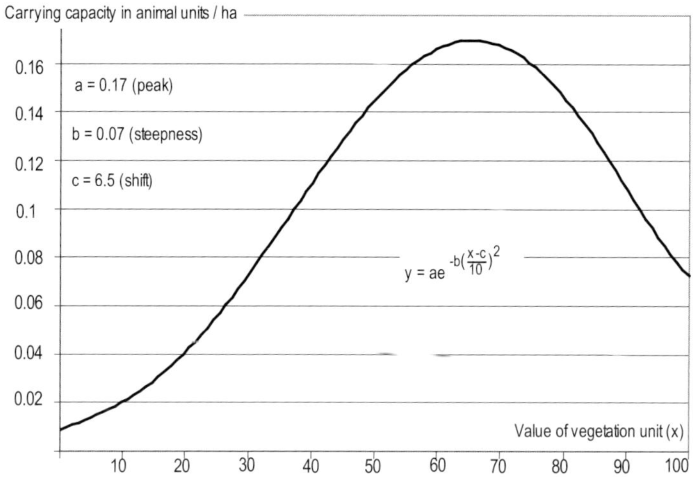

Fig. 7. Example of Gaussian equation curve for carrying capacity of grazing in natural vegetation

In order to exclude any short-term effect from a single model run, a 5-year run was based on sustainability estimations. A feedback loop considered the values of the respective previous year (dotted line on left in Fig. 3). Those areas, aggregated to districts and clustered by greatest value losses, were identified as actual risk areas.

Results and Discussion

The integrated efforts of the German and Brazilian working groups led to a unique GIS database. Approximately 20 thematic maps were created and assembled, including a digital elevation model, a fine-grained, disaggregated soil map (Gaiser et al., this Vol., Oliveira et al., this Vol.), and a land use/land cover classification based on remotely sensed data. The MOSDEL database also included many detailed assessments and research results that cannot be discussed in this article. Oliveira et al. (this Vol.) give some examples.

The data structure of MOSDEL was designed in such a way that sustainability outputs can be (dis)aggregated for each of the eight administrative districts of Tauá, the four landscape types and each of the eight farm types. Thus, a detailed pattern of different sustainability issues can be spatially represented.

Economic Sustainability and Risk Areas in Tauá

Figure 8 gives an example of the income (per capita and year) for market-oriented farm types in different landscape units and in years with contrasting precipitation types, grouped by the eight districts of the municipality of Tauá. The black column represents the reference year 1996 (total rainfall of 734 mm), the grey column a second consecutive run with the same precipitation data to evaluate yield decline through soil fertility loss. The white column is for a separate run for a „dry" year (total rainfall of 157 mm). In all of the districts, a similar ranking can be observed: Farms in the landscape unit Pediplain east were most successful, whereas the farms in the mountainous and alluvial areas were less favored in the three model runs.

However, any further analysis and potential action should carefully consider the complexity of the system and the relative influence that specific parameters may have on the results. In the more fertile Pediplain west, e.g., population density is much higher than in other units; thus, the monetary gains have to be divided by more people. In this case, higher yields/ha, paradoxically, are associated with lower per capita income. The distribution was always carried out equally without regarding social stigmas created through disparity. This was done for two reasons: no data or algorithms were available and discussion of this sensitive point in a socially polarized environment (Oliveira et al., this vol.) tends to heat political arguments. Another reason for disparity is market accessibility. The city of Tauá which represents the market place (0% loss through transportation costs) is situated in the Pediplain east, whereas mountainous areas are peripheral to Tauá and have to bear up to 20% of loss.

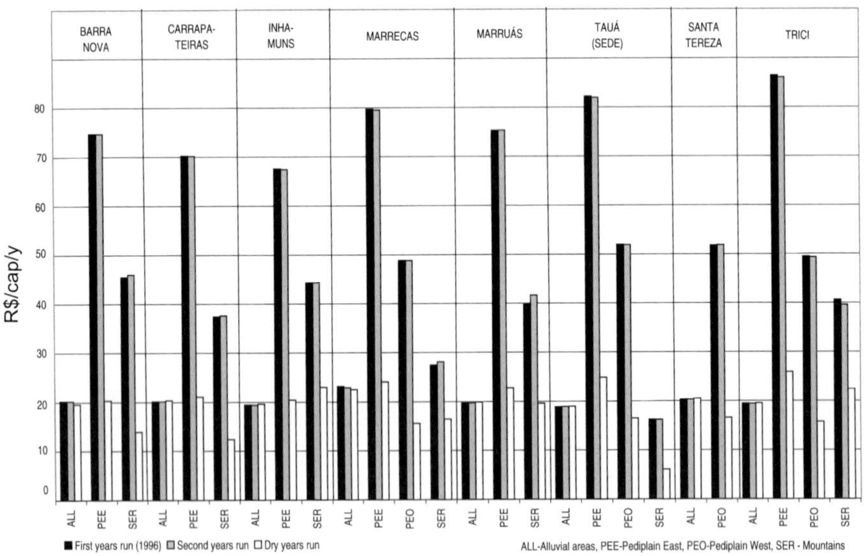

Fig. 8. Income per capita/year of market oriented farm types in districts and landscape units

Apart from these discrepancies, the low level of income is alarming, affecting all aspects of livelihood. The per capita income turned out to be far below the „minimum income“ established by the Brazilian government (approx. R$ 4.5/day/capita in the reference year 1996). The mountain areas of Tauá constitute an extreme case: average income is calculated to be less than R$ 20 per capita and year, and in a dry year it falls below R$ 10. However, it should also be noted that monetary gains through animal husbandry are not considered yet (even though in 1996 the economic value of this activity was approximately two-thirds of the total farm income in Tauá; IBGE 1998b). Further, case studies of the socioeconomic working group (Fuhr et al., this Vol.) showed that non-agricultural income and monetary transfer from other regions (e.g., through emigrated family members and pensions) often have a disproportionately large importance in rural areas. In both cases, no detailed data were available for implementation into the model.

A comparison of consecutive run series did not show significant differences, which indicates that there is enough area within a district's particular landscape unit to practice crop rotation and allow fallow fields to recuperate. In some cases a slight increase is observed, which can be attributed to random distribution (selection of better soils). A detailed investigation of a series of several runs can indicate areas where agricultural uses could be increased or decreased.

The main purpose of the economic sustainability/risk assessment is to study the effect of different precipitation regimes, especially drought, on income. During the drought year (Fig. 8, white columns), all land scape units suffer drastic yield declines, except for alluvial areas; here, water availability allows for the mitigation of drought effects through irrigation.

Zones with the highest declines were assigned to be those with the highest economic risk. The respective areas are marked with dark colors in Fig. 9.

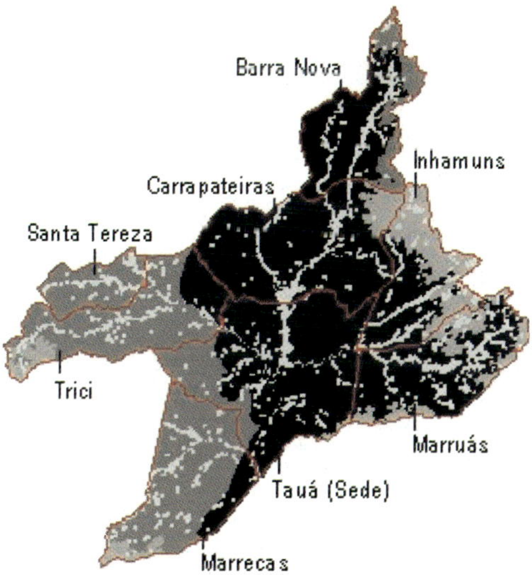

Fig. 9. Economic sustainability/risk areas

Hydrological Sustainability and Risk Areas in Tauá

Figure 10 shows the water balance (available water-demand) for the main user groups (in m^3/year) in the landscape units and districts of Tauá in the reference year 1996. The model was programmed in a way that surplus of water with higher quality could be transferred within the respective landscape units and thus satisfy demands of user groups with lower requirements of water quality. Thus, a total water balance is performed for each landscape unit (TOTAL, white column). Figure 10 shows that in 1996 no problem with respect to water availability for human consumption existed in any district; this is surprising and in contrast to traditionally severe water supply problems for the rural population. The results may be attributed to the fact that in 1996 the amount and distribution of rainfall were favorable. Furthermore, due to lack of data, ownership and distribution of water were not considered. Therefore, in years with regular rainfall, the supply of drinking water is merely a matter of distribution.

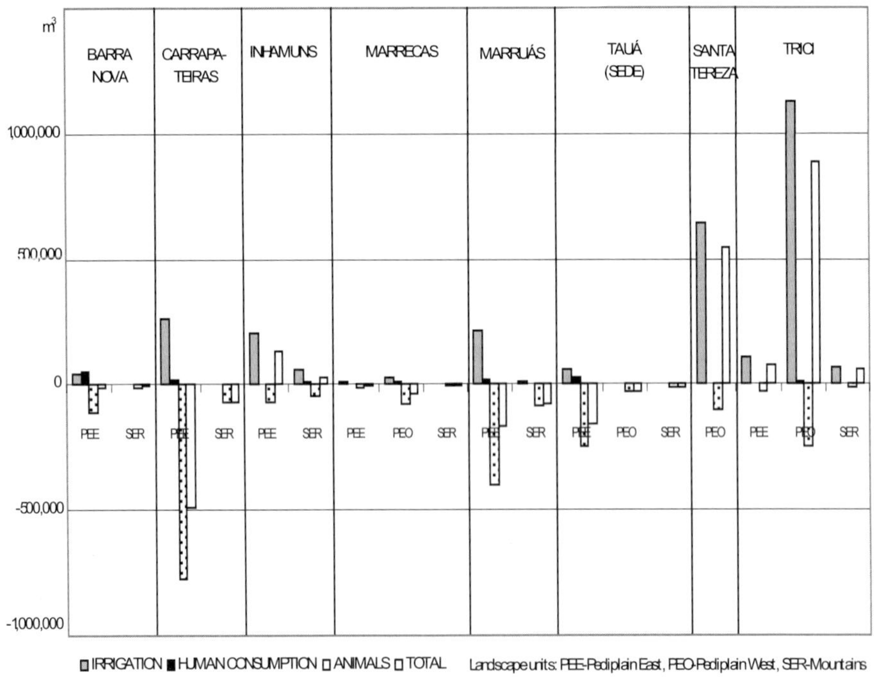

Fig. 10. Water balance for different user groups by landscape units and districts of Tauá (reference year 1996)

Quantitatively, human consumption is the smallest user group. In spite of the fact that 1996 was a year with regular rainfall, a negative water balance could be observed especially in the district of Carrapateiras, due to the water demands for animal husbandry.

Irrigation of small households was not considered at the model's scale. The user group irrigation in Fig. 10 always shows positive values, as irrigation was excluded from landscape units others than „alluvial areas“ (as it is in reality). For better graphical comprehension, the landscape unit „alluvial areas“ is not shown in Fig. 10, as values reach dimensions outside of the graph.

The water balance for irrigation in all alluvial areas was highly negative. Due to the high risk of soil and groundwater salinization, only 50% of the estimated recharge has been considered as a base for sustainable exploitation of alluvial groundwater. In addition, the recharge model does not fully account for indirect recharge to the alluvium by transmission losses. Hence, recharge rates may be underestimated. Nevertheless, even if the potential of alluvial groundwater bodies are underestimated and exploitation is even higher in some areas, low recharge rates of the hot and dry climate in Tauá will not allow for extended large-scale irrigation.

Water balance was calculated for all landscape units, including alluvial areas and irrigation demands. Zones with the highest deficits were assigned to be most at-risk. They were finally grouped into four classes (Fig. 11 – darker shade signifies higher risk).

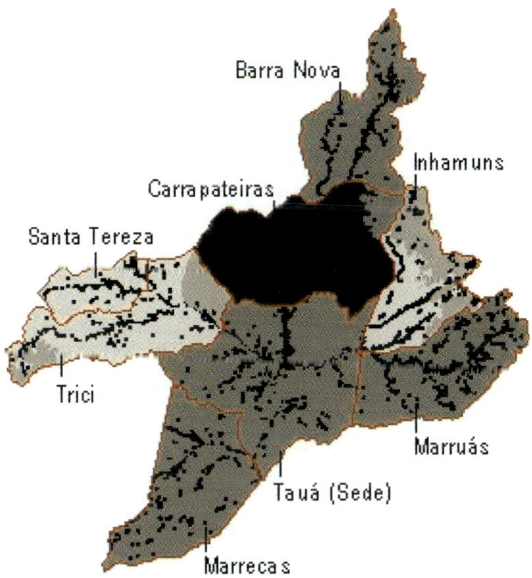

Fig. 11. Hydrological sustainability/risk areas

Ecological Sustainability and Risk Areas in Tauá

In order to correctly model grazing effects, only a time series of several years can show reliable trends. To shorten computing time, we decided to take a 5-year run series for comparing impact effects. As effects of climate change on vegetation in northeastern Brazil are still unknown or have not been modeled, no respective algorithm was implemented. Thus, the selection of precipitation types had only an indirect impact on the results of this part of the model through changing availability of crop residues. As described in the methods section, several parameters had to be implemented, and the systematic testing of the effects of changing values of uncertain variables was the focus. Some results of the sensitivity analysis are presented here. They are based on the Gaussian equation given in Fig. 7. Intensity of grazing in all runs was kept at the same level and correspond to the number of animals in the 1996 agrarian census (IBGE 1998a).

To quantify grazing impact, all grid cells with respective vegetation units were counted before (status quo, obtained from satellite image) and after running the model. Results are given in the form of numbers (ASCII format) and additionally the new status can be stored as a map (grid). Thus, effects can rapidly be analyzed both visually and statistically.

Figure 12 depicts results as sample histograms. Black columns represent the reference situation 1996, the white ones the status after a 5-year run. Column height shows the number of counted cells, corresponding to the area in hectares. Vegetation classes are ordered by decreasing vegetation values (Table 1). The example shows the total municipality of Tauá and two sample districts with accentuated degradation. The first two vegetation units represent the most vigorous state (dense arborous Caatinga: 1 and 2), while the last two correspond to more or less bare soil (10 and 11). A general shift of the vegetation towards the „middle" classes 4 and 5 and the „bare soil" class 11 is shown. In general, the loss of area with dense vegetation (class 1) is partially compensated by a slight increase of class 2.

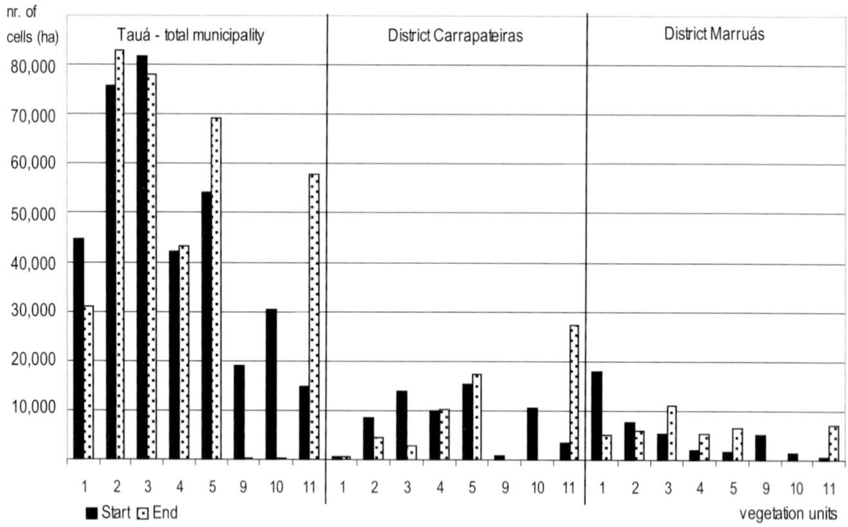

Fig. 12. Change of coverage (number of cells = ha) of vegetation units (Table 1) after 5 years of grazing impact

In the case of the two districts with highest degradation susceptibility, the two classes with vigorous vegetation lose both. Classes 9 and 10 almost disappeared due to the low vegetation value (corresponding to short recuperation or degradation resistance, Table 1). The most degraded class (11) increases remarkably, mainly in the two sample districts. It should also be mentioned that in other districts (e.g., Barra Nova and Marrecas) partial recuperation effects were observed.

The first two and last two classes (most mature and most degraded) were grouped (summarized) to identify sustainability/risk areas with respect to land degradation through grazing. The difference between the start and end status of the respective groups was calculated and the results were clustered into four sustainability classes. As data for the decisive input variable density of „animal units" only was available for districts, in this case results were not given per landscape units but per districts (Fig. 13).

The modeled vegetative balance strongly depended on the variables of the Gaussian curve. The shift of the peak of the curve had a decisive influence on the modeled redistribution of vegetation types. However, the general shift towards greater degradation was remarkable for all runs even with changing shape of the curve. This suggests that the number of grazing animals in 1996 was probably above the carrying capacity of the natural vegetation. If further degradation aggravation is to be prevented, current animal husbandry practices have to change. Unfortunately, because of the above-mentioned lack of data, no reliable thresholds for sustainable carrying capacity have been derived yet. An increased nutritional productivity by herb management and planting of fodder trees, accompanied by carefully limited grazing of the „cultivated" Caatinga, could be a sustainable solution for large areas of the semiarid Sertão. Once data are available for these management methods, their effectiveness could be tested through the model. Furthermore, the model could accompany future long-term research. Investigations to simulate vegetation status in the past could also be supported.

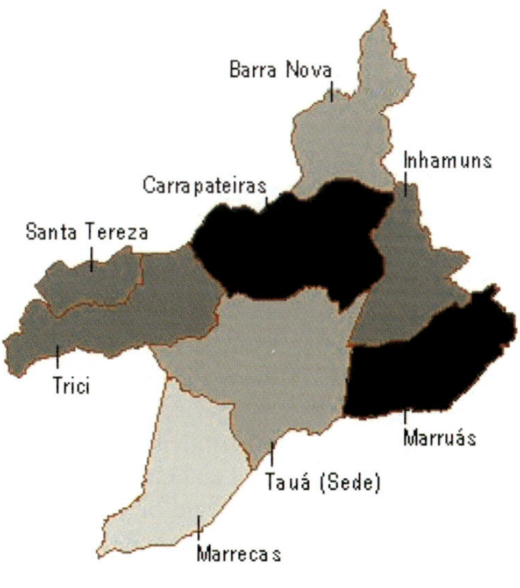

Fig. 13. Ecological sustainability/risk areas

Synopsis of Sustainability and Risk Areas in Tauá

Each cited sustainability/risk area was given a value (1 to 4), and a superposition for a synopsis of the sustainability analysis for the reference year 1996 was made. The result is represented in Fig. 14, where darker color indicates increased risk. Alluvial areas show mostly higher sustainability status in relation to their surroundings. High hydrological risks by irrigation are compensated by highly stabile economic production conditions. In contrast, accumulated risks occur mainly in the lowlands in the landscape unit Pediplain east. There, overgrazing is combined with soil with low productivity (stony Luvisols and Planosols with high sodium saturation) and precipitation values. In the districts of Carrapateiras and Marruás the situation is most aggravated. Whereas overgrazing is the main problem for Carrapateiras, the district of Marruás suffers (apart from other deficiencies) from poor market accessibility. The population in these zones of accumulated risk level are most threatened by droughts. If no intervention is started, the highest rates of emigration are expected in these areas.

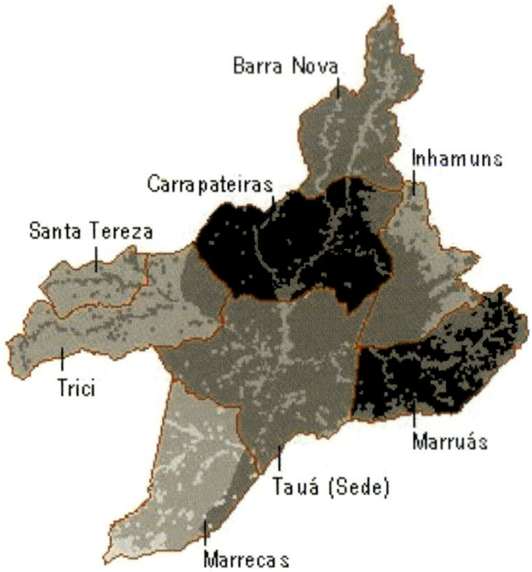

Fig. 14. Synopsis of sustainability/risk areas

Conclusions

It has been shown that the integrative approach of MOSDEL can lead to a new dimension of land use analysis and planning in northeastern Brazil. The great advantages of a GIS-based simulation model are high resolution and easily identifiable and understandable outputs. As the database is increased, numerous land use queries, optimized for both singular and multidimensional constraints, could be performed and the results presented in high resolution, despite the large area of a municipality like Tauá. Through MOSDEL, the ecological and economic consequences of different land use options for the entire region could be simulated at different levels. Current ecological conditions, future climate change, and effects of human intervention can be considered. Resulting sustainability and risk potential could then be easily identified and allocated. Due to its capability, MOSDEL could, potentially, become a major planning and decision-making tool in rural areas, by meeting the needs of local and state institutions to develop monitoring techniques and to assess the impacts of alternative development scenarios or intervention programs.

It should further be stressed that an improvement in the database (e.g., by systematically surveyed data on distribution of cattle) would lead to a higher precision and reliability of the model results. The facility of aggregation of different sustainability issues (especially for the synopsis) should not prevent a model user from having a closer look at the details. Intentionally, the authors of the model did not implement any preconceived decisions or priorities for any kind of sustainability or risk (weighting each sustainability level equally). MOSDEL, as an integrated planning tool, should simply illustrate the effects that specific interventions will produce on three sustainability levels. The decision of priority for sustainable development will ultimately remain with the planners and politicians.

The application of MOSDEL was supported through discussions that took place during a series of three workshops, under the auspices of SEPLAN (Secretaria de Planejamento do Estado do Ceará/Upper Planning Secretary of Ceará State), with representatives from important state ministries discussing the potential use of the model. It also served as a guide for a special task force program: „Coexistence with the drought" („Conviver com a seca"), in which the government expressed intentions of initiating a program of actions, especially geared towards those areas most affected by drought.

Although the model is designed for adaptability and ease of implementation, in order to achieve its goal, a certain degree of technical training is necessary. It will also be necessary to establish a long-term integrative cooperation process. A support consultancy, including official representatives and scientific staff, needs to be established. GIS and remote sensing departments should participate in this effort (especially since almost no reliable digital data are available at this scale in northeastern Brazil). For example, a small group of trained specialists successfully transferred the model within 1 year to a new municipality. In order to continue to extend the model to the 184 municipalities of Ceará, an efficient framework for data collection, data management and adaptation would be of great importance.

Two important side effects of MOSDEL and the WAVES program on administrative and scientific representatives should be mentioned: During the period of cooperation, we felt a growing consciousness for the needs of systematic data surveys and integrated approaches. It was realized that these results could only be enabled by an extensive and consistent database that was linked carefully. WAVES (and especially MOSDEL) was an incentive for many partners to experience new challenges beyond traditional sectoral thinking and communication.

During the modeling process, many research necessities emerged concerning the integration of different disciplines into the process. Here only one example of pertinent landscape ecology issues is given as a key to the future viability of the model: Long-term research on carrying capacity is urgently needed for further model implementation. Stress sensibility and the recuperative ability of native Caatinga and Cerrado vegetation demands further research efforts. The development of sustainable, semi-natural grazing management schemes as potential buffers in the semiarid environment is of great interest, not only for preserving its biodiversity, but because of its (socio-)economic value as well.

References

De Araújo Filho J[a], de Carvalho FC (1997) Desenvolvimento Sustentado da Caatinga. Embrapa, Circular técnico no 13; Sobral-CE

Haber W (1993) Menschliche Einflüsse auf Hochgebirgsökosysteme. Biologie in unserer Zeit. Volume 23.

Hörmann G (1999a) The SIMPEL soil water spreadsheets: defining the low end of hydrologic computing. Ecosystem Research Centre Kiel, Germany, http://www.pz-oekosys.uni-kiel.de

Hörmann G (1999b) SIMPEL – Speichermodelle zum Bodenwasserhaushalt; Ecosystem Research Center Kiel, Germany

IBGE - Fundação Instituto Brasileiro de Geografia e Estatística (1998a) „Censo Agropecuário 1995-1996, no 9. Ceará", Rio de Janeiro

IBGE - Fundação Instituto Brasileiro de Geografia e Estatística (1998b) Base de informações municipais (1996) [CD-ROM], Rio de Janeiro

Smith RE, Williams JR (1980) Simulation of the surface water hydrology. In: Knisel WG (ed) CREAMS: a field-scale model for chemicals, runoff, and erosion from agricultural management systems. US Department of Agriculture. Conserv Res Rep 26:13–35

USDA (1972) SCS National engineering handbook, Sect 4: Hydrology. USDA, Washington. Reprinted March 1985

Williams JR, Jones CA, Dyke PT (1984) EPIC-erosion/productivity impact calculator. 1. Model documentation. USDA ARS. Technical bulletin number 1768

Part 2: Natural and Socio-Economic Conditions in the States of Piauí and Ceará

The Climate of Piauí and Ceará

Peter C. Werner and Friedrich-Wilhelm Gerstengarbe

Potsdam Institute for Climate Impact Research, P.O. Box 60 12 03, 14412 Potsdam, Germany, pcwerner@pik-potsdam.de

The northeast of Brazil is a semiarid region for which precipitation is the most important climatological variable, so that even slight changes in this variable can lead to climate impacts that considerably influence water availability and thus vegetation, water management, agriculture, industry and finally the society as a whole.

The present situation of the region can climatologically be described as follows:

The area under investigation (Piauí and Ceará) is defined by the trade wind region with the summery maritime weather of the trade wind limit. This climate is most importantly characterized by a long winter period with little precipitation. Due to the adjacent boundary climates, the mountain and coastal regions show modified climate characteristics. The annual cycle of rainy and drought periods is controlled by large-scale circulation. This cycle is schematically represented in Fig. 1.

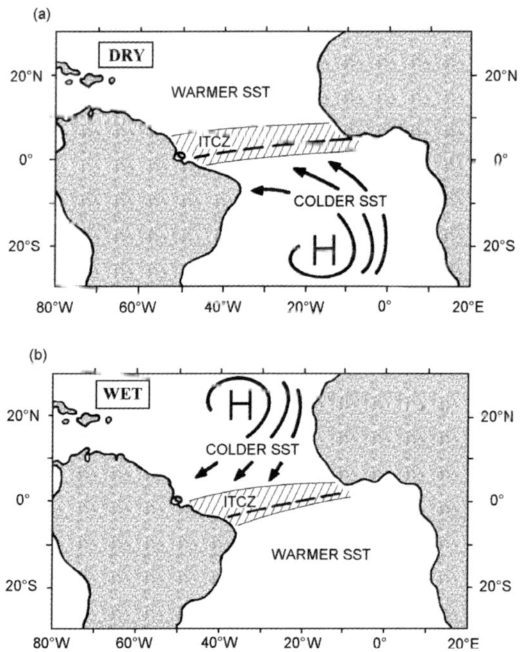

Fig. 1. Scheme of the large-scale circulation patterns over the tropical Atlantic

It is clear that the two seasons are determined by the position of the intertropical convergence zone (ITCZ) that is also determined by the position of the North- and/or South-Atlantic high-pressure area. The position of the high-pressure areas depends on the present sea surface temperature (SST). If, for example, the ITCZ does not shift far enough to the south in one year due to these conditions, this can lead to a weakening of the rain period that can cause a drought (see, e.g., Hastenrath 1991).

Another phenomenon that sustainably influences the climatic conditions in this region are the El Niño events which up to now in most cases have not only led to increased drought periods, but, in some cases, also to an increase in the rainy period (Parry et al. 1988).

The spatial structure of the precipitation distribution is shown in Fig. 2 and has the following important characteristics:

- Precipitation is distinctly decreasing from the coast to the interior of the country.
- Precipitation is higher in the west of the region than in the east.
- In mountain regions and especially at their weather sides, precipitation is usually higher than in other regions. The maximum values are between 1900 and 2000 mm (Serra da Ibiapaba).
- A large area with little rain (500–600 mm) extends from the south of Serra do Machado to the northwest edge of Chapada do Araripe (focus area Tauá).

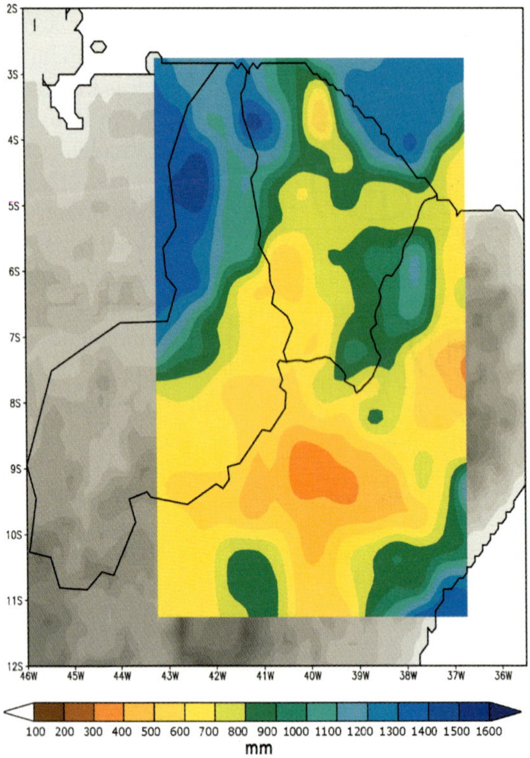

Fig. 2. Spatial distribution of precipitation: mean values for the period 1921–1980

The mean temporal distribution for a region that is characterized by a rainy and dry period is shown in Fig. 3. During the rainy period (December–May), the structure of the spatial distribution of precipitation in the individual months corresponds with that of the annual distribution of precipitation. During the dry period, these structures almost completely disappear so that in September, even in mountainous regions, only precipitation sums from 10–20 mm can be measured. Another characteristic of the temporal development of precipitation distribution is that one can observe a distinctly earlier set-in of precipitation after the drought period in the central and western parts of Piauí compared with the other regions. This occurs especially in the months of October and November for which, in the affected area, monthly sums up to 80 mm and/or 180 mm can be measured in contrast to the coastal region, for example, with values around 10 mm.

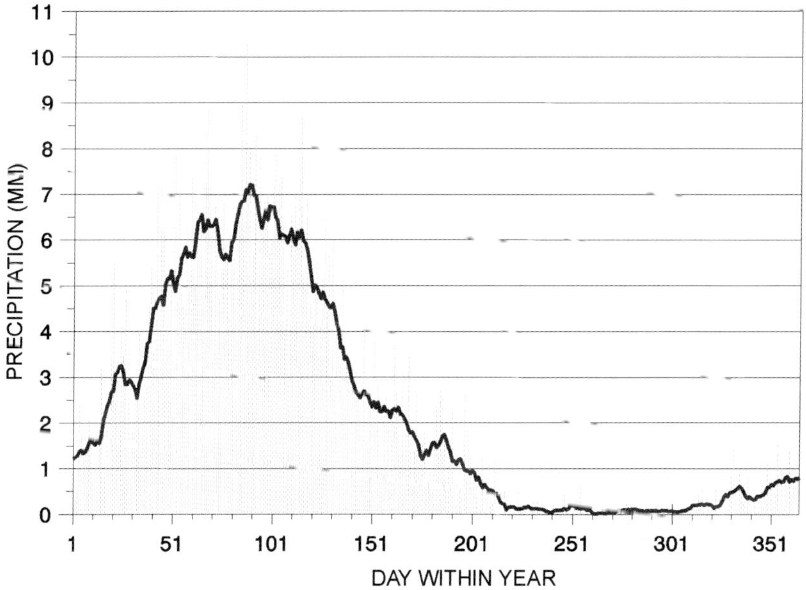

Fig. 3. Mean annual course of precipitation for the station Cedro, period 1921–1980

Taking temperature as an example, it can be demonstrated that the variation width of the other meteorological parameters available is much smaller in space and time than the variation width of precipitation. One can see from the spatial distribution of the mean temperature conditions (Fig. 4) that within Piauí and Ceará, including the mountainous regions, maximum variation widths of 7 K occur.

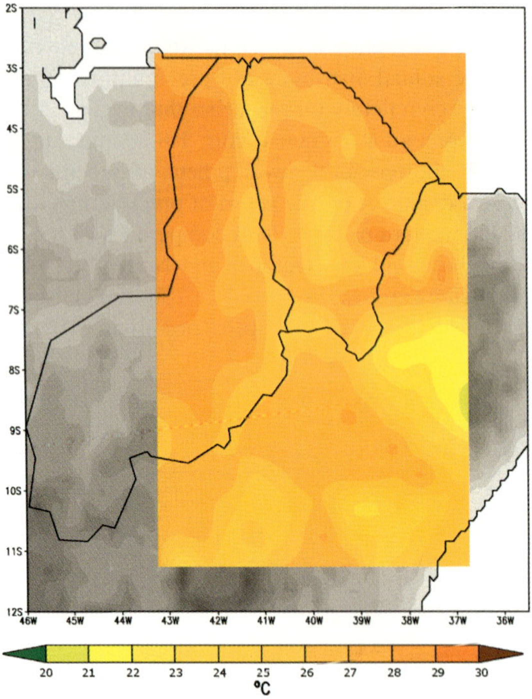

Fig. 4. Spatial distribution of temperature: mean values for the period 1921–1980

The same is almost true for the annual course of temperature (see Fig. 5), which does not exceed an amplitude of less than 3 K on average. Moreover, it is remarkable that the interdiurnal changeability is much larger in the rainy period than in the dry period. The reference areas that were determined by the disciplines at the beginning of the project are in one line from Gilbués in the south of Piauí over Projeto Piloto and Picos to Tauá in the southwest of Ceará. Along an imaginary line, the annual sums of precipitation are decreasing from approx. 900 mm (Gilbués) to approx. 600 mm in the area of Tauá. Moreover, the beginning/end of the rainy and drought periods are shifting. The rainy period starts first in Gilbués and last in Tauá. The length of the drought period is increasing in the same direction. Although the station of Tauá has the lowest annual sum of precipitation, it has the highest monthly sum (March); this means that, from the climatological point of view, an unfavorable hydrological situation can be assumed.

The area under investigation shows a very high mean natural variability of precipitation. As can be seen in Fig. 6, the area mean of precipitation for this area varies between 400 (1915) and 1500 mm (1985).

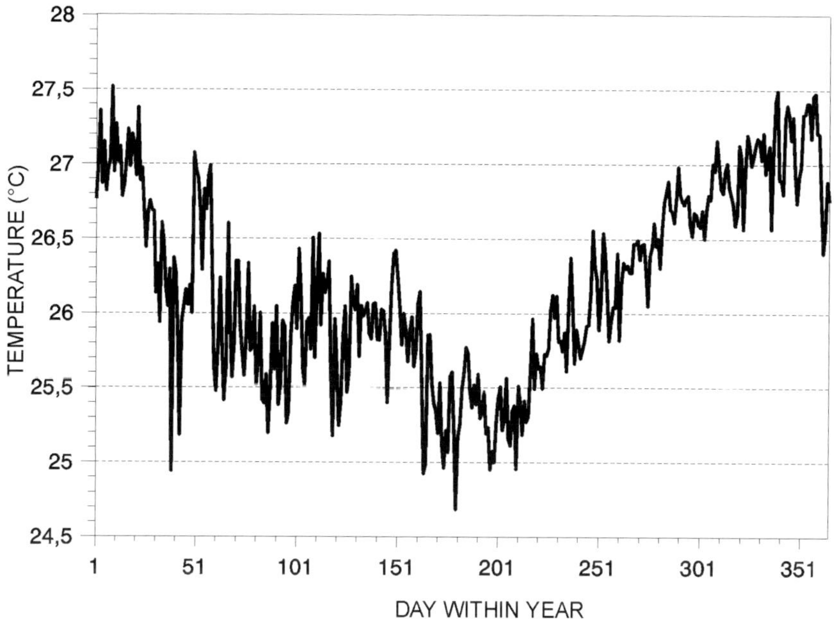

Fig. 5. Mean annual course of temperature for the station Cedro, period 1921—1980

This variability is accompanied by strong fluctuations in the length of rainy and drought periods and by developments over longer periods. Figure 7 is a typical example for this. The strong fluctuations in the duration of the dry period (between approx. 50 and 270 days) and the trend towards a later beginning, and thus a shortening of the dry period, can be easily recognized.

References

Hastenrath S (1991) Climate dynamics of the tropics. Kluwer Academic Publ, Dordrecht

Parry M, Carter TR, Konijn NT (eds) (1988) The impact of climate variations on agriculture. Kluwer Academic Publishers, Dordrecht

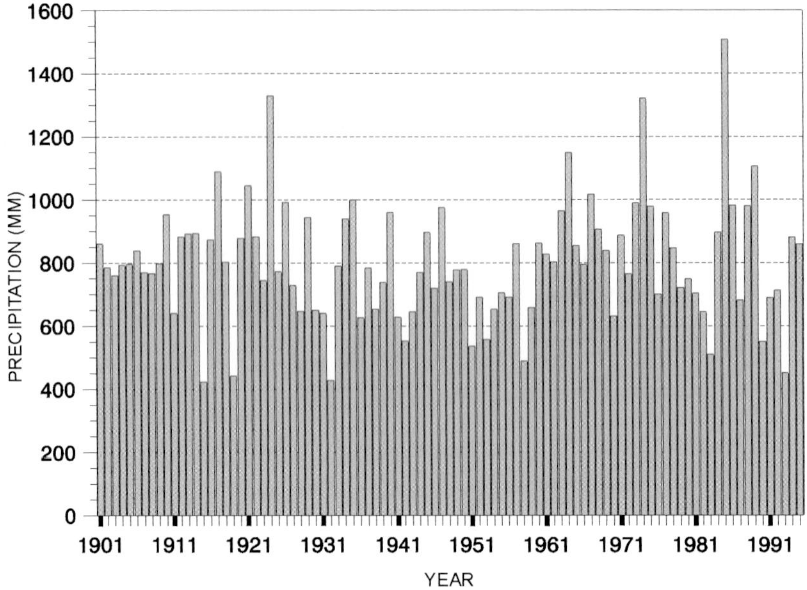

Fig. 6. Temporal course of the mean precipitation sum for the investigation area, period 1901—1995

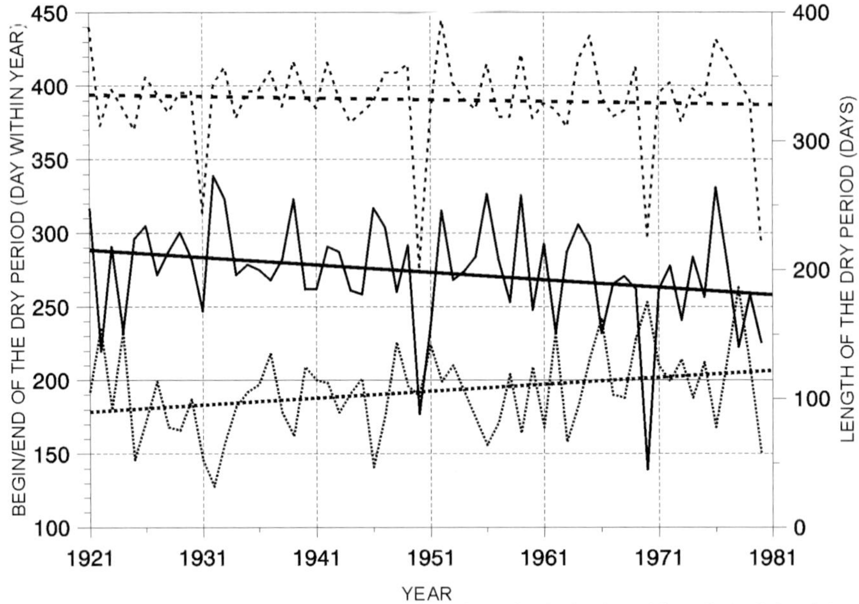

Abb. 7. Temporal course of the beginning (*dotted*), end (*dashed*) and length (*solid*) of the dry period at the station Cedro, period 1921–1980

Water Resources of Ceará and Piauí

Horst Frischkorn, José Carlos Araújo, Maria Marlúcia F. Santiago

Universidade Federal do Ceará, Departamento de Hidráulica e Ambiental, CEP 60451-970, Fortaleza – Ceará, Brazil, cariri@ufc.br

Climate Basics of Ceará and Piauí

Most parts of Ceará and southeastern Piauí are part of the „drought polygon", where potential evaporation by far exceeds precipitation. A mean annual rainfall of about 900 mm competes with a potential evaporation of 2200 mm powered by 3000 h of sunshine. Real evapotranspiration is of the order of 700 mm (SUDENE 1980; corresponding to 78% of rainfall), leaving only about 120 mm (13%) for runoff and 80 mm (9%) for percolation. Specific runoff in the region is of the order of 4 L/s/km^2 to be compared with 21 L/s/km^2 for all of Brazil (Barth et al. 1987). Mean annual temperature ranges from 19 to 29 °C. Köppen/Guettner-Kutzbach climate classification is „BS" for the focus areas of WAVES (Picus and Tauá) whereas northern Piauí and northwestern Ceará are „AW". Climate details, areal and temporal distribution of parameters may be found in Werner and Gerstengarbe (this Vol.) and Gerstengarbe and Werner (this Vol.). Here, we shall restrict our description to basic climatic features that define the general framework.

However, as life depends on minimum rather than on average conditions, in this semiarid region, long-term mean values can hide the decisive characteristic of its climate, which is irregularity of precipitation in time and space.

In this trade-wind belt climate, rainfall is basically limited to a 4-month period (February–May). Whereas parts of the region may enjoy abundant rainfall, others can, at the same time, suffer a drought. Even within the area of the same township, variation can be considerable. [In 1994, two rain gauges in Tauá measured yearly totals of about 770 mm (town) and 370 mm (Favelas Dam), respectively.]

Inter-annual variability is extremely high. For a series of 27 years (1974–2000) from Tauá we obtained a mean value of 564 mm exhibiting a standard deviation of 289 mm (Queiroz 2000)! These variations make it difficult to plan and dimension dams. Campos et al. (1996) showed that a series of 691 years would be necessary in order to achieve the same statistical error in the planning of dams as in the northeast of the United States based on a series of 30 years. Equally, intra-seasonal variations are extreme. Especially during dry years, rainfall is limited often to a few isolated, torrential events. It is common that half of a year's rainfall is concentrated in 1 month during which, again, one half of that pours down during a single event (Duque 1980). On the 24th of April 1997, Fortaleza experienced

270 mm of rain in about 11 hours, limited practically to the urban area, after what seemed to be a „dry“ rainy season.

Recurrent droughts, „secas“, with yearly rainfall totals far below average, castigate the region. History has registered ten such events during the past century. Whether they are periodic or not is subject to further research. So far, a strong correlation with the „El Niño“ phenomenon has been proved (Popelewski and Halpert 1987). The bar chart for annual rainfall totals (Fig. 6 in Werner and Gerstengarbe, this Vol.) clearly identifies these droughts. (On the other hand, equally evident are years of extremely high rainfall.)

It was the severe drought of 1877–1879, the most catastrophic of all, that struck Ceará after a period of 32 years without major pluviometric problems, that made Dom Pedro II pronounce the famous cited declaration of intention: „I shall sell the last stone of my crown in order to prevent the death of a north-easterner“. The drought claimed possibly half a million lives. It took 27 years before the first big dam that could resist pluri-annual droughts was completed (Cedro in 1906). Since then, programs of reservoir construction and well drilling alternate with each other; however, without really alleviating the vulnerability of the region.

Whereas the climate of the reference areas of WAVES in Piauí and Ceará is similar, hydrogeology is fundamentally different. Southeastern Piauí disposes of rich groundwater resources in the Serra Grande and Cabeças aquifers of the sedimentary Maranhão-Piauí Basin. In contrast, Tauá lies in an area of outcropping basement rock with very little vocation for groundwater.

Surface Water in Ceará

In the State of Ceará all rivers are naturally intermittent due to low (and highly irregular) precipitation rates, high evaporation rates (water deficit in relation to atmosphere) and shallow soils on top of the crystalline basement. Even the Jaguaribe River, the largest in the state, with a catchment area of 74,000 km^2, is intermittent (Ceará 1992). It is commonly locally referred to as „the largest dry river of the earth“. Due to water scarcity in rivers in the dry season and to the low groundwater reserves, the only way to have water available with reasonable confidence for social and economical use is to store the excess of water during the rainy season. This policy, historically known as the „hydraulic approach“, directs all efforts towards the construction of artificial reservoirs (the so-called *açudes*), starting in the year 1885 with the construction of the Cedro Dam in the township of Quixadá (Araújo 1992).

Although the policy of constructing dams has been continuously maintained, water planners began to seriously consider optimization of water use in the early 1970s, when irrigation projects were designed and operated by DNOCS, the Brazilian National Department of Works Against Droughts. Up to that time, water had been kept in the reservoirs basically for drinking purposes. In the year of 1987, with the creation of the Water Resources Secretariat of Ceará and with and the formulation of the State Water Management Plan (Ceará 1992), a new approach towards water management started.

After a century of engineering interventions, the State of Ceará alone has more than 7000 dams, which means a density of about one dam every 20 km^2, or about one dam for every 1000 inhabitants. However, while the total water storage capacity of the dams in the state is 12,500 million m^3, the average storage level at the end of the rainy season usually ranges from about 50 to 70% due to hydro-climatic constraints. In the relatively dry year of 2001, e.g., storage volume at the end of the rainy season was little less than 40% of storage capacity (COGERH 2001). At the end of the dry season, effective storage volume of the reservoirs usually ranges from 30–40%.

In spite of the enormous number of dams, there is a high spatial water availability concentration: the 113 largest reservoirs alone (1.6%) can store over 10,400 million m^3, i.e., more than 83% of the state's whole capacity. The dams in Ceará have been mostly designed, built and operated by DNOCS (Araújo 1992). Yet, since the creation of COGERH, the Water Resources Management Company of the State of Ceará, in 1994, the strategic dams have been operated together by DNOCS and COGERH.

Most of the small dams are not able to stand a drought, even if it lasts some months only, as they run dry due to the high evaporation losses. The very small ones usually run dry within 3 months, i.e., even before the regular dry season comes to an end. On the other hand, the large (volume greater than 50 million m^3) and middle-sized dams (volume between 10 and 50 million m^3) are designed so that they dry out only in long droughts, i.e., in droughts that last over 2 years.

According to Campos (1996), well-designed reservoirs in the semiarid northeast can yield approximately 40% of the yearly inflow with 90% confidence. Consequently, the reservoirs in Ceará allow 2560 km of rivers to flow all year round (COGERH 2001), i.e., for every km^2 in the state there are 17 m of continuous flowing river due to reservoirs. The reservoir water yield is computed as a function of supply confidence based on evaporation rate, on topographic characteristics, on average inflow, as well as on the coefficient of variation of annual inflow (Campos 1996). The usual long-term planning yield is calculated for 90% yearly confidence (Q_{90}), which, today, represent some 3100 million m^3 per year in Ceará.

In spite of the dams, the population still suffers intensely the consequences of the frequent droughts, when water-tank trucks (so-called *carro-pipa*) are the only water supply source for small cities, villages and rural communities at the beginning of the third millenium. Water quality of such services is usually poor and, when delivery service is private, every cubic meter may cost up to 20.00 US$, whereas average good quality tap water in Ceará costs less than 0.40 US$/m^3.

Another problem concerning water supply and management is the poor spatial distribution of water facilities: infrastructure is usually only available in the larger urban areas, but absent in almost every village and in the rural areas. The state government admits the urgent need for drinking water facilities for about 30 municipalities out of 184 in Ceará. Possible solutions to the problem are the construction of an extensive net of pipelines, home-tank storage of rainwater and, principally, an improved exploitation of groundwater.

Groundwater in Ceará

Ceará has an extension of 146,000 km^2, 93% of which are included in the so-called drought polygon. In 75% of that area, crystalline Precambrian basement rock crops out. Shallow soils and a reduced vegetative cover enhance the problem of poor groundwater resources.

Sedimentary basins form, with the exception of the interior Iguatú Basin, only the fringes of the state: the Araripe/Cariri Basin in the very south, on the border with Pernambuco, the Apodi Plateau in the east, bordering Rio Grande do Norte, the littoral stripe in the north, and the Ibiapaba Plateau in the west. However, due to higher rainfall in these areas and more favorable hydrogeological characteristics, despite making up only 25% of the territory, they detain about 77% of the state's exploitable groundwater resources, which are estimated to amount to 1.6×10^9 m^3/year (Ceará 1992).

A groundwater „census" by CPRM (1999) registered 13,296 „points" of groundwater exploitation, 11,889 of which are tubular wells, with depths scattering closely around a mean of 60 m, 67% in crystalline and 33% in sedimentary environment. (It is noteworthy that Fortaleza, the capital of the state of Ceará, is not represented in this registry because of lacking data.) Nevertheless, because of the higher mean yield of sedimentary wells (6.7 m^3/h) in comparison to crystalline (2.4 m^3/h), groundwater extraction from sediments exceeds that from crystalline rock.

According to Barbosa (2000), the disposable yield (the capacity of all existing wells, including inactivated and not installed ones) is 147.4 hm^3/year for the crystalline and 350.4 hm^3/year for sediments, whereas the exploited yields (the 12-h pumping capacity of actually exploited wells) are 44.1 and 146.1 hm^3/year only. These large differences reveal a great extravagance with respect to groundwater, which often treats the drilling of wells as electoral free gifts without any future care for maintenance. In addition, frequently, elevated salinity of groundwater in crystalline areas induces the population to disdain groundwater during periods of regular pluviosity. In general, ≈50% only of crystalline water exhibits salinity lower than 2900 mg/L and still ≈10% has more than 10,000 mg/L, with chloride reaching 3000 mg/L in about 25% of wells.

It is noteworthy that there is no correlation between salinity and ^{14}C age for crystalline waters. In general they are surprisingly young. From a compound of 55 wells, 29 produced „modern" waters (pmC >100), 23 exhibited residence times of few hundreds of years and only 3 had around 5000 years (Rebouças et al. 1984; Frischkorn et al. 1989).

In recent years, small scale (1 m^3/h) desalination by reverse osmosis has been used in order to make such waters „acceptable" to a population that, during drought periods without access to nearby surface water, would have attended its basic water needs through expensive tank lorries. However, it turned out that the lack of technical knowledge in such communities is a severe obstacle to a rational and continuous operation of the equipment. If, in spite of these difficulties, the program continues, we expect a reduction in salt concentration in the wells intensively pumped as a consequence of increased circulation and renovation of the water stored in fractures, improving, that way, chemical quality simply by continuous exploitation.

Furthermore, improved location of wells in crystalline rock, based on geoelectric and electromagnetic methods and on fracture tracing rather than on visual indicators for surface near groundwater, may reduce the salinity of the water produced, as the latter criterion makes a systematically erroneous choice through selecting locations of groundwater discharge, where salinization happens through evaporation from water accumulated in depression areas (with invariably clayey soils, which favor capillarity) during the rainy season and after capillary rise during drought periods. Bischoff (1969) already recommended that crystalline fractures should be exploited near recharge rather than discharge areas when reporting that groundwater quality at the foot of „inselbergs“ is much better than at some distance.

In spite of the large fraction of crystalline area in Ceará and the high number of wells (about 8000) in this environment, so far no systematic observations have been performed in order to understand the functioning of fractures as aquifer, their recharge and discharge behavior.

From environmental isotope (^{14}C, ^{18}O) measurements (Frischkorn et al. 1989), we know that crystalline waters exploited (through wells with about 60 m in depth) are young, normally some hundreds of years, that their ^{18}O is that of rainfall, without showing major signs of previous evaporation which could justify the high salt content, and that fractures are normally individuals, without evident hydraulic connection with neighboring ones. There are no signs for a regional flow net.

Strong changes in electric conductivity at the beginning of the rainy season indicate an efficient and rapid renovation of water in wells that are well located. During the 5-year drought period at the beginning of the 1980s, we observed 29 crystalline wells around the Araras Dam in the northwest of Ceará. Whereas the static level of wells in sedimentary environment, near Fortaleza, dropped by about 15 m, none of the basement rock wells observed fell dry, indicating recharge even during isolated rainfall events during a period of very reduced pluviosity. This is understandable because sparse vegetation and shallow soils lead to strong surface runoff, funnelled into fractures due to the fact that surface drainage follows mainly tectonic fractures. In addition, reduced discharge seems to protect crystalline wells naturally against excessive exploitation.

Thus, we believe that wells in crystalline regions, despite their obvious natural restraints, detain an equally natural vocation for supplying good quality water, even during periods of drought, however at reduced discharge.

Likewise, shallow wells tapping baseflow in alluvial deposits in beds of intermittent rivers may have surprising social impact at very low cost. The „Pingo d'água“ (water droplet) project in Quixeramobim, central Ceará, hand-drilled about 200 such wells with depths ranging from only 7 to 12 m at a cost of a modest 50 USD per unit. The mean yield is 5, the maximum 90 m^3/h! These wells provide supply for human consumption and small-scale irrigation ($\approx$1 ha) of fruits, guaranteeing a mean monthly income of about 250 USD to their owners (O POVO 2001).

This way, although it is obvious that groundwater in Ceará, with a disposable capacity of 498 hm^3/year, which corresponds to only 8.7% of the regulated discharge (at 90% of guarantee, Q_{90}) from dams, cannot respond to industrial and large-scale irrigation demands, and for that reason is not attractive to governmental planners, who are used to „thinking big", it offers localized solutions (adequate to the wide dispersion of rural population in the northeast of Brazil), with scaled investments and minimal environmental damage.

Water Resources in Piauí

Whereas in Ceará 75% consists of crystalline area, in Piauí 83% is part of the sedimentary Maranhão/ Piauí Basin, where sandstone aquifers store rich groundwater resources and where, due to large recession time and groundwater discharge, rivers are mostly perennial.

Surface water resources are represented by the Parnaiba River and its tributaries to the right, with an annual discharge of 20×10^9 m^3 and a dry season minimum flow of 300 m^3/s in its lower part.

According to PLIRHINE (Plano de Aproveitamento Integrado dos Recursos Hídricos do Nordeste, SUDENE 1980), the total installed capacity of dams in Piauí is 5511 hm^3, 5085 hm^3 of which, however, refer to the Boa Esperança Dam alone, used exclusively for hydroelectric generation and fish-breeding. Thus, a volume of only 426 hm^3 is used for regulated discharge and irrigation purposes. Whereas the national agency DNOCS (Departamento Nacional de Obras Contra as Secas) constructed 539 dams in Ceará, the number for Piauí is only 16 during the same period (up to 1995).

This disparity is due to the fact that Piauí can rely on extended groundwater reserves in the sediments of the Maranhão/Piauí Basin, which cover an area of 450,000 km^2. Permanent reserves of the basin are estimated to amount to 2×10^{12} m^3, with a sustainable yield of 3×10^9 m^3/a. An exploitable yield for a 50-year period (consuming about 20% of permanent reserves) of 10^{10} m^3/a is estimated.

The main aquifers are the Serra Grande and Cabeças formations. Outcropping in narrow strips of some kilometers only, a series of sedimentary deposits, aquifers alternating with aquitards in the southeast of Piauí (in the Picos region) slopes gently northwest towards the Parnaiba river (Fig. 1). The Serra Grande aquifer, reposing on the Precambrian crystalline basement, is confined by the aquitard Pimenteiras; the following aquifer Cabeças, is confined by the aquitard Longá. This sequence promotes artesianism with flowing wells in large parts of the southeast of Piauí. At the Violeto well, in the township of Cristino Castro, in the central part of the Basin, the sedimentary package reaches a thickness of nearly 1000 m, with the Serra Grande, Pimenteiras, and Cabeças formations contributing respectively 350, 320, and ≈250 m. ^{14}C ages of the water exploited reach more than 30 ka. Noble gas and ^{18}O measurements prove that these groundwater resources formed during a Pleistocene cold period (≈5 °C colder than today; Frischkorn et al., this Vol.). Their exploitation has to be considered as „mining" and ought to be done in the most rational way. However, producing water at no pumping costs is seductive and easily leads to misuse. From a sample of 85 flowing wells in the township of Bom Jesus (in the Gurgeia valley), 46 were without any use whatsoever, 27 were partially used, and the water from only 12 wells (14%) was used adequately (Barjud Filho 2001).

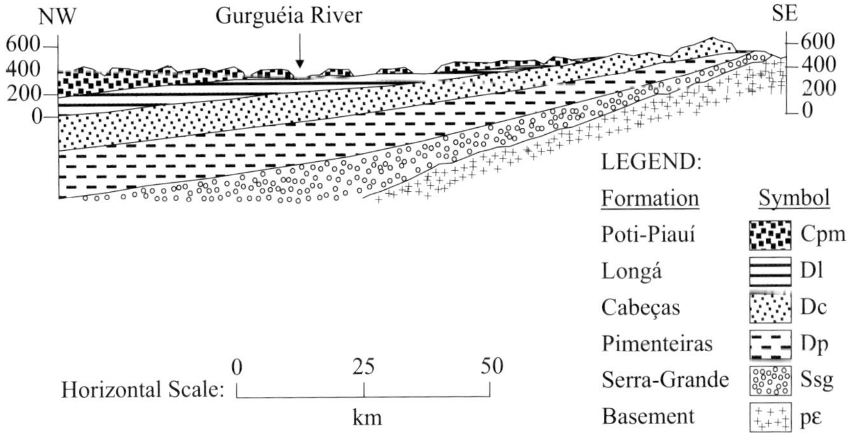

Fig. 1. Hydrogeological draft of central/southeastern Piauí (after COTEP 1976)

References

Araújo JAA (1992) Dams in the Northeast of Brazil. DNOCS, Fortaleza/BR

Barbosa CP (2000) Avaliação dos custos de água subterrânea e de reuso de efluentes no Estado do Ceará. M Sc Thesis, Universidade Federal do Ceará, Centro de Tecnologia, Fortaleza/Brazil

Barjud Filho M (2001) O maior lençol freático do Brasil pede socorro. Revista ABASTECE, (Associação Brasileira de Águas Subterrâneas – ABAS – Ceará, Fortaleza/BR) ano 2, no 7, p 12

Barth FT, Pompeu CT, Fill HD, Tucci CEM, Kelman J, Braga Jr BPF (1987) Modelos para gerenciamento de recursos hídricos. Nobel/ ABRH, São Paulo/BR

Bischoff G (1969) Die entwicklungspolitische Wirksamkeit der Deutschen Geologischen Mission in Brasilien. Infrastruktur Beratungs- und Forschungsgesellschaft – IBF, Bonn

Campos JNB (1996) Dimensionamento de Reservatórios: O Método do Diagrama Triangular de Regularização. UFC, Fortaleza

Campos JNB, Filho FAS, Araújo JC (1996) O processo de estocagem de águas em reservatórios: o papel da variabilidade dos deflúvios. In: 3° Simpósio de Recursos

Hídricos do Nordeste. Associação Brasileira de Recursos Hídricos – ABRH, Salvador/BR

Ceará (1992) Plano Estadual de Recursos Hídricos. Secretaria de Recursos Hídricos, Fortaleza/BR

COGERH (2001). Website http://www.cogerh.com.br. Fortaleza/BR

COTEP (1976) Consorcio OESA/ TECNOSOLO/ EPTISA: Estudo de Viabilidade do Vale do Gurgéia. Teresina/BR

CPRM (1999) Companhia de Pesquisa de Recursos Minerais. Atlas digital dos recursos hídricos subterrâneos do Ceará. Fortaleza/BR

Duque FG (1980) Solo e água no poligono das secas. Coleção Mossoroense v. CXL II, Mossoró/BR

Frischkorn H, Santiago MMF, Serejo ANC (1989) Isotope study of wells in crystalline rock of the semi-aride northeast of Brazil. IAEA, Agência Internacional de Energia Atômica, Isotope Hidrology Investigations in Latin America, Vienna, pp 73–90

O POVO Jornal (2001) Fortaleza/BR. 06.05.2001, p 25

Popelewski CF, Halpert MS (1987) Global and regional scale precipitation patterns associated with the El Niño/southern oscillations. Mon Wea Ver, Boston, Ma 115:1606–1626

Queiroz GH (2000) Salinização em açudes no município de Tauá-Ceará. MSc thesis, Universidade Federal do Ceará, Centro de Tecnologia, Fortaleza/Brazil

Rebouças AC, Santiago MF, Frischkorn H, Torquato JR (1984) Aplicação de C-14 na avaliação da recarga de zonas aqüíferas nas rochas cristalinas no Ceará. In: 3° Congresso Brasileiro de Águas Subterrâneas, Fortaleza/ BR, vol 2, pp 52–62

SUDENE (1980) PLIRHINE – Plano de Aproveitamento Integrado de Recursos Hídricos do Nordeste do Brasil. Recife/BR

Physical and Environmental Context in the State of Ceará

Marcos José Nogueira de Souza and Vladia Pinto Vidal de Oliveira

State University of Ceará (UECE), Department of Geosciences, Fortaleza – CE, Brazil

Introduction

This work presents the fundamental aspects of the physical and environmental context of the state of Ceará, located at the northern portion of the northwest region of Brazil. It comprises an area of 148.016 km^2 between the coordinates 3° 46′ 30" and 7° 52′ 15" of latitude south and between the coordinates 37° 14′ 54" and 41° 24′ 55" of longitude west (Fig. 1). Within the state territory, 136.335 km^2 comprise areas of semiarid climatic conditions.

This research was conducted for the identification and brief characterization of the geo-environmental compartmentation, the theoretical and methodological basis of which has geo-systemic fundamentals (Bertrand 1969; Sotchawa 1977). It should be considered that the physical and biotic systems are integrated through various elements that maintain mutual relations between themselves, and are simultaneously submitted to matter and energy flows. In each one of the systems, there is a harmonic relationship between its components, and they are also endowed with a capacity for support that individualizes them from the point of view of potentialities and limitation of natural resources.

The criterion adopted to delimitate the geo-systemic units is of geomorphologic nature, synthesizing the physical-environmental characteristics of the context. This way, the limits of the relief and the features of the landscape are subject to a precise delimitation through the use of orbital images.

Geo-morphological compartmentation derives from a slow and integrated evolution of nature through the Quaternary. Each resulting compartment and the morpho-dynamic influences have, this way, their own standards in the characteristics of the network of superficial drainage, associated soils and vegetal coverage.

Fig. 1. Map of the localization of the state of Ceará (source: http://www.ceara.br)

Methodological Procedures

The study produced results from a revision of the systematic gathering of information, from analysis and interpretation of orbital images, besides many field trips through different quadrants of the state.

The thematic analyses were conducted through perception of the systemic relationship between the geo-environmental components. The functional dependency between environmental variables was highlighted through geological--geomorphological conditions, hydro-climatic characteristics, soil distribution and vegetal coverage patterns. This sequence presents a chain of partial products that expose a relationship of dependency between the variables, in such a way that each one of them is based on the previous ones and acts as a foundation for those that follow. This way, the geo-systemic variables were identified as a result of the specific combinations between the variables.

The analysis to be performed for each geo-environmental unit synthesizes the geo-environmental context and indicates its capacity for support.

Geo-environmental Context of Ceará

In the context of northeastern Brazil, Ceará comprises a certain diversity of natural and landscape domains. Their geological conditions vary, in spite of the primacy of pre-Cambrian terrains of crystalline basement. In its adjacent areas with other states, the Palaeo-Mezozoic sedimentary basin, present as plateaus, assumes peculiar characteristics. In littoral areas, disposed in discordance over the basement, we find the occurrence of detrital sedimentary coverage, forming pre-coastal barren tableland that was deposited throughout the recent geological history of the territory.

The Cearense relief has a very significant predominance of lands situated under the 200-m level. Crystalline residual massifs and sedimentary plateaus above the level of 700 m have restricted extensions. At the littoral, besides the dune fields modeled as actual sediments, the older deposits are shaped incipiently by surface drainage, isolating interfluvial plateaus that represent pre-coastal plateaus *(tabuleiros pré-litorâneos)*.

The regional climate, in spite of the predominance of the semiarid -- with rainy seasons irregular in time and space -- also presents differentiations. Humid spots, with rainfall totals above 900 mm, comprise the coast. Here, a reasonable level of humidity is preserved for 6--7 months of the year. The semiarid, strictly speaking, with water shortage for the most part of the year, presents a pronounced character in the interior depressions such as Inhamuns, Irauçuba and Médio-Jaguaribe (Soares et al. 1995). It attenuates at the foothills, at the low crystalline residual mountain and in the *Sertões* (hinterlands) near the littoral. Generally, the rainy period comprehends summer and autumn; an average temperature above 24 °C configures a hot climate or megathermal.

Under the influence of annual and inter-annual irregularities of rainfall, the Cearense territory is subject to eventual periods of calamitous droughts that generate severe social and economic problems.

Water resources, both superficial and subterranean, depend on the morpho-structural and climatic conditions already related. In the crystalline rock semiarid hinterland (*Sorties semi-radios*), there is a high density of rivers and creeks with intermittent flow. In the littoral sedimentary areas and plateaus, water shortage is attenuated through a higher potential of groundwater.

Soils have a very complex mosaic, derived from different combinations of factors and processes of formation. In the hinterland areas (*areas sertanejas*), shallow soils, often interspersed with gravel and outcropping basement rock, define the typical substratum for the semiarid vegetation, like the tropical xerophytic deciduous broadleaf (*Caatingas*). There is, on the other hand, the occurrence of expressive soil spots endowed with a natural fertility that varies from medium to high. The same is true for soils derived from the calcareous rocks that cover the Chapada do Apodi (Apodi Plateau). On sedimentary plateaus, such as Ibiapaba and Araripe, in spite of a larger thickness, soils are acid and have a low natural fertility.

From these conditions the vegetal cover emerges as the environmental component that best reflects relations between the other natural components. The most representative component of vegetation is the tropical xerophytic deciduous broadleaf (*caatinga*), which, however, exhibits various physiognomic and floristic standards. Generally, they are subordinated to the soils' hydric potentialities and disposability. Forests on humid hills and cilliary forests (*matas ciliares*) that cover the fluvial plains with alluvial soils are restricted to small exceptional areas.

We should emphasize that, in Ceará, ecosystems and/or geosystems originating from self-organization of the biosphere inside a physical environment are presently almost nonexistent. Transformation of the landscape and its degradation by man, through rupturing the original equilibrium of nature, are evident.

Anthropism is identified by various manifestations which include among others: indiscriminate deforestation, inadequate management of water resources, predatory hunting, acceleration of erosive processes with a consequent thinning of the soils and the intensification of the silting up of rivers and sluices, as well as the disappearance of perennial and seasonable fonts.

The impact on nature caused by the presence of man as a populator and as a predator verged through time has provoked environmental disequilibria, the intensity of which is distinct in the various geosystemic units. It is a fact that man, when populating a territory, seeks to exploit to a maximum the available natural resources. In order to do so, he uses techniques that are very often improper, as they depend on his cultural level as well as the development of society. Rudimentary technology, besides retarding the development process, contributes to accelerate or reactivate the degradation processes whose effects are felt in the short, medium and long term.

In Ceará, as in other underdeveloped regions of the world, the diversification of the natural framework and the potentialities of its resources have always constituted themselves as conditioning factors for land use. The structuring of subregional frameworks, through the localization and the variety of economic activities, has been and is dependent on the natural resources available. When we deal, more specifically, with primary activities, these conditions assume a clearer character.

The conditions imposed on the utilization of natural resources are, for the state of Ceará, of relevant character. In general, they have been considered as unfavorable as a function of the large semiarid portion of space. Here, precipitation is irregular and uncertain. These characteristics are considered the major natural obstacle for the performance of rural activities in the semiarid and are the main cause of its vulnerability.

On the other hand, there are a small and dispersed proportion of areas in which soil and topographic conditions are favorable. Examples are the Apodi Plateau (Chapada do Apodi) and the alluvial hinterland plains (*Planícies aluviais sertanejas*). Equally dispersed and in small proportions are sectors with favorable climatic and soils, but where the degree of relief accidents is very strong. This is the case in the humid and subhumid crystalline residual massifs.

Altogether, allying the poor natural resources to the rudimentary techniques for exploitation, there is a tendency for the environmental imbalance to be accentuated. With an accelerated demographic growth -- in spite of the strong populational mobility through migratory movements -- the pressure on the natural resources is amplified. It is recognized that, traditionally, the productive process of combined agriculture and stock-rearing has been successful more through the incorporation of new land than because of increases in productivity. It is to be expected, therefore, that human action on the environment will increase further in the future causing severe impacts on ecosystems.

Results

The following map deals with the geo-environmental compartmentation of Ceará (Fig. 2). The characterization of the units is organized in a matrix format (Synopsis, Table 1), which includes the following aspects of geo-environmental units: (a) Geo-environmental unit; (b) Dominant natural features; (c) Support capacity (referring to the potentialities and limitations of each unit); and (d) The main risks of occupation and the impacts produced.

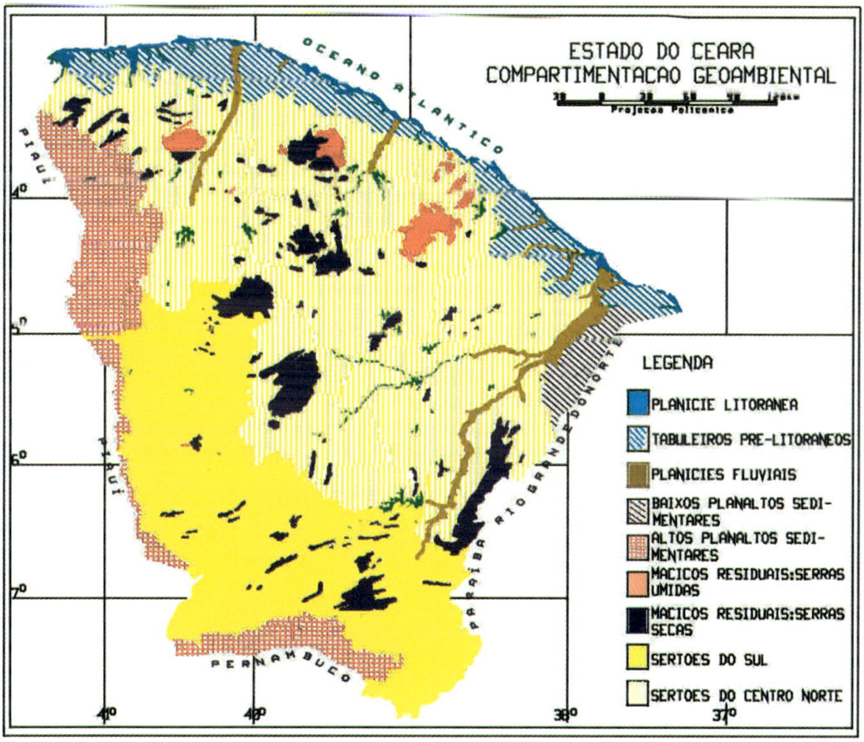

Fig. 2. Geo-environmental compartmentation of Ceará

Final Considerations

Due to their homogeneity, it is possible to define delimitations of geo-environmental units for the state of Ceará and establish general directions for an orderly occupation of the territory. It is found that some units have many common characteristics from the point of view of potentialities and limitations of use. Eco-dynamics present very differentiated conditions. The greater fragility and instability of the littoral geo-environmental unit makes it more vulnerable to anthropic occupation than others. Impacts and risks from a disorderly occupation are identified for each unit, suggesting directions for planning and land use.

In other geo-environmental units, generally located in the interior, some common problems were identified, which also occur at the littoral:

1. Deficiency of surface and groundwater.
2. Acidity and low fertility of soils for agricultural use.
3. Generalized deforestation with evident implications in devastation of fauna and flora.
4. Environmental vulnerability increasing the risks of expansion of degradation of natural resources.

Solutions for the problems indicated require the implementation of public policies and the cooperation of governmental and non-governmental organizations in participative planning, in search of sustainable development.

References

Bertrand G (1969) Paisagem e Geografia Física Global-Esboço Metodológico. IGEOG-USP. Caderno de Ciências da terra 13, São Paulo

Soares AML, Leite FRB, Lemos JJS, Martins MLR, Mera RDM, Oliveira VPV (1995) Áreas degradadas suscetíveis aos processos de desertificação no Ceará. In: Gomes GM, Souza HR, Magalhães AR (eds) Desenvolvimento Sustentável no Nordeste. IPEA, Brasília

Sotchava VB (1977) O estudo dos geossistemas. IGEOG-USP. Métodos em Questão 16, São Paulo

Souza MJN (1998) Análise Geoambiental e Ecodinâmica e das Paisagens do Estado do Ceará. Tese para Concurso de Professor Titular. UECE, Fortaleza

Table 1. State of Ceará: geo-environmental compartmentation (synopsis)

Geo-environmental unit	Dominant natural features	Support capacity		Occupational risks
		Potentialities	Limitations	
Littoral Plain (Planície Litorânea)	Surface made up with terrains of neo-formation, submitted to sea, wind, fluvial and pluvial influences, composed of large beach areas, fields of dunes with different generations, mangroves, lines of cliffs, lacustrine plains and accumulation and floodable areas; Climatic conditions varying from semiarid to subhumid and with annual rainfall varying from 700--1200 mm; frequent existence of estuaries and good potential of groundwater; Aeolian sands and mangrove soils covered by the vegetal complex of the littoral.	Tourist attraction; Groundwater resources; Ecological tourism; Controlled extractivism; Aquifer recharge; Legal restrictions; Road construction; Residential development.	Poor soil conditions for construction; Unfavorable eco-dynamics; Restrictions to mining; Soils with low natural fertility; Perturbation or interruption of sediment transport because of irregular occupation; Disequilibria in the littoral sedimentological balance.	Pollution of water resources; Loss of tourist attractions; Dismounting of dunes through deforestation; Retaking the process of erosion and natural sedimentation; mining in the sands surrounding the littoral lakes can lead to ruptures in the hydric equilibrium causing floods, destruction of the fluvial channels and changes in water flow
2. Pré-Littoral Plateau (Tabuleiros Pré-litorâneos)	Surface with smooth topographic slope towards the coastal line, developed in Plio-Quartenary sediments of barrier formation, dissected by interfluvial plateaus. The plateaus present quartzose and podzolic (yellow-red) sand, dense, covered with plateau vegetation, littoral stunted sparse forest and eventual spots of open pasture with patches of stunted vegetation. The primary vegetation is strongly characterized by land use (mainly mixed cattle breeding/farming).	Agrarian vegetable extractivism; Urban expansion; Controlled mining; Construction materials; Enhanced cattle raising; Road construction; Groundwater	Low soil fertility; Hydric deficiency during droughts.	Erosive processes in degraded areas; Risks of pollution of groundwater and soils; Uncontrolled mining; Urban expansion and impermeabilization compromising aquifer recharge.

Table 1. (Continued)

Geo-environmental unit	Dominant natural features	Support capacity		Occupational risks
		Potentialities	Limitations	
3. Fluvial Plains (Planícies Fluviais)	Plain areas resulting from fluvial deposits that border the river channels, reaching considerable enlargement in the low valleys of the rivers Jaguaribe, Banabuiú, Acaraú, Coreaú, Curú, Choró, Pacoti, among others. Occurrence of deep alluvial soils, poorly drained and with eventual problems of salinization. They are associated with other types of soils and are covered with cilliary carnaúba woods.	Agrarian extrativism; Water resources; Agriculture and stock raising; Controlled mining; Irrigated agriculture; Fishing; Tourism and leisure.	Legal restrictions for cilliary woods (Mata cilia); Poor drainage of soils, salinization and periodic flooding; Uncontrolled mining; Urban expansion on low levels of the fluvial terrace.	Degradation of cilliary woods unleashing erosive processes and silting up the rivers; Water resources pollution; Salinization of soils; Floods and inundation.
4. Low Sedimentary Plateaus (Baixos Planaltos Sedimentares)	Plateaus with altimetric levels under 100 m in the potiguar mesozoic basin, covered by calcareous rock (Jandaíra formation) superimposed by sandstone from the Açú formation. Hot semiarid climate with average rainfall between 650--700 mm; scarcity of surface water; with soils of high fertility (as cambisoils), covered by tropical xerophytic deciduous broadleaf (caatingas).	High natural fertility of soils; favorable topography; good groundwater potential; efficient drainage; sedimentary calcareous deposit.	Low potential for surface water; limitations of recharge; great depth of water table; low and irregular rainfall; low potential for dam construction.	Exploration of calcareous minerals is uncontrolled with damage to environment.

Table 1. (Continued)

Geo-environmental unit	Dominant natural features	Support capacity		Occupational risks
		Potentialities	Limitations	
5. High Sedimentary Plateaus (Altos Planaltos Sedimentares)	Plain surfaces or smoothly skewed, conditioned by geological structures of Paleozoic sedimentary basins (Bacia do Parnaíba) or Mesozoic basins (Bacia do Araripe). Altimetric levels between 750--900 m, with average precipitations between 900--1200 mm. Soils of high permeability provide fast infiltration and result in scarcity of surface water. Vegetation varies from tropical xerophytic deciduous broadleaf closed scrub (caatinga arbustiva densa) and brushwood (carrasco) to fluvial foggy woods and tropical xeromorphic semideciduous broadleaf of cerrado; areas for agricultural use and extractivism	Plain relief, smoothly undulated deep soils; regular to high pluviometry with well-distributed rainfall.	Soils of low natural fertility. Scarcity or inexistence of surface water. Elevated depth of water table.	Indiscriminate deforestation leads to an increase in lixiviation and impoverishment of soils; Risks of contamination of soils and water resources by agricultural toxins; Deforestation has led to a decline of primary biodiversity.
6. Crystalline Residual Massifs (Maciços Residuais: Serras Úmidas e Serras Secas)	Hilly areas dispersed over the inland depressions and near the littoral, with altimetric levels between 600--900 m. Reliefs strongly dissected into hills, crests and lengthened ridges of crystalline base rock. Dense fluvial network of dendritic pattern and eventually dendritic retangular, carving profound valleys in V format. On ridges and crests, yellow and red podzolic soils predominate, and in the smooth sectors, alveolar plains with alluvial soils of average to high natural fertility developed. On steep slopes and on ridges, shallow soils occur (litholic) and rock formations emerge. Agricultural use with various cultures, prevailing coffee, bananas, horticulture. Dry hills dispersed in the hinterland (sertões), have intermediary characteristics between humid hills and semiarid hinterland (Sertões).	Favorable hydroclimatic conditions; Medium to high fertility of soils. Presence of fractured rock aquifers. Ecotourism.	Areas protected by environmental legislation because of high declivity of slopes. No possibility of mechanization. High susceptibility to erosion.	Accelerated erosion due to deforestation and the use of rudimentary agricultural techniques. Impoverishment of biodiversity. Deficient sanitary conditions in urban areas. Silting up of valleys. Contamination of water bodies through agricultural toxins.

Table 1. (Continued)

Geo-environmental unit	Dominant natural features	Support capacity		Occupational risks
		Potentialities	Limitations	
7. Hinterland (Sertões: Sul e Centro-Norte)	Eroded surface in the hinterland's (sertões) semiarid or subhumid depressions. Truncated rocks of the crystalline complex at a level between 250 and 400 m. Dissected reliefs with hills alternate with open valleys with flat bottoms. Superficial drainage system with sub-dendritic standards and eventually controlled by structural features. Predominancc of seasonably intermittent fluvial regimes. Podzolic and dark non-calcic sands at the top and steep slopes of hinterland hills (colinas sertanejas), covered primarily with tropical xerophytic deciduous broadleaf scrub. On weak slopes, there is the occurrence of litholic soils and rocky emergence, covered with tropical xerophytic deciduous broadleaf scrub. At the bottom of the valleys, the fluvial plains are covered with cilliary forests (Mata cilia), with the predominance of the optician; areas of extensive agricultural-extractive use and with the predominance of extensive cattle rearing; areas that are strongly degraded as the Inhauls and Medio-Jaguaribe, evidence of desertification.	Average fertility of soils. Extensive cattle rearng. Mining (ornamental rocks). Vegetal exploitation (linoleous plants from stunted sparse forest). Relief favorable to cattle rearing and agriculture. Environmental recovery. Silviculture.	Irregular and scarce rainfall. Susceptibility to soil erosion. Vulnerability to desertification. Vulnerability to salinization of soils and water bodies. Shallow, stony soils. Outcropping of rock.	Unleashment of erosion in degraded areas. Salinization of the soils in lowlands; Pollution of water resources in urban areas. Impoverishment of biodiversity and susceptibility to degradation of soils and to desertification

Survey of the Vegetation in the State of Ceará

Carlos Lineu Frota Bezerra, Eliedir Ribeiro da Cunha Trigueiro, Antonio Alberto Jorge Farias Castro and Antônio Sérgio Farias Castro

Departamento de Biologia, Universidade Federal do Ceará, Bloco 906, Centro de Ciências, CEP 60455-970 Fortaleza – Ceará, Brazil, lineu@ufc.br

Introduction

From an ecological point of view, the vegetal coverage in the state of Ceará spans from a native vegetation climax to secondary degraded vegetation. Among the vegetational types, we may find: vegetação da planície litorânea (vegetation of the littoral plain), mata de tabuleiro (forest from the tableland), mata de dunas (forest from the dunes), manguezal (mangroves), caatinga (stunted sparse forest), cerrado (open pasture with patches of stunted vegetation), mata ciliar (forest bordering rivers, ponds and lakes), and matas serranas (forest in the mountains). They may present subtypes according to the environmental conditions and human influence.

Human action causes serious impacts, modifying the ecological conditions of the native vegetal communities. Frequently, an ecological imbalance occurs because of the deprivation of environmental characteristics manifested by the erosive processes of the soils, evidenced in the climate, in the vegetation and the fauna.

The advance of „development" is causing sensible effects on the areas of native vegetation, especially at the duna (dunes), at the terrenos de tabuleiro (tablelands) and at the serras (mountains).

This chapter has the objective of describing the vegetal coverage of Ceará from its different aspects (physiognomic, phytogeographic and floristic).

Methodological Procedures

The working methodology was based on a revision of the literature of the following authors: Fernandes (1990b); Figueiredo (1990) and Lima et al. (2000), who studied the vegetal coverage of the state.

Results and Discussion

Based on the observations and analyses of the authors that classified the vegetal coverage of the state of Ceará, i.e., Fernandes (1990a), Figueiredo (1990), Lima et al. (2000), we arrived at the following results that are discussed below:

The vegetative setting encompasses, as expressed by Fernandes (1990a), three environments: o Litoral (the Littoral), o Sertão (the Hinterland), and Serranias (the Hills). In each of these environments, the authors identified different vegetational types: (1) Litoral – vegetação das planícies litorâneas (coastal plains), vegetação de dunas (dune vegetation), vegetação dos tabuleiros (tableland vegetation), vegetação das planícies flúvio-marinhas (river/sea plain vegetation), aquáticas das lagoas (aquatic vegetation); (2) Sertão – vegetação das caatingas altas e baixas (high and low stunted sparse forest), cerrado (open pasture with patches of stunted vegetation), campestre (campaestrian), carnaubal (carnauba grove); and (3) Serranias – vegetação higrófila (hygrophilous), mesófila (mesophilous), xerófila (xerophilous), subxerófila (subxerophilous), escleromorfa (scleromorphic).

Figueiredo (1990) classified 10 phytoecological units and 8 subtypes. For the phytoecological units the classification received the following denominations: complexo vegetacional da zona litorânea (vegetational complex of the coastal zone); floresta subperinifólia tropical nebular (subperennial foliage tropical cloudy forest); floresta subcaducifólia tropical fluvial (subdeciduous foliage tropical river forest); floresta caducifólia espinosa (deciduous foliage thorny forest); caatinga arbustiva densa (dense stunted sparse forest); caatinga arbustiva aberta (open shrub stunted sparse forest); carrasco (brushwood); floresta paludosa marítima (mangrove forest); floresta mista dicótilo-palmacea (mixed dicotilo-palmacea forest), and floresta subcaducifólia tropical xeromorfa (drymorphic subdeciduous foliage tropical forest).

Lima et. al. (2000) classified the vegetal coverage in terms of phytoecological units based on the geo-environmental units of the state of Ceará such as: complexo vegetal da planície litorânea (coastal plain vegetal complex); vegetação de tabuleiro (tableland vegetation); caatingas das depressões sertanejas (stunted sparse forest of the hinterland depression); matas ciliares (gallery forest); enclaves de mata pluvionares (remnants of rain forest); and cerrados (open pasture with patches of stunted vegetation).

Based on the results from the cited authors, we have elaborated a table (Table 1) containing the geographic regions of the state of Ceará, the vegetation types, their subtypes and the species found most frequently, having as reference the collections in the Prisco Bezerra Herbarium – EAC of the Federal Universtiy of Ceará (UFC).

Table 1. Geographic regions in the state of Ceará with their vegetation types and subtypes and major species

Geographic region	Vegetation type	Subtype	Major species
Litoral (Littoral)	Vegetação das Planícies Litorâneas (Vegetation of Coastal Plains)	V. do Berma (Berm Vegetation)	*Paspalum vaginatum* Sw. *Sesuvium portulacastrum* L. *Ipomoea pes-caprae* Sweet *Sophora tomentosa* L. *Caesalpinia bonduc* (L.) Roxb.
		V. Flúviolacustre (Fluvial-lacustrine Vegetation)	*Montrichardia linifera* (Arruda) Schott *Nymphoides indica* (L.) Kuntze *Ammania latifolia* L. *Stemodia maritima* L. *Clitoria guianensis* (Aubl.) Benth.
	Vegetação de Dunas (Dune Vegetation)	Dunas Semi-Fixas	*Cyperus maritimus* Poir. *Canavalia rosea* (Sw.) DC. *Paspalum maritimum* Trin. *Macroptilium panduratum* (Mart. Ex Benth.) Maréchal & Baudet *Chamaecrista hispidula* (Vahl.) Irwin & Barneby
		Dunas Fixas	*Chiococca alba* (L.) Hitch. *Tocoyena sellowiana* K. Schum. *Chloroleucon foliolosum* (Benth.) G.P. Lewis *Byrsonima crassifolia* Kunth *Ziziphus platyphylla* Reissek

Table 1. (Continued)

Geographic region	Vegetation type	Subtype	Major species
Litoral (Littoral)	Vegetação à Sotavento das Dunas (Leeward Dune Vegetation)		*Macclura tinctoria* (L.) Engl. *Commiphora leptophloeos* (Mart.) Gillet. *Copaifera cearensis var. arenicola* Ducke *Vitex flavens* Humb., Bonpl. & Kunth
	Vegetação de Tabuleiro (Table and Vegetation)		*Andira surinamensis* (Bondt.) Splitz ex Pulle *Cochlospermum vitifolium* Spreng. *Protium heptaphyllum* (Aubl.) March. *Tapirira guianensis* Aubl. *Zanthoxylum syncarpum* Tul.
	Vegetação de Mangue (Mangrove Vegetation)		*Rhizophora mangle* L. *Avicennia germinans* (L.) Stearn. *Avicennia schaueriana* Stapf. & Leechman *Laguncularia racemosa* C.F. Gaertn. *Conocarpus erectus* L.
Sertão (Hinterland)	Caatinga (Stunted Sparse Forest)	Arbustiva (with shrubs)	*Croton sonderianus* Müll. Arg. *Combretum leprosum* Mart. *Jatropha mollissima* (Pohl.) Baill. *Mimosa tenuiflora* (Willd.) Poir. *Piptadenia stipulaceae* (Benth.) Ducke
		Arbórea (with trees)	*Auxemma oncocalyx* (Allemão) Taub. *Myracrodruon urundeuva* Allemão *Schinopsis glabra* F.A. Barkley & T.Mey *Tabebuia impetiginosa* (Mart. ex DC.) Standl. *Anadenanthera colubrina var. cebil* (Griseb.) Altschul

Table 1. (Continued)

Geographic region	Vegetation type	Subtype	Major species
Sertão (Hinterland)	Mata Ciliar (Forest bordering rivers, lakes and ponds) (Gallery Forest)	Com Carnaúba (Gallery forest with carnauba)	*Copernicia prunifera* (Mill.) Moore *Erythrina velutina* Willd. *Licania rigida* Benth. *Albizia inundata* (Mart.) Barn. & Grimes *Ziziphus joazeiro* Mart.
		Sem Carnaúba (Gallery forest without carnauba)	*Trema micrantha* Blume *Triplaris gardneriana* Wedd. *Guazuma ulmifolia* Lam. *Sapindus saponaria* L. *Lonchocarpus sericeus* (poir.) Kunth.
	Cerrado (Open pasture with patches of stunted vegetati-on)		*Curatella americana* L. *Acosmium dasycarpum* (Vogel) Yaklov. *Qualea parviflora* Mart., *Qualea grandiflora* Mart. *Simarouba versicolor* A. St.Hil. *Schefflera morototoni* (Aubl.) Maguire, Steyerm & Fredin *Cupania oblongifolia* Mart.
Serras (Hills)	Mata Úmida (Humid Forest)		*Garcinia gardneriana* (Planch. & Triana) Zappi *Pera glabrata* (Schott) Baill. *Myrcia prunifolia* Cambess. *Ceiba glaziovii* (Kuntze) K.Schum. *Astronium fraxinifolium* Schott.
	Mata Seca (Dry Forest)		*Cordia trichotoma* (Vell.) Arráb. ex Steud. *Spondias mombim* Jacq. *Tabebuia serratifolia* (Vahl) G.Nicholson

Table 1. (Continued)

Geographic region	Vegetation type	Subtype	Major species
Serras (Hills)	Cerradão („Cerradão“ de Cerrado in the Northeast, according to Castro and Martins 1999)		*Parkia platycephala* Benth. *Dimophandra gardneriana* Tul. *Caryocar coriaceum* Wittm. *Lonchocarpus araripensis* Benth. *Magonia pubescens* A St.Hil. *Salvertia convallariodora A. St.-Hil.* *Bowdichia virgilioides* Kunth
	Carrasco (Brushwood)		*Mimosa verrucosa* Benth. *Piptadenia moniliformis* Benth. *Memora candolleana* Miers. *Hymenaea eriogyne* Benth. *Lindackeria ovata* (Benth.) Gilg.

Vegetação das Planícies Litorâneas (Vegetation of Coastal Plains)

Vegetação do Berma (Ante-Duna, Alta Praia) (Berm Vegetation – Pre-dune, High Beach)

Occupies the sandy plains, pre-dunes, psammophilous vegetation adapted to an environment of high salinity, poor nutrients and instability of the substratum. The herbaceous species predominate, with emergence of a few shrubs.

Vegetação Fluviolacustre (Fluvial-lacustrine Vegetation)

Occupies the low portions between dunes, or occurs near lakes and rivers; including aquatic vegetation, hydrophilous, hygrophilous and heliophilous plants.

Vegetação das Dunas (Dune Vegetation)

According to Fernandes (1998), the dune vegetation covers an area that begins in the primary dunes, including the ones already stabilized, and ends at the border of the tableland. According to the ecological conditions, type of soils, relief and exposition to the wind, the vegetation presents itself in a variety of species. The species remain green during the whole year without leaf fall (perennial leaves). We have two types of environmental situation: in function of larger (windward) or smaller (leeside) exposition to solar radiation, directions and intensity of wind. At windward expositions, a shrub vegetation predominates and, at leeside, a vegetation of arboreous nature is developed, the dune forest (mata dunar).

Vegetação de Tabuleiro (Tableland Vegetation)

According to Fernandes (1990b), this type of vegetation occurs on the leeside slopes of the secondary dunes and in between them, and can even be found in their interior, reaching the limit of the caatinga (stunted sparse forest). It presents two types of behavior: one is semipermanent (semiperenifólio; with 90% of the leaves keeping their green color) and the other with most of the leaves falling (caducifólio). This is evidenced by the ligneous plants, which show this adaptation due to soil and microclimate conditions. The falling of the leaves is a survival strategy of the plants due to the limitations of water availability in unfavorable microclimatic conditions (low soil humidity).

Vegetação de Mangue (Mangrove Vegetation)

From the physiognomical point of view, it presents itself as a dense arboreous environment, floristically poor in vegetal species, but very interesting regarding the morpho-physiological adaptations that these species exhibit for fixation, breathing and reproduction. They are adapted to an environment deficient of oxygen with salinity and variation, streams, having soils that are halomorphic, sulfuric, poorly drained and subject to the influences of the tides. In the littoral of Ceará, the manguezal (mangrove) is found at the marginal parts of estuaries of its most important rivers (Coreaú, Acaraú, Ceará, Maranguapinho, Pacoti, Mundaú, Cocó, Curu e Jaguaribe), in coves and around coastal lagoons (Fernandes 1990b).

Vegetação da Caatinga (Stunted Sparse Forest)

The area that encompasses the geological contact between the sediments of the Barreiras formation and the rocks of the crystalline basement, with the conditions of the Sertão (hinterland), propitiates a landscape with the emergence of a type of vegetation well adapted to the conditions of water deficiency. It is called the caatinga. This vegetation type can present many physiognomies, which are observed depending on the ecological conditions such as: geology, climate, relief and soil. It reflects physiognomic patterns that are different regarding stratification, size and density. It can be classified as caatinga alta-arbórea-densa (high-arboreous-dense), alta-arbórea-aberta (high-arboreous-open), baixa-arbustiva-densa (low-shrubby-dense), and baixa-arbustiva-aberta (low-shrubby-open). From a physiognomic and floristic point of view, it can be separated into two basic types: the caatinga arbórea (arboreous caatinga) or the caatinga arbustiva (shrub caatinga). The caatinga arbórea (arboreous) presents three well-defined strata: one arboreous with trees 8–15 m in height, one shrubby/sub-shrubby with a height of 2–5 m, and a herbaceous with annual and ephemeral character. The latter is characteristic for plain or ondulated surfaces and occurs on the hillsides in isolated granitic elevations (inselbergs – isolated hills) and in dry slopes of hills.

Mata Ciliar (Forest Bordering Rivers, Lakes and Ponds=Gallery Forest)

Along the margins of the rivers that allow access to littoral or coastal zones occurs the vegetation type Mata ciliar, a vegetal community with perennial leaves that, due to the proximity of rivers, is favored by humid soil conditions. Its distribution is well evidenced along the rivers and the tributaries. There are two subtypes: (1) Mata ciliar com carnaúba (Mata ciliar with carnaúba); and (2) Mata ciliar sem carnaúba (Mata ciliar without carnaúba). The *mata ciliar com carnaúba* presents arboreous physiognomy, 10–12 m in height including shrubs in the lower stratum, occurring along the tributaries, where the alluvial soils are developed and the presence of carnauba is very striking. The carnaúbas form stripes of various kilometers in width, and are subject to flooding. The presence of saline soils is frequent, especially at the lower course of the rivers. The *mata ciliar sem carnaúba* is found at the margins of smaller and narrow channels, and does not have the conditions for the existence of carnaúbas, although its species may also exist in the carnaubal (carnaúba forest).

Cerrado (Open Pasture with Patches of Stunted Vegetation)

There are enclaves of scleromorphic vegetation with significant dimensions in the region that comprehend the middle and the southern part of the state. It has two strata, one ligneous shrubby-arboreous, with low density, interposed by the herbaceous stratum, with abundance of gramineous plants. Since the report by Castro (1994), the northeastern cerrados are now called „marginal“. Notwithstanding, the connotation of the term „marginal“ is not because a flora is predominantly composed of „accessory“ species (sensu Rizzini), but because they are distributed in the margins of a geographical space occupied by the Brazilian cerrado. There does not need to be any necessary connection with the quality of the flora or of the species, i.e., they have oreadic aspects. There is the idea of the existence, in the northeast, of one of the supercenters of biodiversity of the cerrados, along with the other two, cerrados of the southern southeast and those in the central plateau which are separated by two great climatic barriers: the „frost polygon“ in the south, and the „drought polygon“ in the northeast.

Mata Úmida and Mata Seca (Humid and Dry Forest)

These two vegetation types are distributed in a hill system of humid/subhumid hills and subhumid/dry hills.

Fernandes (1990a) divides the humid and dry forest into three types, according to the distance from the sea and the pluviometric regime, correlating them with a greater or smaller degree of humidity: „hygrophilous, mesophyllous and xerophilous vegetation“. The hygrophilous vegetation corresponds to the humid and subhumid hills and is distinguished in the municipalities of Baturité, Pacatuba and Maranguape as being of perennial character (perenifolio). It begins at 100 m altitude and extends to the upper limit of the hills.

According to Lima et al. (2000), the hills of Pacatuba and Maranguape are inserted in the residual massifs that comprehend the humid and subhumid hills where we find the humid forest. These are areas of strongly dissected relief, endowed with good edaphic and climatic conditions, with intense agricultural utilization, remnants of humid rain cloudy forest and high ecological instability. The crests and slopes of the dry hills and groupings of inselbergs are areas endowed with a dry climate that, as a consequence of the semiarid conditions, constitute a habitat for the development of dry forest (subdeciduous plant communities=permanent leaf fall during the dry season). In the residual massifs, the windward conditions favor the existence of humid forest. They develop at an altitude above 700 m, having species with a height of 8–10 m, sometimes reaching 20 m in deep and fertile soils. The conditions at the leeside favor the presence of dry forest with species of 8–12 m height, sometimes reaching 16 m in fertile and deep soils spreading into the limits of the littoral tablelands, in the hills close to the coast.

Cerradão

This possesses a type of scleromorphic vegetation with species reaching 8–15 m in height, occurring on the Chapada do Araripe (on the plateau) between the humid forest of the hill slopes and the carrasco (brushwood) in the south on red-yellow dystrophic Latosols. It differs from the cerrado especially by its denser forest physiognomy.

Carrasco (Brushwood)

This is characterized by a xerophilous vegetation, closed, branched, with thin stalks and rigid leafs, with many creeping plants and few herbs, in sedimentary terrains with sandy soils at an altitude of 600–750 m

Conclusions

As a consequence of the environment in Ceará, we have the predominance of the vegetation type that is most characteristic for northeastern Brazil: the caatinga. It is adapted to the semiarid conditions that are prevalent in the hinterland, called „Sertão". However, there is a variety of other vegetation types like the Dune and Tableland vegetation that is dominant in the coastal plains or the dry and humid forests that are characteristic of the mountain regions (Table 2).

Table 2. Characteristic vegetation of Ceará state with total area with forest cover and total anthropic area during years 1992 and 1998

	Total area (1992)		Area with forest cover (1998)		Anthropic area (1998)	
Caracteristic Vegetation	km^2	%	km^2	%	km^2	%
Complexo Vegetacional Litorâneo	22 175	15,10	4 295	19,36	17 881	80,64
Caatinga*	114 810	78,20	56 442	49,16	58 366	50,84
Matas Úmi-das (Humid Forest)	2 642	1,80	1 055	39,94	1 587	60,06
Cerrado	2 247	1,53	341	15,16	1 907	84,84
Cerradão	278	0,19	278	100,00	-	-
Carrasco	2 991	2,04	1 746	58,38	1 245	41,62
Permanent Preservation	1 675	1,14	1 674	100,00	-	
Total	146 818	100,00	65 831	44,84	80 986	55,16

* Type of vegetation predominant in Ceará.
Total area=area with forest cover+anthropic area.
Source: SEMACE SDU(1998)

References

Braga R (1960) Plantas do Nordeste, especialmente do Ceará. Imprensa Universitária, Fortaleza

Brummitt RK, Powell CE (1992) Authors of plant names. Royal Botanic Gardens, Kew, London

CASTRO, A.A.J.F., 1994. Comparação florístico-geográfica (Brasil) e fitossociológica (Piauí - São Paulo) de amostras de cerrado. Campinas, UNICAMP.

Castro AAJF, Martins FR (1999) Cerrados do Brasil e do Nordeste: caracterização, área de ocupação e considerações sobre a sua fitodiversidade. Pesq Foco 9:147–178

De Barbosa MRB, Mayo SJ, Castro AAJF, de Freitas GL, Pereira M, Do S, da Gadelha Neto PC, Moreira HM (1996) Checklist Preliminar das Angiospermas. In: Sampaio EV de SB, Mayo SJ, Barbosa MRV (eds) Pesquisa Botânica Nordestina: Perspectivas. SBB/SRP, Recife, pp 253–415

Ducke A (1979) Estudos Botânicos no Ceará. Mossoró, RN. ESAM, Coleção Mossoroense

Fernandes A (1990a) Estudo Fitogeográfico do Brasil. Editora Stylus Comunicações, Fortaleza

Fernandes A (1990b) Temas Fitogeograficos. I. Deriva Continental; II. Conjunto Vegetacional Cearense; III. Manguezais Cearenses. Editora Stylus Comunicações, Fortaleza

Fernandes A (1998) Fitogeografia Brasileira. Multigraf, Fortaleza

Figueiredo AG (1990) Vegetação. In: Atlas do Ceará. IPLANCE, Fortaleza

FUNCEME, Ceará (1993) Cobertura Vegetal e Antropismo do Estado do Ceará: Uma Abordagem Utilizando Imagens T.M – Landsat. Fortaleza

Lima LC, Morais JO, Souza MJN (2000) Compartimentação Territorial e Gestão Regional do Ceará. Editora Funeci, Fortaleza

Oliveira Filho AL (2000) O Estudo da Comunidade Vegetal da Mata. Semiperenifolia do Tabuleiro do Campus do Pici. Monografia de Bacharelado em Ciências Biológicas, UFC, Fortaleza – CE

Oliveira Filho AL, Rêgo PS, Bezerra CLF, Souza EB (1998) Observações Preliminares da Composição Florística no Processo de Sucessão Ecológica da Mata Semiperenifolia de tabuleiro, Campus Universitário do Pici, UFC, Fortaleza - CE. In: 40 Congresso de Ecologia do Brasil, Resumos, Faculdade de Ciências Agrárias do Pará, Belém- PA, p 409

Rêgo PS, Oliveira Filho AL, Bezerra CLF, Souza EB (1998) Considerações Preliminares sobre um Estádio Seral da Sucessão Ecológica em Mata Semiperenifolia de Tabuleiro Pró Litorâneo do Campus do Pici/ Universidade Federal do Ceará. In: 40 Congresso de Ecologia do Brasil, Resumos, Faculdade de Ciências Agrárias do Pará, Belém - PA, p 409

Rodrigues RR, Leitão-Filho H (2000) Matas ciliares: conservação e recuperação. Editora Universidade de São Paulo, Fapesp, São Paulo

Sales MF, Mayo SJ, Rodal MJN (1998) Plantas Vasculares das Florestas Serranas de Pernambuco – Um Checklist da Flora Ameaçada dos Brejos de Altitude. UFRPe/Royal Botanic Gardens, Kew/CNPq/Darwin Initiative for the Survival of Species/PNE, 130 pp

SEMACE SDU (1998) Diagnóstico e Macrozoneamento Ambiental do Estado do Ceará. Convênio FCPC/SEMACE, Fortaleza

Silva EV (1987) Modelo de Aprovechemento y Preservatión de los Manglares de Mariscos y Barro Preto - Aquiraz, CE. Instituto Agronómico Mediterraneo de Zaragoza/Espanha (EXCLUIR)

Silveira ARM (1993) Estudos Estruturais e Comparativos em Bosques de Mangues, Aquiraz - CE. Monografia de Bacharelado em Biologia, UFC. Fortaleza – CE

Souza ES (1997) Diagnóstico Geoambiental: Uso e Ocupação do Campus do Pici da Universidade Federal do Ceará. Monografia de Bacharelado em Geografia, UFC, Fortaleza – CE.

Souza MJN (1998) Análise Geoambiental e Ecodinâmica e das Paisagens do Estado do Ceará. Tese para Concurso de Professor Titular. UECE. Fortaleza.

Survey of the Vegetation in the State of Piaui

Antonio Alberto Jorge Farias Castro

Universidade Federal do Piauí (UFPI), Núcleo de Referência em Ciências Ambientais do Trópico, Eonal do Nordeste (TROPEN), CEP 64049-550 Teresina – Piauí, Brazil, aajfcastro@uol.com.br

Summary

This paper introduces a general vision of the main vegetal formations of the state of Piauí in terms of estimates of occupation areas with comments on the form as this has been considered. The current estimates reveal, in spite of the superpositions, that few of the vegetal formations of the state are being protected, not guaranteeing the necessary minimum for the conservation of the tropical biota. The caatinga and the cerrado are the great phytogeographic domains of Piauí. The transition areas are significant and they have a character of singularity due to the ecotones that have established in these areas. Quantification of transition areas depends on further studies and mapping in large scales because of the high spatial heterogeneity and of the mosaic distribution patterns of the vegetation. The caatingas of Piauí are marginal, as well as the cerrados. Current data on their characteristic physiognomies and principal species are updated.

Main Vegetal Formations of Piauí

In Piauí, the delimitation of the vegetal formations needs a wide revision because the sole publication on this issue is still incomplete and refers to all of them on a very small scale (1:6,000,000; CEPRO 1990, 1992a; Fig. 1). Emperaire (1985) presents the main vegetal formations, but as an exercise of subdivision in which she delimitates the Piauí semiarid area.

There are no direct estimations of the coverage of the vegetal formations of the state of Piauí. Usually, what exists in the literature refers only to the vegetation types and the name of the municipalities in which they appear. This information does not allow a good estimation, because the areas of the municipalities were modified due to a recent process of subdivision among them, which increased their number from 114 to 222 „municípios".

When one crosses and associates data reported by Emperaire (1985) and CEPRO (1992b), the main vegetal formations are the caatinga with 34.7%, the cerrado with 20.8%, the semideciduous forest with 14.9%, and the coastal vegetation with 1.0%. There are still two transitional areas, the caatinga/cerrado with 22.1% and cerrado/semideciduous forest with 6.5%.

In terms of isolated formations, the „carnaubais“ (populations of *Copernicia prunifera*), both native and cultivated, occupy an area of about 100,000 ha (CEPRO 1992c) and the coastal strip area occupies an area of 46,000 ha (CEPRO 1996), approximately.

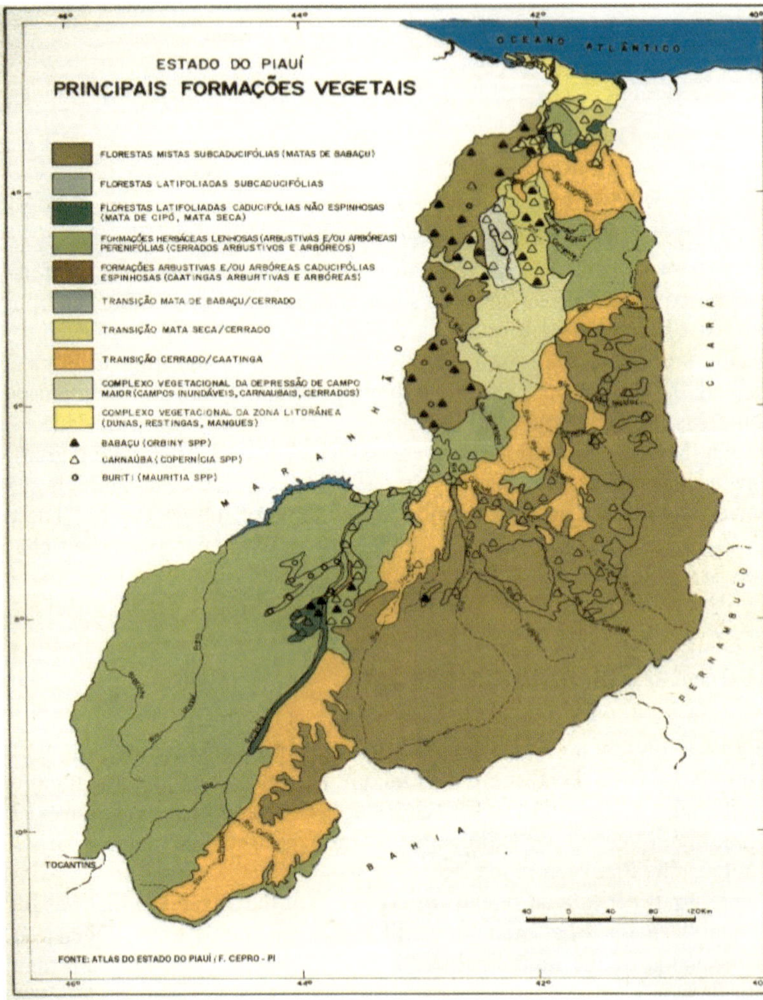

Fig. 1. Main vegetal formations of the state of Piauí according to CEPRO (1990, 1992a)

For Oliveira et al. (1997), the cerrado area corresponds to the one that was referred by CEPRO (1992a); however, for the caatinga areas and the transition area, the numbers differ with 37.3 and 18.9%, respectively.

Again, when one crosses and associates data from CEPRO (1992b) and IBGE (1996), the Piauí caatinga occupies 21.0%, the cerrado 28.4%, the transition 51.1%, and the mangrove/sandbank as a whole only 0.1%.

Tables 1, 2 and 3 show the comparative estimation of the areas occupied by the caatinga, cerrado, and transitional vegetation, respectively. These figures are calculated, in most cases, based on indirect data of some references.

Table 1. Comparison of estimations found in the literature with respect to the coverage of the caatinga vegetation in the state of Piauí

Reference	Occupation area (km^2)	% of total area *
Emperaire (1985)	87.451	34.8
CEPRO (1992b); IBGE (1996)	52.664	21.0
Oliveira et al. (1997)	93.600	37.3

*Base: 250.934 km^2

Table 2. Comparison of estimations found in the literature with respect to the cerrado vegetation

Reference	Occupation area (km^2)	% of total area *
Emperaire (1985)	52.276	20.8
CEPRO (1992a)	83.498	33.3
CEPRO (1992c); IBGE (1996)	71.359	28.4
Oliveira et al. (1997)	83.500	33.3

*Base: 250.934 km^2

According to CEPRO (1990), the vegetation in Piauí is divided into two large domains: caatinga and cerrado.

Caatinga

The caatingas, which are exclusive to the northeast and the north of Minas Gerais, according to Eiten (1972), in general present shallow soils, originated from limestone or crystalline rocks, which are fertile, when compared with the soils of the forest and of the cerrado zone. The annual average precipitation is less than 300 mm in its interior, 750–800 mm at the western border and up to 1000 mm at the eastern border. The number of dry months varies between 5 and 8, reaching up to 11 months. It does not possess a permanent water-table and the storage of plant-available water in the soil during the dry season, in typical conditions, is zero. The xeromorphic character of the leaves is rare, in spite of the considerable hydric deficiencies that the soil-plant-atmosphere system is submitted to.

Table 3. Piauí transitional vegetation: estimation of occupation area

Vegetation type	Reference	Occupation area (km^2)	% of total area *
Cerrado/caatinga transition	Emperaire (1985)	55.666	2,2
Cerrado/semideciduous forest transition (Babaçu Forest included)	Emperaire (1985)	16.347	6,5
Cerrado/other vegetational formations	CEPRO (1992a)	35.071	14,0
Several vegetational formations	CEPRO (1992c) and IBGE (1996)	128.233	51,1
Caatinga/cerrado/carrasco /semideciduous forest	Oliveira et al. (1997)	47.519	18,9

*Basc: 250.934 km^2

The physiognomy of the undisturbed vegetation can be of arboreal caatinga with low canopy, varying from closed to open; arboreal–shrubby caatinga with closed shrubby stratum, with or without low trees of 5–8 m height, dispersed on the shrubby layer, or of open shrubby caatinga, with 2–5m height (Eiten 1972). However, today, the caatinga is highly disturbed as a consequence of the itinerant pattern of agricultural exploitation during the last centuries. Despite the fact that the Piauí caatingas are still the most conserved of northeastern Brazil, mainly because of the low population density, in today's caatingas secondary vegetation caatingas frequently occur with an age of up to 20 years (Castro 2000, pers. inform.).

According to Fernandes (1982), the caatinga frequently represents a deciduous community, thorny, full of boughs, occupying, preferably, the inland depression. In Piauí, however, they are considered marginal, as long as they are distributed at the western borders of the drought polygon and lots of them on sedimentary soils.

Andrade-Lima (1981) characterizes the Piauí caatingas in two types. First, the low arboreal with trees of 5–7 m in height, on deep soils. Secondly, the arboreal type, open, ciliary, located on alluvial soils on the riverbanks with streams heading to the north. The more common species of the first type are: *Poeppigia procera*, *Dalbergia cearensis*, *Piptadenia moniliformis*, *Commiphera leptophloeos*, *Anadenanthera colubrina* var. *cebil*, and *Acacia bahiensis*. In the second type, the predominant species are *Copernicia prunifera*, *Licania rigida* and *Geoffroea spinosa*.

According to Fernandes (1982), the most common species of the caatingas in Piauí are: *Bursera leptophloeos*, *Mimosa acutistipula*, *Miracrodruon urundeuva*, *Caesalpinia bracteosa*, *C. pyramidalis*, *C. ferrea*, *C. microphylla*, *Combretum leprosum*, *Croton sonderianus*, *Aspidosperma pyrifolium*, *Tabebuia avelanedae*, *Luetzelburgia auriculata*, *Calliandra depauperata*, *Pilosocereus gounelli*, *P. glaucescens*, *P. pihauiensis*, and *Cereus jamacaru*. Emperaire (1982) adds among others: *Acacia glomerosa*, *Bocoa mollis*, *Thiloa glaucocarpa*, *Ziziphus joazeiro*, *Mimosa caesalpiniifolia*, *Cratylia mollis*, *Croton argyrophylloides*, and *C. hemiargyreus*.

Cerrado

According to Eiten (1972), the cerrado soils are almost always deeper than 10 m, with poor nutrient status, acid soils, and a low content of organic matter.

The annual precipitations vary from 1217–1709 mm, the average precipitation of the driest month varies from 5–30 mm. The number of dry months is between 3 and 5. In the interfluvial areas, the depth of the water-table is permanent at 10–35 m. In relation to the storage of water in the soil during the dry season, in typical conditions only the upper 2 m are dry. The scleromorphic character is very frequent, because, in conditions of light abundance, given by the latitude and relief, and abundance of water, in comparison with the caatinga, the carbohydrate production is low, producing thick membranes, developed veins and abundant sclerenchymas, according to Arens (1958).

It is a mesophyte vegetation, semideciduous, composed of two layers, an arboreal–shrubby (woody layer) with irregular canopy and, a herbaceous shrubby (low layer). In the areas of contact with the caatinga or the carrasco it has a deciduous form. The sympodial growth is frequent, stimulated by the death of the terminal yolks, provoked by fire or herbivorous insects and birds, translated by the irregular and tortuous ramification of the branches. The suberosy of the logs and the frequency of cartaceous and coriaceous leaves are also common.

It is a structural physiognomic continuation of the cerrados of the Central Plateau, but is floristically distinguished by the substitution of species provoked by three types of factors. First, the annual hydric deficiency of the soils, which grows in the direction Central Plateau-Northeast; second, because of the low altitude compared with the Central Plateau; and the third factor is related to the characteristics of an areal flora that answers to lati-altitudinal patterns associated with high levels of spatial heterogeneity, according to Castro (1994) and Castro and Martins (1999).

Fernandes (1982) points out that the cerrado vegetation dominates the Parnaíba river basin, covering plateaus, plains and a great part of the sedimentary pediplane. Locally, it is called „chapada“, when located in the southwest area of the state, and „agreste“, when located in the north-central area, due to the covering of the rural grass (*Diectomis fastigiata*). The „chapada" is located on very deep soils and the „agreste" on shallow soils, even Leptosols.

Among the most frequent woody species can be mentioned: *Salvertia convallariodora, Curatella americana, Qualea grandiflora, Q. parviflora, Annona coriacea, Magonia pubescens, Dimorphandra gardneriana, Caryocar coriaceum, Stryphnodendron coriaceum, Hymenaea stigonocarpa, Parkia platycephala, Sclerolobium paniculatum, Plathymenia reticulata, Anacardium occidentale, Byrsonima crassifolia, B. sericea, Vatairea macrocarpa, Bowdichia virgilioides, Terminalia fagifolia*, and *Hancornia speciosa.*

Among the species of restricted distribution can be mentioned: *Copaifera coriacea, C. luetzelburgii, Dimorphandra gardneriana, Hymenaea maranhensis, H. stigonocarpa* var. *pubescens, H. velutina, Caryocar coriaceum, Davilla cearensis, Diospyros brasiliensis, D. coccolobaeifolia, Lonchocarpus araripensis, Byrsonima blanchetiana, B. vacciniifolia, Calliandra abbreviata, C. fernandesii, Mimosa verrucosa, Stryphnodendron coriaceum, Virola sessilis, Allophylus quercifolius, Solanum baturitense, Callisthene microphyla*, and *Vochysia gardneri*, according to Castro et al. (1998).

Transitional Vegetation Types

The babaçu (*Orbygnia phalerata*) as a spontaneous invader plant occupies large portions of the areas of Teresina towards the north when associated with „capoeiras" (secondary regrowth). Towards the south, the babaçu forest maintains more contact with areas of cerrado and caatinga. Because of these associations, the dicótilo-palmácea forest, as documented by Fernandes (1982), has been more and more disturbed by the anthropic action, progressively losing the status of the third most important natural floristic–vegetational domain of Piauí.

The transitional character of the Piauí vegetation, which allows the appearance of ecotone areas, is provided mainly by the contact between cerrado and caatinga. Other contacts also exist enriching these domains with „carrasco" vegetation, deciduous drought forest, seasonal semideciduous forest, babaçu forest, carnaúba forest, and gallery forest. Another contact is with the vegetation complex, denominated „Campo Maior complex", that is formed by a mosaic of composed vegetation including „campo cerrado, a savannah of *Copernicia prunifera*, periodically flooded grassland and/or parkland" vegetation (Castro et al. 1998). Finally, it should be emphasized that most of these transition areas are found in conditions of low altitude and of small topographical unevenness, which has been the object of study of the „Núcleo de Referência em Ciências Ambientais do Trópico Ecotonal do Nordeste" (TROPEN) since its foundation in 1996.

Very little is known about these transition areas, mainly in floristic and phytosociological terms. In these areas, the limits of the ecosystems are often superposed and the lack of large-scale maps hinders the identification of the contact areas. These limits are influenced by the high heterogeneity of the landscape, which is reflected directly by the spatial floristic heterogeneities.

Conclusions

The main vegetation types found in the state of Piauí are the cerrado and the caatinga vegetation with a high proportion of transitional vegetation units like cerrado/caatinga cerrado/semideciduous forest transition.

Due to the lack of studies at a scale greater than 1:600.000, there are contrasting reports about the coverage of the vegetation units. However, all authors emphasize the high frequency of transitional ecotones, which makes it worth developing a conservation strategy for the state of Piauí.

References

Andrade-Lima D (1981) The caatingas dominium. Revta Brasil Bot 4:149–153

Arens K (1958) Considerações sobre as causas do xeromorfismo foliar. Botânica 15:25–56

Castro AAJF (1994) Comparação florístico-geográfica (Brasil) e fitossociológica (Piauí - São Paulo) de amostras de cerrado. Campinas, UNICAMP

Castro AAJF, Martins FR (1999) Cerrados do Brasil e do Nordeste: caracterização, área de ocupação e considerações sobre a sua fitodiversidade. Pesq Foco 9:147–178

Castro AAJF, Martins FR, Fernandes AG (1998) The woody flora of cerrado vegetation in the state of Piauí, northeastern Brazil. Edinb J Bot 3:455–472

CEPRO (Fundação Centro de Pesquisas Econômicas e Sociais do Piauí) (1985) Zoneamento ecológico do babaçu no estado do Piauí. CEPRO, Teresina

CEPRO (Fundação Centro de Pesquisas Econômicas e Sociais do Piauí) (1990) Atlas do estado do Piauí. CEPRO/IBGE, Teresina

CEPRO (Fundação Centro de Pesquisas Econômicas e Sociais do Piauí) (1992a) Cerrados piauienses; estudo preliminar de suas potencialidades. CEPRO, Teresina

CEPRO (Fundação Centro de Pesquisas Econômicas e Sociais do Piauí) (1992b) Perfil dos municípios piauiensis. CEPRO, Teresina

CEPRO (Fundação Centro de Pesquisas Econômicas e Sociais do Piauí) (1992c) Piauí; visão sumária. CEPRO, Teresina

CEPRO (Fundação Centro de Pesquisas Econômicas e Sociais do Piauí) (1996) Macrozoneamento costeiro do estado do Piauí; relatório geoambiental e sócio-econômico. CEPRO, Teresina

Eiten G (1972) The cerrado vegetation of Brazil. Bot Rev 2:201–341

Emperaire L (1982) Delimitação e regionalização do Brasil semi-árido; relatório da área de vegetação. Teresina (mimeographed)

Emperaire L (1985) Végétation de l'État du Piauí (Brésil). CR Soc Biogéogr 4:151–163

Fernandes A (1982) A vegetação do Piauí. In: Castro AAJF (coord) Congresso Nacional de Botânica, vol 32, pp 313–318. Universidade Federal do Piauí, Teresina, Brazil

IBGE (1996) Cidades e vilas do Piauí. Instituto Brasileiro de Geografia e Estatistica, Teresina, Brazil

Oliveira MEA, Sampaio EVSB, Castro AAJF, Rodal MJN (1997) Flora e fitossociologia de uma área de transição carrasco-caatinga de areia em Padre Marcos, Piauí. Naturalia 22:131–150

Degradation Factors in a Risk-Prone Area: the Semiarid Northeast of Brazil

Hartmut Gaese

Institute for Technology in the Tropics, University of Applied Sciences Cologne, Betzdorfer Str. 2, 50679 Cologne, Germany, Hartmut Gaese [bg075@dvz.fh-koeln.de]

The Portuguese settlement of northeast Brazil of the seventeenth century was initially restricted to the favorable locations along the coastal area („zona da mata") and to the neighboring sub-humid hilly region („zona do agreste"). Later, the settlement extended to the limited spots of favorable conditions located within the area of the semiarid savanna, the „Sertão" („brejos" and the humid „serras"). Droughts of varying duration and intensity have severely and repeatedly affected the interior of northeast Brazil, the „drought polygon", ever since the first settlement by the Portuguese. The consequences of droughts are damage to agricultural production and shortage in the supply of water to humans and animals. Cropping systems are affected by delay of rainfall, dry spells in the initial phase of the cropping cycle, or even a total lack of rainfall for one or several years. During the most destructive drought of the nineteenth century in Ceará, half a million people (at that time half of the population of Ceará) and almost all animals died.

Aspects from Settlement Hstory

It is difficult to explain why such an unfavorable area as the interior of northeast Brazil was settled at all. In the old world, such climatic zones were traditionally dominated by flexible nomadic systems. Concerning the land use systems in northeast Brazil in precolonial times, little is known apart from a few exceptions. It is most likely that locally fixed agriculture was exclusively practiced along rivers, e.g., the Rio São Francisco. It is assumed that these agricultural activities were limited to provide supplementary food to fishing and hunting activities (Boland 1997, p. 40). The Indians passed through the Sertão only sporadically during their hunting and foraging endeavors, and they settled at the „oasis" of favorable conditions. They applied gentle extensive techniques such as a slash-and-burn method covering wide areas and preserving soil fertility. A population pressure that leads to unsustainable slash-and-burn practices was not yet prevalent. Looking further into the past and considering climatic change, 10,000 years ago, the conditions in the region were semihumid or even humid.

According to Dickerson (quoted in Boland 1997, p. 41), the Portuguese learned how to cultivate a wide range of indigenous annual and perennial crops, e.g., the cultivation and processing of cassava, cashew nuts, palm trees, corn, and other tree crops. The Portuguese also adopted the slash-and-burn practice, without considering the conditions for sustainability, although, that is,. maintaining an appropriate relation of cropping and fallow periods to allow the regeneration of soil fertility after the land use cycle. Thus commenced a successive degradation of land according to the degree of the sensitivity of the respective ecosystem. Extensive pasture systems evolved as the predominant systems of the Sertão in response to the demand for meat for the labor force of the sugar plantations of the coastal areas and the miners in Minas Gerais. Following the collapse of the sugar industry on the eastern coast in the seventeenth century, labor force was released, and partly drifted to the interior of the country.

Principally, pasture systems can be adapted to the local conditions of such marginal zones as the Sertão. The sustainability, though, is assessed conversely in the literature, similarly to the conception of the climax vegetation. The felling of trees for ship construction and the production of charcoal after the appearance of the Portuguese surely contributed equally to the progressive degradation of the region as inappropriate pasture management. The effects of the degradation appear especially fatal with drought events. Considering that husbandry contributes undeniably to the degradation or even desertification of fragile regions, future research activities should take special account of the various animal production systems playing an important role in small-scale farming as integrated crop-animal production systems, as well as in large-scale extensive pasture management.

Agricultural Constitution

The agricultural constitution, i.e., the property rights and labor constitution, is still a further hampering factor for the development of the human ecological system of northeast Brazil. As a measure for inequality often applied in this context, the application of the indicator farm size structure is misleading. While pasture system farms certainly require substantial land resources (up to 30 ha/animal unit or even more), a few hectares of farm land in valleys of perennial rivers can guarantee the livelihood of a family, even without irrigation. Still, the farm size structure can reflect to some degree the extreme land distribution: More than 70% of all farms operate less than 10 ha, covering a total share of the agricultural land of 5.4%. On the other hand, farms of more than 1000 ha, which represent only 0.4% of all agricultural holdings, cover 32% of the farmland area. Thus, the land distribution gap in northeast Brazil exceeds, by far, the land distribution inequality of Brazil as a whole (IBGE 1998: Agricultural census 1995/1996). The agricultural constitution is a major factor for the low income in the agricultural sector and the unequal income distribution. Northeast Brazil is still among the poorest regions of the world: Three-quarters of the persons engaged in agriculture do not even achieve the income level defined as the „existence minimum“. The least-favored quarter of the agricultural labor force disposes of almost zero income, the second quarter must make a living on half of the minimum income or less, the third quarter finally achieves as a maximum the minimum income, which still is insufficient to make a decent living (Gaese 1996).

Macro and Regional Policy

For decades, the predominant economic policy of Brazil has been a policy of import substitution and the establishment of industries protected from international competition, resulting in a substantial distortion of factor and product markets. Only since the beginning of the 1990s has this protective policy been abandoned in favor of a rather market-oriented policy. The consequence of the long period of distorted product and factor markets, for the northeast, was a continuation of the development towards the regional, technological and economic dualism, initiated already in the nineteenth century. While the southeast and south of Brazil have been characterized by a boosting economy, in both urban as well as in rural areas, the development in the northeast has increased disparities among rural and urban areas, among the agricultural and the industrial sector, and among social levels. Since technical progress has been induced, the distortion of the markets has been transferred to technological development and application (Gaese 1997).

The financing of the development has to a large degree been provided by the primary sector, specifically the agricultural sector. This is also a reason to specially promote the export-oriented production systems within the sector. Research and development of technologies concentrate on those products important for export, such as sugarcane, coffee, soybean, and citrus. Success and progress have been impressive for these products, resulting in „technology jumps“ as, for instance, achieved for sugarcane, benefiting from the alcohol program. Contrary to the „modern“ export products and the secondary sector, the traditional agricultural sector has been disadvantaged. This has resulted in a misallocation of production factors (shift of comparative cost advantages), such as an out-migration of labor and capital from the agricultural sector and the rural areas (Gaese 1994).

This dualism, related to the difference in returns on the production factors, is particularly dominant in northeastern Brazil, featuring a prevailing „traditional“ agricultural structure and agricultural constitution, hampered in development by the risky natural environmental conditions.

Droughts, Emergency Programs and „Industria da Seca”

Natural production risks in the northeast are extremely high. Due to climatic variability, agricultural production suffers extreme yield insecurity, up to consecutive years of total losses. This insecurity also affects the industry and trade sectors, since the processing of agrarian primary inputs and the production of food play a major role in the economy of northeastern Brazil. This group of products suffers from a disproportionate fluctuation in prices while the price elasticity of demand is low. The degradation of large areas, erosion caused by inappropriate land use (population growth), intensity too high, extension of the agricultural frontier to fragile areas, etc., lead to an exponential amplification of the effects of drought events. Nowadays, 30% of the land use area of the drought polygon is considered strongly degraded.

Groundwater resources are said to be statistically speaking sufficient: 4300 m^3 per capita have been calculated (source?) which is more then twofold of what is considered sufficient according to UN standards (2000 m^3/capita). The common statement that drought problems in northeastern Brazil are merely a problem of distribution is a misinterpretation rooted in simplification of the drought problem. Comparing the Sertão to Israel and California, with the result that the Sertão could be turned into a green area using the available groundwater (Der Spiegel 24/1998), is definitely wrong. Still, drought events are usually linked to unexplained actions and inequalities which have led to the introduction of the term „indústria da seca", meaning drought-related business from which landlords, members of parliament, mayors and governors make profit by distracting funds meant for the needy population.

A survey by SUDENE (1999, p. 67) concerning the drought programs reveals that the drought-affected population – small-scale farmers, farm workers, day-laborers, settlers, etc. – are aware of the problems and have commonly experienced inequalities in receiving their assigned benefit. Undoubtedly, the Brazilian government has so far not succeeded in overcoming the evolving drought calamities.

Interaction of Factors Influencing the Agricultural Sector and the Farming System Development

The essential problem in farm development potentials, and thus for the development of the agricultural sector, is the risk, bound to the location, which limits the production and income potentials substantially. The adoption of technical progress is hampered and improvement investments are a gamble. Farmers respond to environmental conditions by adopting a risk-aversion strategy that is rational under the insecure production circumstances. The rationality of the farmers' behavior comprises also that farmers' risk aversion is proportional to the variance of climate and weather conditions. The more difficult it is to generate individual capital stocks and to secure subsistence production, the more dominant is the risk-aversion behavior of farmers. Thus, the objective knowledge of the expected production function (yields in response to production conditions at the time axis) and the probability distribution (variance) are of immense importance for the decision making of farmers, which actually is soundly based on experience.

In semiarid regions, the potential extent of field crops is restrained by the available family labor in periods of peak requirements (field preparation, seeding, harvest), as long as no mechanization of time constraint production processes are introduced. Small-scale farms do not dispose of the required capital, though, and, due to the absence of collateral for loans in these farms (agricultural constitution), this situation is not likely to change.

Additionally, the income generated by farming activities in small units does not allow the employment of permanent labor. Seasonal employment of labor is the most common strategy to overcome peak labor requirements, but, in view of the low cash income generated by the large majority of farmers in the region, household members themselves may need to generate cash income through off-farm employment in seasonal agricultural activities. Thus, farmers neglect their own plantations, experiencing difficulties in finalizing important field activities within the only possible time frame in which weather conditions allow the development of crops.

On the other hand, actual developments show a trend of population growth within the group of small-scale farmers. In the least advantaged group of farms, production conditions tend to be worse then average, and alternative income opportunities for cash income generation are too remote to be accessed. So the available labor force for this group has grown, which has led to a reduction in the income per person engaged in the agricultural sector. Risk potentials for further degradation of semiarid regions are basically located with this group of farms, who have no economic alternative to extracting as much as possible from the limited production factors available.

During recent decades, a substantial extension of the land use area in northeast Brazil has been observed (1959: ca. 58.34 Mio ha; 1985: ca.85.2 Mio ha). Simultaneously, the number of farms has tripled, basically among the group of small-scale farms. During this period, yields of the traditional production (beans, corn, cassava, cotton) dominating the small-scale farm sector have decreased, while those crops produced in the large-scale farm sector, cashew, cacao and sugarcane, show exactly the opposite trend. One reason for this development is that the small-scale farm sector had to abandon part of the favorable production locations in favor of the farming enterprises „with development potentials". Small farmers had to encroach marginal land, leading to a reduction of fallow area in relation to the cropping area and a shortening of the field rotation cycles. The result of these developments is a „poverty induced degradation". Also, research has done little yet to counteract these effects, which are becoming rather more severe due to the continuous increase of population pressure in the traditional sector. In contrast, research and technological development have concentrated on export-oriented production (as may be seen from the research goals of the national organization for agricultural research, EMBRAPA), putting the traditional sector once more at disadvantage. In addition, the political conditions of credit lines and low prices for food in favor of export-oriented agricultural products contribute to the comparative cost advantages of the „modern" agricultural sector and of the southeast, in comparison to the northeast, due to the different structure of agricultural production in the two regions.

Intro-regional Locational Factors of the Northeast

The introduction of modern technologies and, thus, the development of modern agriculture is, besides the expansion of the input and output processing sectors, highly dependent on the availability of water. This fact limits „modern" agriculture in the semiarid areas of the northeast to very few locations, where modern agriculture is almost exclusively found under irrigated conditions. Capital intensive irrigation, up to present predominantly under the regime of national agencies, accounts for high opportunity costs, since they are mostly located in those locations, i.e., river valley (várzeas), where the climate risk is reduced and rain-fed agriculture at a comparatively high intensity level is also possible. In these locations, irrigation systems have been constructed that yield unsatisfactory levels of output, caused by low water use efficiency and inappropriate irrigation management (water use efficiency 30%, 30% of the irrigation areas are severely affected by salinization).

Within the northeast, locational differences are considerable. The Sertão areas will continue to be the net importing areas for food. In response to the low production potentials, production costs are comparatively high. Thus, comparatively, cost advantages are only relevant for local markets, while more favorable areas have high potentials in this respect. The Meio-Norte, for instance, or the humid coastal areas, have a high export potential, exporting not only abroad, but also inter-regionally, for important primary products (rice, beans, corn) and for a wide range of tropical fruits (perennial bushes and trees as well as annual crops). The northeast has already achieved a level of self-sufficiency in the production of vegetables. For this product group there is a marketing potential to produce for seasonally varying markets in other regions of Brazil or the MERCOSUR. Considering the comparatively high-income elasticity of demand in northeastern Brazil, an increase in income can be expected to induce a substantial increase in the demand for food. The capacity of the regions with expressed Sertão characteristics will remain low, even if land is distributed more equally. A strategy of fixing people in the countryside (fixação do homem no campo) has to be considered problematic, as migration to areas of higher carrying potential is rather advantageous.

Political and Economic Environment and Its Effects on Development

Since the early 1990s, the political and economic environment of the northeast has undergone notable changes on a national level as well as on a state level. In the course of continuous liberalization measures, a pressure to adjust has evolved, which has partly been responded to on the level of MERCOSUR. Presently, the effects of liberalization on the agricultural sector in the semiarid areas of the northeast are ambivalent.

Concerning the goals for the predominant farming system, the Brazilian government has taken a considerable turn towards sustainable family farms

(Ministério do Desenvolvimento Agrário 2000). These goals are rooted in recognizing that family farms – as long as they are equipped with a minimum stock of resources – are much more flexible and, thus, less dependent on cyclic fluctuations of the economy than commercial enterprises. Besides, considering the aspect of welfare distribution, a farm structure dominated by family farms is much more sustainable in social terms.

Conclusions and Suggestions for Research

To assess the potential for further development of farming systems in the light of further increasing population pressure, the following crucial questions have to the raised:

Will the extension of the agricultural area be profitable? The national government demands an extension of the agricultural frontier in order to guarantee the supply of food and currency. This raises the question for the existence and location of unused land resources. It must be considered that the high potential land is already under production and that the encroachment of marginal land is linked to enormous transport costs, if market orientation is to be made a profitable endeavor in those regions. The intensity levels of production systems follow the development of infrastructure and the growth of decentralized subcenters.

Which would be the consequences for the agro-ecological environment if the spatial expansions were realized in the diverse locations? Which areas would suffer most from „poverty induced degradation“ if more and more marginal land were to be farmed by the increasing population? An increase in agricultural production under the condition of land scarcity – as happened in the historical settlement area „agreste“ – must inevitably lead to an increase in the optimum specific intensity through the application of improved technologies, which also induces an increase in the farming risk.

- What will be the role of the family farms with growth potentials? How can the new goals of the government be implemented into the agricultural constitution?
- Which will be the role of the presently expensive irrigation technology and what are the required conditions?
- Which are the optimal technologies for which location, and which land use systems can contribute to achieving sustainable land use (preservation of the carrying capacity of soils, long-term stability of the farms)?

The quantitative assessment of these interactions is complex. Economic models developed to include ecological restrictions have to be simplified. Literature in this context (Kutcher and Scandizzo 1981; BNB 1990) supports the recent approach of the Brazilian government to strengthen the family-farming sector. Family farms hold by number the largest share of farms in all regions of the northeast apart from the dry zones, which are dominated by extensive pasture systems. In the course of the ongoing structural change in the agricultural sector, the family-farming systems can be expected to increase their share related to both the operated land and the production value.

Plausibility considerations, thus, lead to the conclusion that the goal pursuit by the Brazilian government as well as by the state governments in the northeast, which is the „fixação do homen no campo“, cannot be successful, even if the economic environmental conditions were improved. The continuous out-migration should rather be considered as a favorable process in reducing the population in the fragile areas of the northeast. Further, it must be expected that technical progress will release even more labor force in the future. General economic development leads to a higher return on labor in other sectors and regions, which will in the long run increase the opportunity costs for labor in the agricultural sector. The ongoing integration into the framework of MERCOSUR will provoke additional impulse to adjust the production systems and spatial resource allocation to the competition situation for specific markets (e.g., meat, milk and cereals) in response to shifting comparative cost advantages. Resuming, all these factors will lead to accelerate the out-migration of the labor force from the disfavored regions of low production potentials to the preference regions with high development potentials. A large share of the labor force will migrate to the subcenters and centers and find employment there in the expanding processing and service sectors. For the primary production, only the „modern“ sector, which can only evolve in the high potential preference locations, can be expected to increase the demand for labor.

Concluding from the actual developments of the agricultural sector in northeast Brazil, demands for research arise especially concerning the development of appropriate system models to analyze economic-ecological trade-offs and to analyze sustainability-oriented resource management. The WAVES program responds to this demand, but WAVES experience has also shown that further research is required.

References

Banco do Nordeste do Brasil (BNB) (1990) A Unidade de Produção Agropecuária, vols 1–6, Fortaleza

Boland GM (1997) Farmers' organizations, poverty and the environment in the Sertão, North-East Brazil. Nijmwegen Studies in Development and Cultural Change. Verlag für Entwicklungspolitik Saarbrücken GmbH, Saarbrücken

De Melo ML (1978) Regionalização Agrária do Nordeste. Recife SUDENE, CPR

De Melo ML (1997) Estratégia Espacial para o Nordeste, Divisao do Espaço Territorial Nordestino em Subregiões segundo os Sistemas de Atividades e de Uso de Recursos. Recife SUDENE, CPR 1997

Der Spiegel (1998) Lateinamerika-Vertrauen auf Gott, 24/1998, pp 162–163

Gaese H, Hinterthür H (1999) Die Berücksichtigung des Produktionsfaktors Arbeit in einem Agrarsektormodell für einen bevölkerungsreichen semiariden Raum in Entwicklung: das Beispiel Nordost-Brasilien. Göttinger Beiträge zur Land- und Forstwirtschaft in den Tropen und Subtropen, „Stabilisierung und nachhaltige Entwicklung land- und forstwirtschaftlicher Systeme in den Tropen", vol 133, Göttingen

Gaese H (1994) Ursachen und Entwicklung regionaler Disparitäten im Agrarsektor Brasiliens. IKO-Verlag, Frankfurt/Main (ABP, Schriftenreihe des Zentrums Portugiesischsprachige Welt an der Universität zu Köln)

Gaese H (1995) Technischer Fortschritt und Dualismus in der Agrarentwicklung. VDI-Landtechnik

Gaese H (1996a) Standorte und Diversität landwirtschaftlicher Betriebssysteme in Nordost-Brasilien. IKO-Verlag, Frankfurt/Main (ABP, Schriftenreihe des ZPW an der Universität zu Köln, 1/1996)

Gaese H (1996b) Sektorale und räumliche Wirkungen des Technischen Fortschritts: Aspekte der Agrarentwicklung am Beispiel Brasilien. Entwicklungsländerbezogene Forschung - standortgerechte Technologien für den ländlichen Raum. FH Köln, Institut für Tropentechnologie, Cologne

Gaese H (1997) Protecionismo, mercados de fatores de produção e retorno à economia de mercado na América Latina. In: Moniz Bandeira LA, Pinheiro Guimarães S (eds) Instituto de Pesquisa de Relações Internacionais (IPRI): Brasil e Alemanha: a construção do futuro, Brasilia

Hall AL (1978) Drought and irrigation in north-east Brazil. Cambridge University Press, Cambridge

IBGE (Fundação Instituto Brasileiro de Geografia e Estatística) (1998) Censo Agropecuário Brasil 1995/95, Rio de Janeiro

IBGE (Fundação Instituto Brasileiro de Geografia e Estatística) (1990) Geografia do Brasil, vols 1–4, Rio de Janeiro

Kutcher GP, Scandizzo PL (1987) A Economía Agrícola do Nordeste do Brasil, Centro de Pesquisa para o Desenvolvimento – Banco Mundial

Ministério do Desenvolvimento Agrário (Governo Federal) (2000) Novo Retrato da Agricultura Familiar – O Brasil Redescoberto, Brasilia

Rönick V (1986) Regionale Entwicklungspolitik und Massenarmut im ländlichen Raum Nordost-Brasiliens; Ursachen des Elends und Hindernisse bei der Erfüllung der Grundbedürfnisse. Schöningh, Paderborn

SUDENE (1999) A Seca Nordestina de 1998-1999: Da Crise Econômica à Calamidade Social, Recife

Part 3: Climate Scenarios and Climate Modeling

Climate Analysis and Scenarios for Northeast Brazil

Friedrich-Wilhelm Gerstengarbe and Peter C. Werner

Potsdam Institute for Climate Impact Research, P.O. Box 60 12 03, 14412 Potsdam, Germany, gerstengarbe@pik-potsdam.de

Introduction

The climate system of the Earth is an extremely non-linear system. This means that its development cannot (or only in a very restricted way) be predicted (Lorenz 1963). If global climate changes are to be expected, their regional impacts will primarily be of special interest. Since global climate models are presently, and in the near future, not able to provide applicable model results of meteorological parameters at the regional scale (IPCC 1995), one has to find other possibilities for data supply for the estimation of climate developments in such areas. In doing so, consistency should not be violated, either spatially/temporally or between the meteorological parameters.Three fundamental methods for scenario development are presently being discussed:

Within the framework of the first method (the dynamical approach), one attempts to develop a regional climate model that is embedded in a global one and will be supplied by the GCM with the required boundary values (Machenhauer et al. 1996). The advantage of this procedure is the physical connection of large- and small-scale processes. Its disadvantage is that the modelling of connections and processes is presently still too inaccurate so that the model results cannot be used for regional investigations.

The second method uses climate model results, which will be transformed into smaller-scale areas by means of statistical procedures (Zorita et al. 1993). The advantage of this method is that direct access to the model results is possible so that a physical propagation of errors, as is the case with coupled climate models, will not occur. One disadvantage is due to the fact that the errors in the climate model will not be removed by the statistical approach but will directly be entered into the scenario.

Since the first two procedures do not yet provide satisfactory results, a statistical model was developed that led to distinctly improved results compared with those obtained previously (Werner and Gerstengarbe 1997).

The aim of the further investigations is the description of the future climate in northeastern Brazil using the new developed climate scenario model. After presentation of the method the climate development over the next 50 years is discussed in comparison with the results of the climate analysis between 1921/80 (see also Krol et al., this Vol.).

Method

The basis for the construction of a scenario is formed, on the one hand, by observed time series of meteorological parameters and, on the other, by information about the future development of climate (expressed by a selected climate parameter) calculated by a GCM. (For instance, the expected trend of the precipitation within the area under investigation can be defined as such a parameter.) According to these conditions, the expected changes (in general, a trend) are imposed on the observed values of the climate parameter (henceforth called the reference quantity). Using a special algorithm, the other observed meteorological parameters are adapted consistently to these changes. It is important that the statistical characteristics are kept intact. Figure 1 shows the complete scheme of the scenario calculation. It can be outlined as follows:

Step 1 - Selection of information about a meteorological parameter from GCM results (in this case: precipitation trend over northeast Brazil derived by ECHAM-4 calculations; Roeckner et al. 1996).

Step 2 - Making observed data available (for instance, a long-term series of all available meteorological daily data for a number of stations in the region under investigation).

Step 3 - Simulation of the series of the reference quantity using the information from step 1, maintaining the statistical characteristics of the observed data, i.e., inter-annual variability, annual cycle and persistence.

Step 4 - Cluster analysis based on those parameters which describe the reference quantity as exactly as possible (Gerstengarbe and Werner 1997; Gerstengarbe et al. 1999). In general, these are the parameters with the highest degree of information (for instance, daily sum of precipitation, intensity, precipitation sum of the previous day).

Step 5 - Calculation of the same parameter combination as in step 4 for each time step (day) of the simulated series and assignment to the most similar cluster using the Mahalanobis distance.

Step 6 - Each parameter combination of the simulated series calculated in step 5 can be completed with the observed data set which best corresponds to it.

Using this method it is possible to define appropriate simulated series for each of the other stations simply by classifying the given observed data on the basis of each element. Thus, the temporal and spatial consistency of climate conditions is preserved.

Remark: when applying this algorithm, physical laws must be considered as well as the statistical characteristics. This means, for instance, that the simulated time series should contain no abrupt breaks. For more information on the method, see Werner and Gerstengarbe (1997).

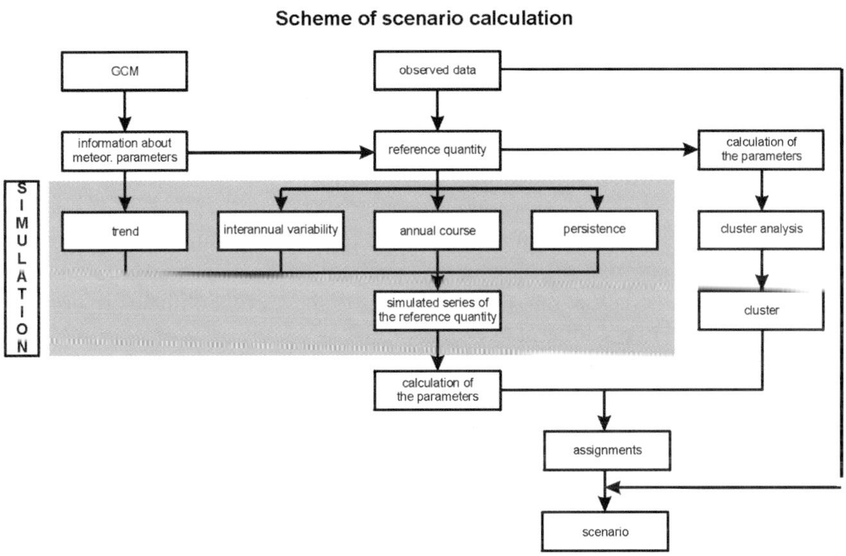

Fig. 1. Scheme of scenario calculation

Results and Discussion

A future scenario of the future (about 2050) climate in northeastern Brazil was developed. It proceeded on the assumption that the presently observed rise in CO_2 will continue until 2050. Based on the model simulations of the Max-Planck-Institute for Meteorology Hamburg, the future precipitation behaviour of the investigation area was estimated. Precipitation was defined as the selective parameter since it is the most important meteorological parameter for this region, and it also refers to the hydrological research questions of the project. On this basis, a climate scenario was calculated for the period 2001–2050.

The results can be interpreted as follows:

Precipitation

Figure 2 shows the mean precipitation conditions for the period 2041–2050 of the regional scenario. The last 10 years of the transient scenario run (2001/50) were selected in order to be able to evaluate the mean conditions at the end of the calculation period. One can see that the principal spatial structure has changed only negligibly over these 50 years (see Fig. 2 in Krol et al., this Vol.).

The situation is different if we look at the map of differences 2041/50–1921/80 (Fig. 3). A substantial decrease in the annual sums of precipitation can be observed for almost the entire investigation area with a span from 0 to 400 mm. The few small areas with a slight increase, however, do not count. This means that the entire area under investigation will be negatively influenced by longer and more intensive dry periods if the greenhouse gas concentration increases further.

Temperature, Relative Air Humidity, Radiation

Figures 4, 5 and 6 show the mean annual values for air temperature, relative humidity and global radiation in their spatial distribution for the period 2041–2050. For all three parameters it is true that the differences to the current climate are negligibly small (see, e.g., Fig. 4 in Krol et al., this Vol.).

Since it is therefore proven that these elements do not have any relevant changes influencing the climate, these results will not further be investigated or interpreted.

Duration of the Dry Period

The spatial distribution of the dry period duration referring to the analysis period is shown in Fig. 7. One can see a distinct spatial structuring varying between 165 and 225 days. Figure 8 shows the mean condition for the last 10 years of the scenario period. Maintaining the spatial structures, the dry period is getting longer in most areas of the states of Piauí and Ceará (Fig. 9). The scale of these changes varies from 0 to 20 days. A distinct decrease in the dry period can be observed in the south of Ceará, a region that is not included in the investigation area. It should also be mentioned that the increased length of the dry period corresponds to a simultaneous decrease in precipitation in this region.

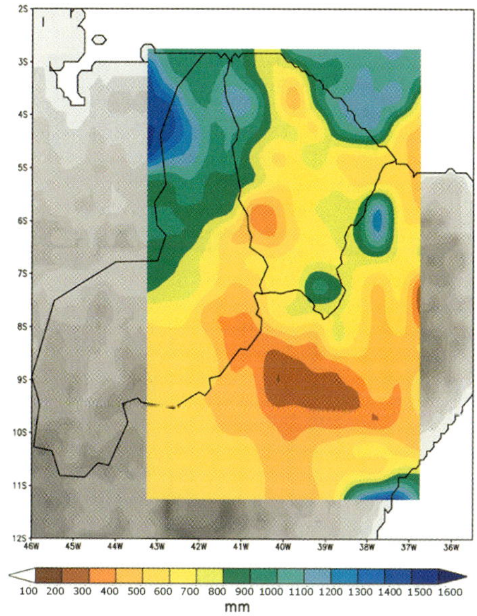

Fig. 2. Spatial distribution of the precipitation, mean values for the period 2041–2050

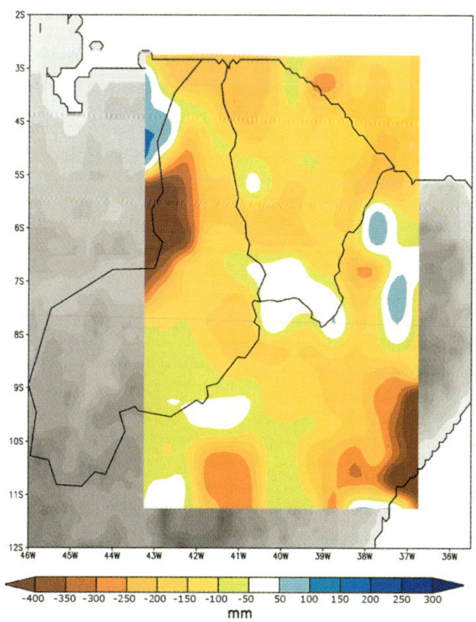

Fig. 3. Spatial distribution of the precipitation differences scenario 2041/50–1921/80

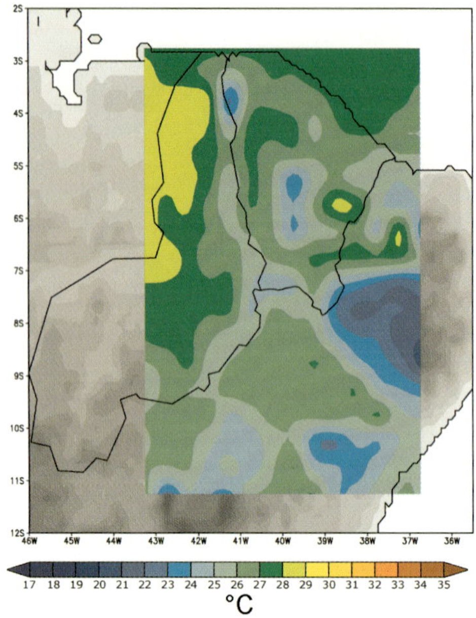

Fig. 4. Spatial distribution of the air temperature, mean values for the period 2041–2050

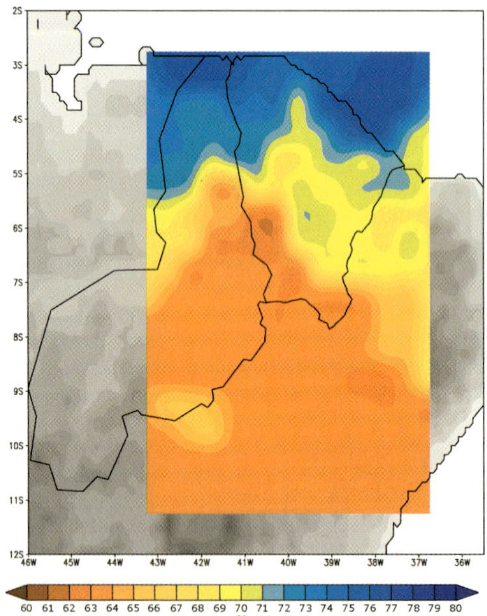

Fig. 5. Spatial distribution of the relative humidity, mean values for the period 2041—2050

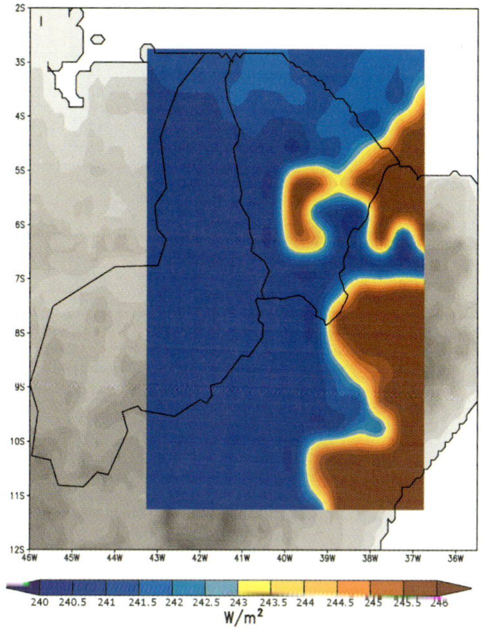

Fig. 6. Spatial distribution of the global radiation, mean values for the period 2041–2050

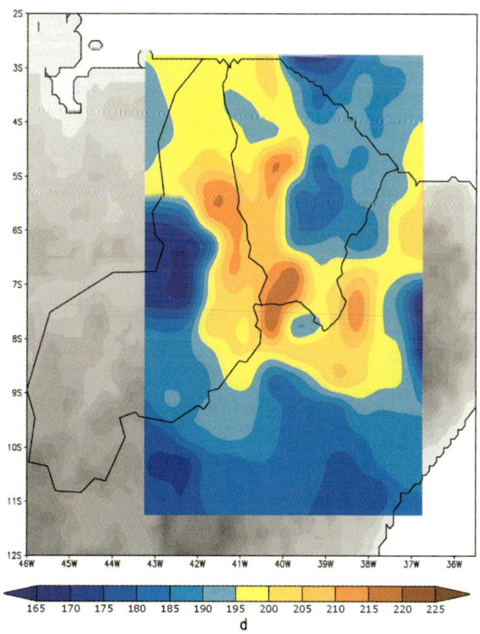

Fig. 7. Spatial distribution of the duration of the dry period, mean values for the period 1921–1980

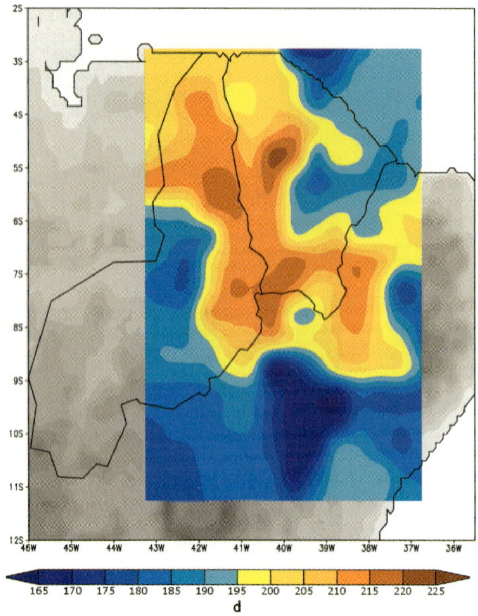

Fig. 8. Spatial distribution of the duration of the dry period, mean values for the period 2041–2050

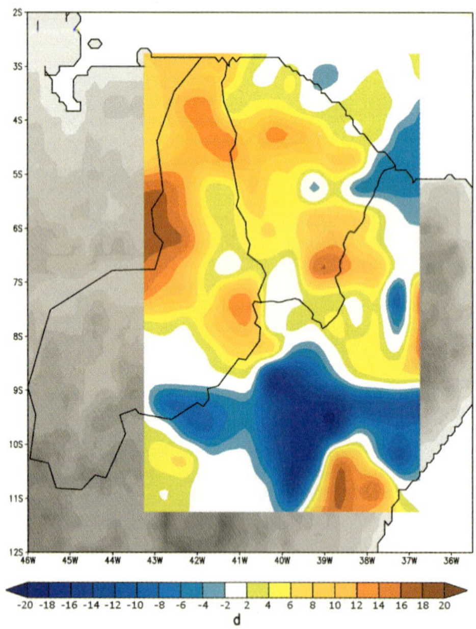

Fig. 9. Spatial distribution of the differences of the length of the dry periods 2041/50–1921/80

Shift of Climate Areas

According to the climate classification of Köppen (1936), the climate types of the investigation area were calculated with a modified variant (better inclusion of vegetation) of Köppen/Trewartha (1980) for the analysis period as well as for the scenario period which helped to determine the different shifts. Figure 10 shows that there are five climate types in the entire region: Af - tropical, permanently moist; Aw - tropical, alternately moist, dry winters; As - tropical, alternately moist, dry summers; BS - steppe climate; BW - desert climate; Cs - subtropical, summer dry, that shift in the period 2041–2050 according to Fig. 11. Two important climate shifts can be observed: On the one hand a shift from Aw to BS (for the entire area 12.4%=95,347 km²), on the other, a shift from BS to BW (for the entire area 19.8%=152,533 km²). The first shift is important for the areas of Piauí and Ceará, the second one for the southern region. The reference area around Tauá (Ceará) is also affected where the steppe climate changes to a desert climate.

Extreme Value Behaviour

To evaluate the extremes, the natural extreme value region NE (Gerstengarbe and Werner 1989) was used. As is generally known, the probability W of $x_1 \leq x_w \leq x_2$ can be defined as the integral of the probability density function between the limits x_1 and x_2:

$$W(x_1 \leq x_w \leq x_2) = \int_{x_1} f(x)\, dx \tag{1}$$

From the equation it is evident that for a constant probability W and a simultaneously increasing probability density f the difference between the limits x_1 and x_2 decreases, and vice versa. Hence, within a sample arranged by the magnitude of the values ($x_1 < x_2, \ldots, < x_w$), the magnitude of the difference between adjacent values is inversely proportional to the probability of the occurrence of the sample value. Therefore, the limit of an extreme value region can be defined as the point beyond which the changes of the differences of a partial region are significantly different from those of the remaining region of the differential series. This is the point where an obvious development showing a tendency starts. So, the definition of an extreme value region is reduced to the estimation of the beginning of an evident tendency. This point can be found by an adequate statistical test (for more information, see Gerstengarbe and Werner 1989).

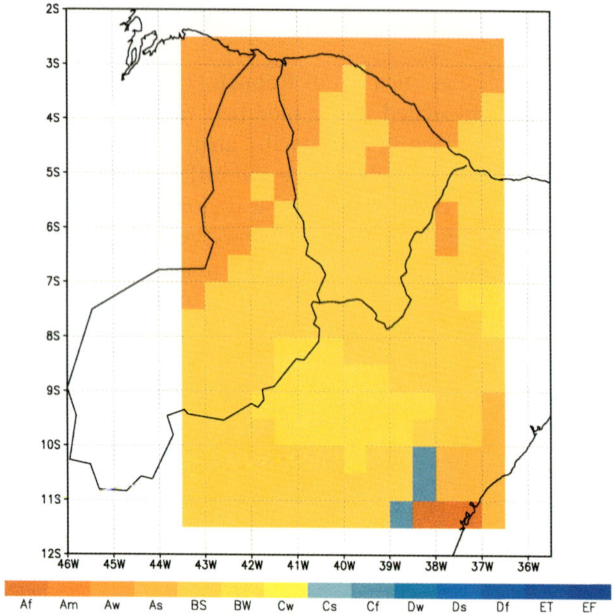

Fig. 10. Spatial distribution of the Köppen climate types, period 1921–1980

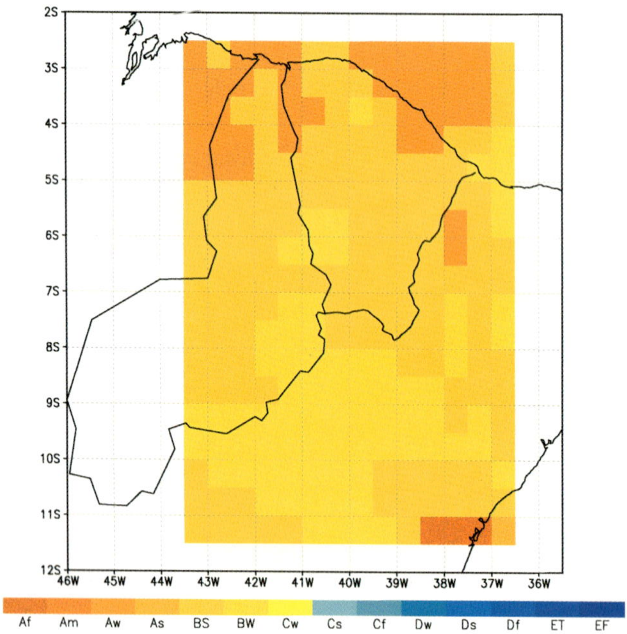

Fig. 11. Spatial distribution of the Köppen climate types, period 2041–2050

Figures 12 and 13 show the values of the NE for the daily sums of precipitation in the periods 1921/80 and 2001/50. One can see that the natural extreme value region in most regions is shifting to distinctly smaller values in the period 2001/50. This corresponds with the decrease in precipitation in these regions that could have already been proven in the scenario calculations. This means that, apart from the decrease in the sum of precipitation, its intensity is also expected to be weaker.

It is remarkable that in the investigation of the NE, no great changes could be observed for the duration of the dry periods in both periods (Figs. 14 and 15).

As an additional example for decreasing intensity, the 100 highest daily sums of precipitation for the station Cedro are shown in Figs. 16 and 17. Also here, one can see that the sums of absolute values per year are lower in the second period than observed so far and that at the same time the frequency of extremes is decreasing at the end of this period.

Thus, it could be proven that the climate changes to be expected have an influence on the extreme value behavior.

Conclusions

To sum up it can be said that a moderate global climate change scenario within the next 50 years will lead to considerable changes in the climate regime in the northeast of Brazil. This is mainly expressed by a decrease in annual precipitation in the entire region connected with an obvious shift of the climate areas.

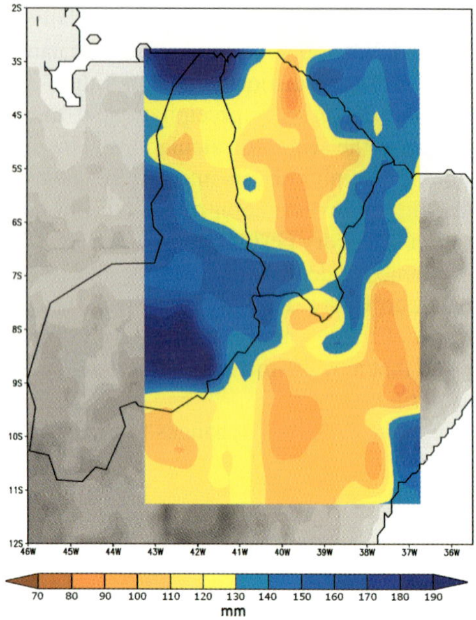

Fig. 12. Spatial distribution of the natural extreme value region for the daily sums of precipitation, period 1921–1980

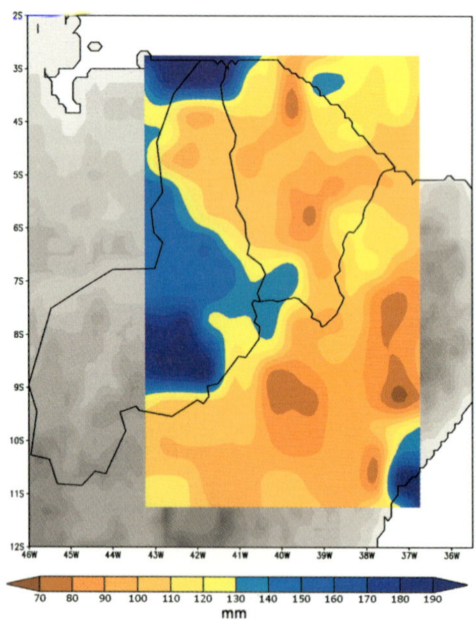

Fig. 13. Spatial distribution of the natural extreme value region for the daily sums of precipitation, period 2001–2050

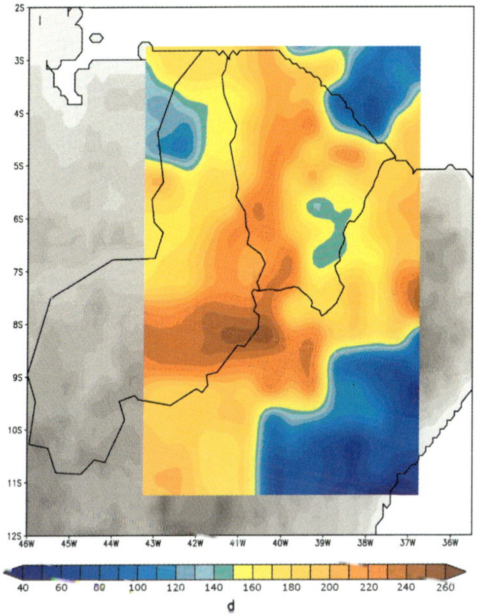

Fig. 14. Spatial distribution of the natural extreme value region for the length of the dry period, period 1921–1980

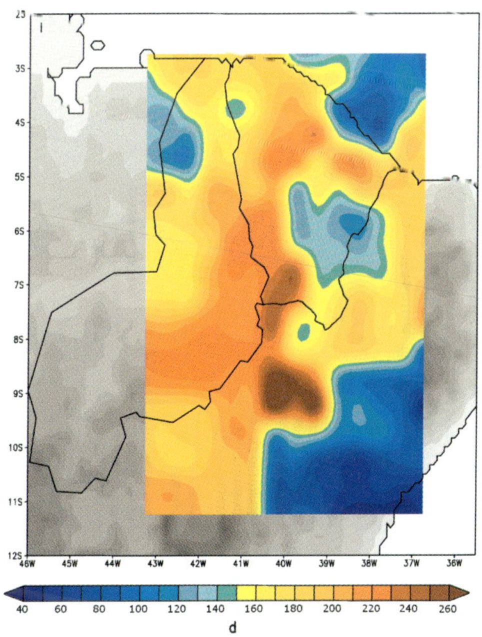

Fig. 15. Spatial distribution of the natural extreme value region for the length of the dry period, period 2001–2050

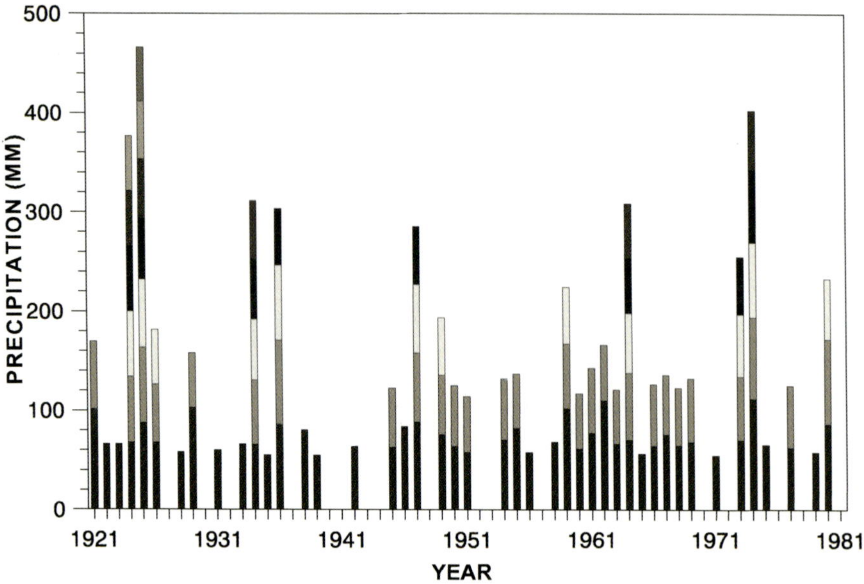

Fig. 16. 100 largest sums of precipitation, Cedro, period 1921–1980. Different events in the same year are plotted with different grey levels

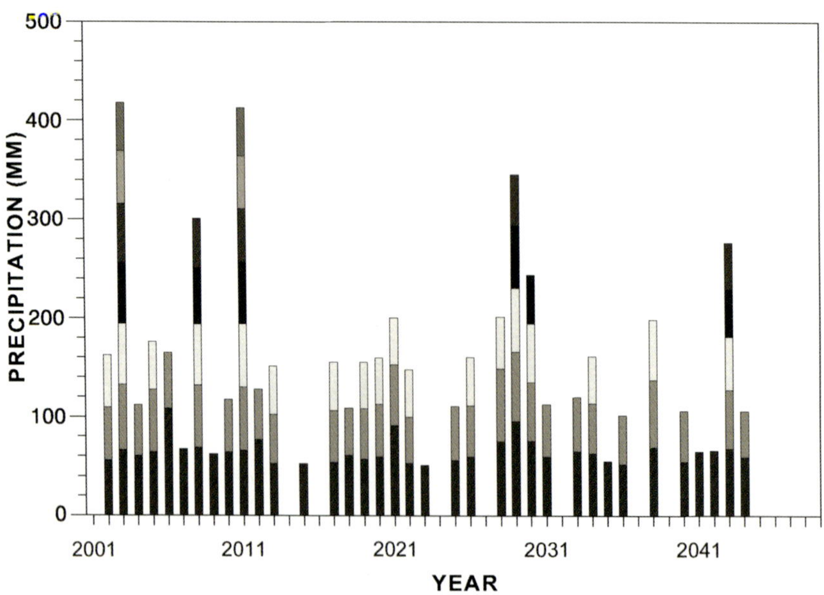

Fig. 17. . 100 largest sums of precipitation, Cedro, period 2001–2050. Different events in the same year are plotted with different grey levels

References

Gerstengarbe F-W, Werner PC (1989) A method for the statistical definition of extreme value regions and their application to meteorological time series. Z Meteorol 39(4):224–226

Gerstengarbe F-W, Werner PC (1997) A method to estimate the statistical confidence of cluster separation. Theor Appl Climatol 57:103–110

Gerstengarbe F-W, Werner PC, Fraedrich K (1999) Applying non-hierarchical cluster analysis algorithms to climate classification: some problems and their solution. Theor Appl Climatol 64:143–150

IPCC (1995) Climate change 1995. Summary for policymakers. Cambridge University Press, Cambridge UK

Köppen W (1936) Das geographische System der Klimate. Bornträger, Berlin, p 388 (Handbuch der Klimatologie, vol 1/C)

Lorenz EN (1963) Deterministic non-periodic flow. J Atmos Sci 20:130–141

Machenhauer B, Windelband M, Potzet M, Jones RG, Déqué M (1996) Validation of present-day regional climate simulations over Europe: Nested LAM and variable resolution global model simulations with observed or mixed layer ocean boundary conditions. Max-Planck-Institut für Meteorologie, Hamburg (MPI report 191)

Roeckner E, Arpe K, Bengtsson L, Christph M, Claussen M, Dümenil L, Esch M, Giorgetta M, Schlese U, Schulzweida U (1996) The atmospheric general circulation model ECHAM-4: model description and simulation of present-day climate. Max-Planck-Institut für Meteorologie, Hamburg (MPI report 218)

Trewartha GT (1980) An introduction to climate, 5th edn. McGraw-Hill, New York, p 437

Werner PC, Gerstengarbe F-W (1997) A proposal for the development of climate scenarios. Climate Res 8:171–182

Zorita E, Hughes JP, Lettemaier DP, von Storch H (1993) Stochastic characterization of regional circulation patterns for climate model diagnosis and estimation of local precipitation. Max-Planck-Institut für Meteorologie, Hamburg (MPI report 109)

Dynamic Regional Climate Modeling and Sensitivity Experiments for the Northeast of Brazil

Uwe Böhm, Friedrich-Wilhelm Gerstengarbe, Detlef Hauffe, Martin Kücken, Hermann Österle and Peter.C. Werner

Universität Potsdam, Institut für Physik, Postfach 60 15 53, 14415 Potsdam, Germany, boehm@pik-potsdam.de

Introduction

This research topic has two main goals The first is to supply spatially and temporally consistent data for a drought period in the northeast of Brazil, generated by a dynamic modeling approach. The second is to elaborate a state-of-the-art regional-scale climate model, which can be applied to simulate scenarios of future climatic developments in the northeast of Brazil. For this, two different types of such models were adapted to the region of interest and were investigated for their ability to reproduce present-day extreme weather conditions. To assess a model's performance here in terms of the specific drought conditions in the WAVES project area under agricultural aspects, a new validation algorithm has been developed and applied. With this method, errors can be quantified on a statistical basis and can then be accumulated to form general quality measures. Possible sources of inaccuracies were identified as being linked to the representation of hydrological soil processes. On the basis of these results, sensitivity experiments were designed and carried out to investigate the effect of implementing various modified descriptions of the soil hydrology on the accuracy of the model simulations. Finally, a regional climate model version was elaborated which is able to reproduce present-day climatic conditions within a range of errors as characteristic for coarser-scale dynamic modeling approaches, but with a better representation of smaller scales. Remarkable improvements were achieved regarding the representation of drought conditions by the dynamic climate model. It is, however, still a challenging task for the future to further reduce the range of errors for regional-scale features in the simulation results to such an order of magnitude as is known from gridded observations. To reach this objective, further investigations on soil–water and energy balances and near-surface vertical exchange processes are required.

Regionalization Strategy and Validation Procedure

To achieve the goal of this WAVES subproject as formulated above, we used a step-by-step strategy. Because no long-term climate simulations for South America using dynamic regionalization techniques have been undertaken so far (IPCC 2001), we first set up a dynamic regional modeling system for our region of interest using a so-called nesting approach. Here, a limited-area fine-scale model is driven by the results of a coarse global model at the lateral boundaries with no feedback to the driving model. This method is therefore also known as one-way nesting technique. Since it is important to estimate the reliability of such a regionalization system as reported, e.g., for Europe by Christensen et al. (1997) or Machenhauer et al. (1998), much effort has been put into the evaluation of the simulation results. In order to assess the accuracy of model results in terms of such climatic parameters that are of interest as inputs for other modeling groups within the WAVES project, we developed a validation algorithm that allows us to evaluate simulation results with respect to such criteria. We applied this technique for a multi-criteria assessment to obtain a general overview of a model's performance. Depending on the results, we subsequently performed a single-criteria quantitative diagnosis to estimate the error amount produced by the individual parameters. Using this information, efforts were made to identify possible sources of the inaccuracies detected. In a next step, we implemented some modifications into the representation of the corresponding processes in the model and performed sensitivity experiments with these configurations. Finally, a re-evaluation of the results obtained gave information on the improvements that had been achieved.

The dynamic regional climate models we applied are described in the following section, and the subsequent two sections describe the general set-up of the simulation runs and the developed validation algorithm and the diagnostic tools used, respectively.

Regional Climate Model Description

In a first step, we used the regional climate model REMO version 1.0, which was developed by the Deutscher Wetterdienst (DWD) and the Max-Planck-Institut für Meteorologie (MPIfM). It is a so-called hydrostatic model, which solves the basic equations for adiabatic atmospheric processes in a terrain-following hybrid coordinate system under the assumption that time derivatives of the vertical component of the momentum can be neglected as insignificant compared with the typical length scales of changes in the horizontal directions. The prognostic variables (surface pressure, horizontal wind components, temperature, water vapor and cloud water content) are computed using second-order central differences in space and a leap-frog time-stepping scheme with semi-implicit corrections. Diabatic sub-grid scale physical processes are parameterized using two different approaches. One of these packages was originally developed for the former operational weather forecast model of the DWD. The second one originates from the global circulation model ECHAM4 of the MPIfM (Roeckner et al. 1996) and was adapted in REMO for use on regional scales. For more details, see Jacob et al. (1995). The REMO version, which comprises the physical package of the DWD (DWD physics), gave the most realistic results in initial test runs for the northeast of Brazil. Therefore, this version was chosen for the investigations.

Further, we adapted the DWD's new operational forecast model for Europe (LM version 2.1) for long-term simulations and applied it within this WAVES research topic for the first time for an integration period longer than 1 week. It is a so-called non-hydrostatic model without the above-mentioned assumptions concerning the time derivatives of the vertical component of the momentum. The prognostic equations for the physical components of wind velocity, perturbation pressure relative to a hydrostatic basic state, temperature, specific humidity, cloud water content and (optionally) cloud ice content are formulated in such a way that they are fully compressible, with no scale approximations, and are solved in a generalized terrain-following coordinate system. For spatial discretization, second-order finite differences are used. A mode-splitting scheme is employed for time integration, which allows a separate treatment of fast sound waves. The parameterizations of sub-grid-scale diabatic processes are formulated in a way similar to those in REMO with DWD physics, but with additional options, e.g., for the computation of evapotranspiration. More details can be found in Doms and Schättler (1999) and Schättler and Doms (2000). We have implemented this LM version on the SP multi-processor computing platform at PIK and modified it to remove technical restrictions for long-term simulations. Furthermore, new post-processing filters and re-run facilities were developed and implemented. A pre-processing algorithm was elaborated to prepare ECMWF re-analyses as initial and boundary values for the LM in a comparable way as for REMO. The LM itself was amended to allow for dynamical boundary condition updates at the surface as required for long-term simulations.

The use of LM in addition to REMO is justified mainly by two aspects. First, it was expected that a non-hydrostatic model would better represent the real atmospheric conditions when steep orographic gradients occur within the model region, as in the case of the Andes. Secondly, it can be applied on a massive parallel computing platform and allows sensitivity experiments to be performed very quickly.

Experimental Design

In accordance with the goal of this research topic, we selected the year 1983 as a well-known period in the past in which extreme dry conditions were reported in the northeast of Brazil. For this episode, the regional models described above were integrated to reproduce the principal atmospheric characteristics in the region of interest. To allow for a model spin-up, an extended integration period was defined from December 1982 to December 1983. We specified a model region of about 9600×6000 km^2 covering a window from 95.75°W to 0.25°E and from 29.75°S to 30.25°N, as shown in Fig. 1. This very large area was required in order to include at least partially the dominant large-scale atmospheric circulation patterns influencing the northeast of Brazil, such as the Atlantic subtropical high-pressure systems and the Pacific El-Niño phenomenon. The model grid was set up with a latitude/longitude resolution of 0.5×0.5° and 20 levels in the vertical direction for both models. This leads consequently to 193×121 horizontal grid points. Initial and boundary conditions for both regional models were generated from ECMWF (European Centre for Medium Range Weather Forecast) re-analyses (Gibson et al. 1997). We initialized the models once for the 1st December 1982 and afterwards we provided boundary conditions only at 6-h intervals. Model results were saved every 12 h for REMO and every 6 h for LM. In addition to a different formulation of the dynamic equations, the use of an alternative approach to compute evapotranspiration represents the major difference between the two model configurations. In REMO, the Budyko method (Schrodin 1995) is applied, whereas in LM we opted to use the algorithm of Dickinson (1984).

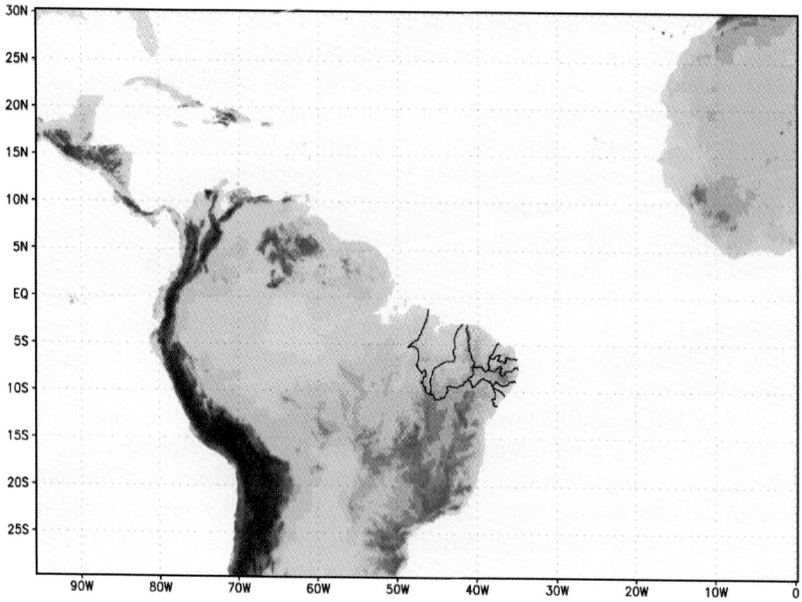

Fig. 1. Model region for REMO and LM used to simulate the period December 1982 to December 1983

Validation Methods and Diagnostic Tools

The central module of our validation approach consists of a technique that was described in detail and extensively tested by Böhm (1999). We applied this method here to evaluate the results of different climate model runs, ECMWF re-analyses and gridded observations against observed reference data at station sites by means of criteria that are capable of describing extreme conditions for agriculture in the northeast of Brazil. The basic structure of this algorithm is shown in Fig. 2. Firstly, criteria suitable for characterizing an agricultural drought in terms of climatic parameters, such as, e.g., the potential total yield loss, were derived for some of the most important crops in the federal states of Piauí and Ceará (maize, cowpea, manioc) in close cooperation with other project groups. Using these criteria, normalized parameters were computed from drought-related climate variables. In our case, monthly precipitation totals were applied. Considering all derived parameters at once, a non-hierarchical cluster analysis (Gerstengarbe and Werner 1999) was performed under multiple aspects, to identify the spatial patterns of an extreme. We applied this technique separately for both either model results or analyzed data that were interpolated to station sites on the one hand, and for observed reference data from station sites on the other. Both sets of patterns were then related to each other. For this purpose, several well-known distribution-free tests were utilized to quantify significant differences between the most similar clusters as derived for both test and reference data. With these results, overall measures were defined to assess a model's performance as comprehensively as possible. For this purpose, we combined the individual test results and distance estimates between comparable clusters in the two data sets that were obtained previously under various interesting aspects.

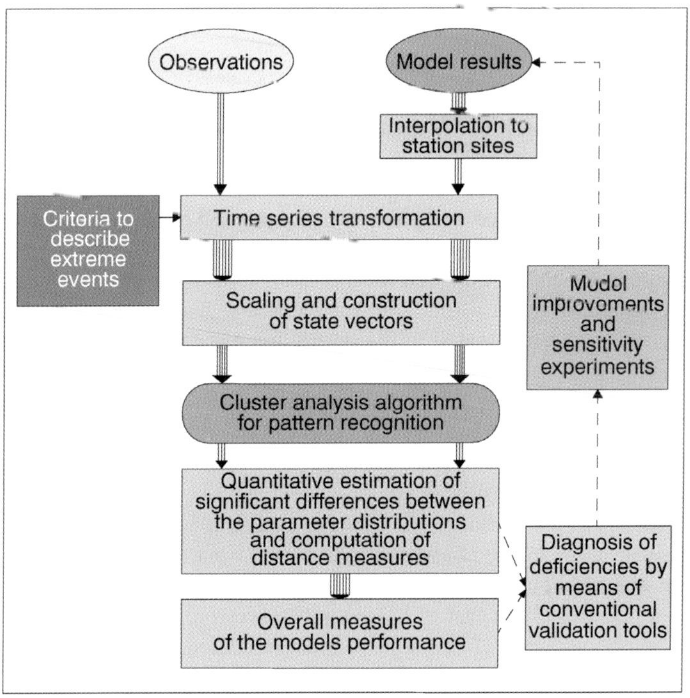

Fig. 2. Underlying structure and basic modules of the developed validation algorithm

If the accuracy of a model is judged to be insufficient according to these generalized error measures, the quantitative test results for the individual parameters can be used to identify those parameters giving the strongest error contributions. For our research within the WAVES project, certain minimum precipitation threshold values are used for the different phases of an idealized vegetation period for the crops investigated. On the basis of this supplementary information, conventional validation tools can be applied to inspect the related climate variables. Then, their representation within the model under investigation can be revised. Finally, a modified formulation of these processes can be implemented and tested as part of further sensitivity experiments.

Results

We assess firstly the performance of REMO and LM with respect to the representation of the extreme climatic conditions of 1983 by means of the validation algorithm described above. We focus here on the investigation of the potential total yield loss for maize, which is the most important agricultural crop in the federal states of Piauí and Ceará. With these results, we then try to identify possible sources of errors in the models. Subsequently, the outcome of some sensitivity experiments is discussed and requirements for further diagnostic investigations and model improvements are derived.

Potential Drought Risk for Maize in the Northeast of Brazil

Fig. 3 shows the outcome of the non-hierarchical cluster analysis for LM and REMO results. For reference, the patterns that were derived from observations are added. The parameters we have utilized describe certain critical precipitation threshold values for maize between the first and fourth month after sowing, from anthesis to the end of grain filling, and for the entire growing season. For REMO, the drought risk is overestimated in the northern and central parts of the diagnosis region. This has proven to be a general tendency in the model results, and the most endangered regions are also shifted to the northwest. Altogether, about one-third more data than for the observations was found in the driest clusters, expressing an over-high risk of agricultural yield loss. With regard mainly to coastal regions, however, the simulation results revealed a tendency to less hazardous conditions compared with the observations. The analysis of LM results gives a similar picture for the regions close to the coast with an underestimation of potential yield loss there. The structural representation of the drought risk, however, is remarkably improved, especially in the east of Piauí and the western parts of Ceará. On average, however, the severity of drought is underestimated. As described above, we computed general quality measures to describe these deviations between the spatial patterns quantitatively.

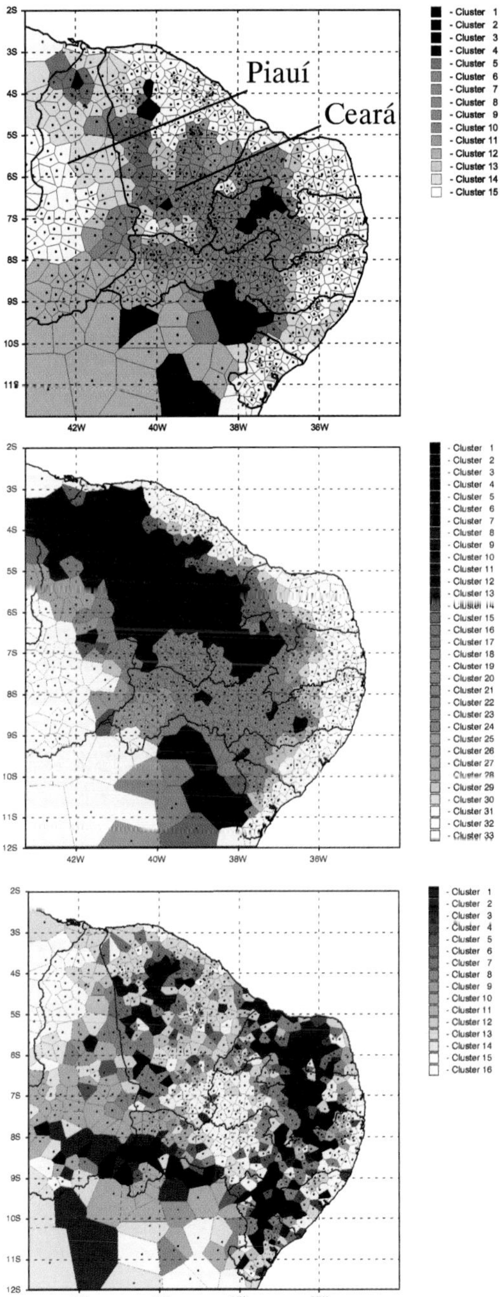

Fig. 3. Potential total yield loss for maize in 1983 for the northeast of Brazil; derived from LM results (upper panel), from REMO results (middle panel) and from observations (lower panel)

In Fig. 4, Q_1 expresses the differences between the most similar clusters of each particular test data set and the reference data. Q_2 describes structural differences. It can be seen that both regional models gave quite similar results for Q_1 in relation to the global circulation model ECHAM4. This leads to the conclusion that no significant improvement in the representation of drought severity has been achieved so far by the regional climate models used. The structural inaccuracies, however, are clearly reduced in the LM results as compared with those from REMO. Nevertheless, Q_2 appears still at the lower boundary of the range that is given by the ECHAM4 results and ERA data, as representatives of global models. Somewhat surprisingly, even the gridded observations that were generated by the Climate Research Unit (CRU) of the University of East Anglia (New et al. 2000) show clear deviations from station observations (an underestimation of the drought severity).

To explain these general validation results for the two regional models, the individual parameters that we used for the cluster analysis were examined. For the REMO results, we obtained the strongest deviations from the observations for those parameters describing the precipitation deficit with regard to the critical threshold values for maize in February and April. Especially in the latter month, all of the most similar clusters when comparing those derived from model results and observations showed significantly different distribution functions for the corresponding parameters. As a consequence, the amount of precipitation during the entire growing season was also underestimated. We likewise diagnosed the inaccuracies for the individual parameters for LM. We found that the total precipitation for January and especially for March was now overestimated, but in no case were the distribution functions for these parameters significantly different in any of the most similar clusters between model results and observations.

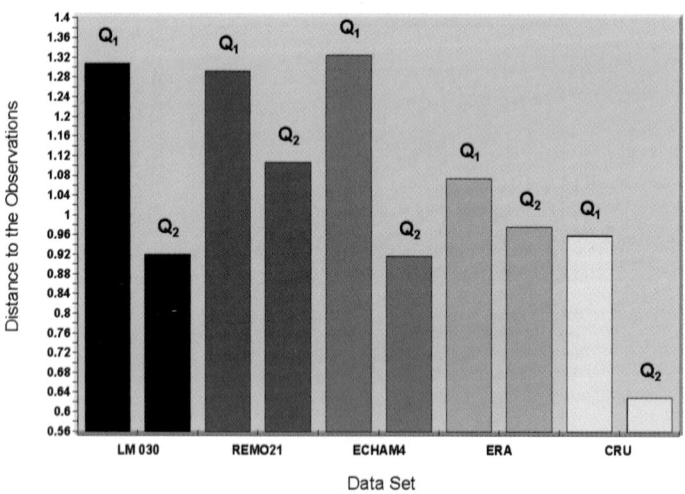

Fig. 4. General quality measures Q_1 and Q_2 for LM and REMO results, ECHAM4 results, ECMWF re-analyses (ERA) and CRU gridded observations

Possible Causes of the Identified Model Deficiencies

The results gathered so far provided evidence of model deficiencies as regards the generation of extreme low rainfall amounts within precipitation distributions. For a further diagnosis, we utilized not only these extremes, but also the entire precipitation distribution for the rainy season. During this period, the major contribution to total annual rainfall occurs. As shown in Fig. 5, both regional models generated more total precipitation in coastal regions. This could be an indication of a deficiency common to both models, regardless of the different formulation of the dynamic and evapotranspiration processes. The problem detected here could probably in this case be attributable to one of the physical parameterizations. There are hints from investigations with LM for the tropical western part of Africa that convection could be a candidate. Focusing on the REMO results, too little rainfall was simulated for the central parts of the northeast of Brazil. The LM results, however, gave a much better representation of precipitation here, especially in terms of the spatial structure. Nevertheless, too much precipitation was now simulated for the entire diagnosis region.

In addition to the spatial representation of typical precipitation patterns during the rainy season, the temporal evolution of precipitation within the region of interest is an important indicator of the model performance. Except for October 1983, REMO generally generated a too-low amount of precipitation for the northeast of Brazil (see Fig. 6). However, the annual cycle is well reproduced. The latter is also true for LM. During the rainy season, however, this model simulated excessive precipitation. For March 1983 in particular, the total amount of rainfall for the northeast of Brazil was strongly overestimated. However, according to the spatial rainfall representation in Fig. 5, this problem seems to be restricted to the regions close to the coastline, whereas large parts of the northeast of Brazil show a reasonable good agreement between model results and observations.

Qualitatively, the same result is obtained when computing the spatial mean for all land grid points within the entire model area (not shown). For REMO, the difference between modeled and observed precipitation is now reduced during the last few months of the simulation period. This indicates that the model is able to recover from over-dry conditions and may simulate subtropical low-pressure systems, which propagate northward during October and November 1983 in a more realistic way than the preceding weather situations.

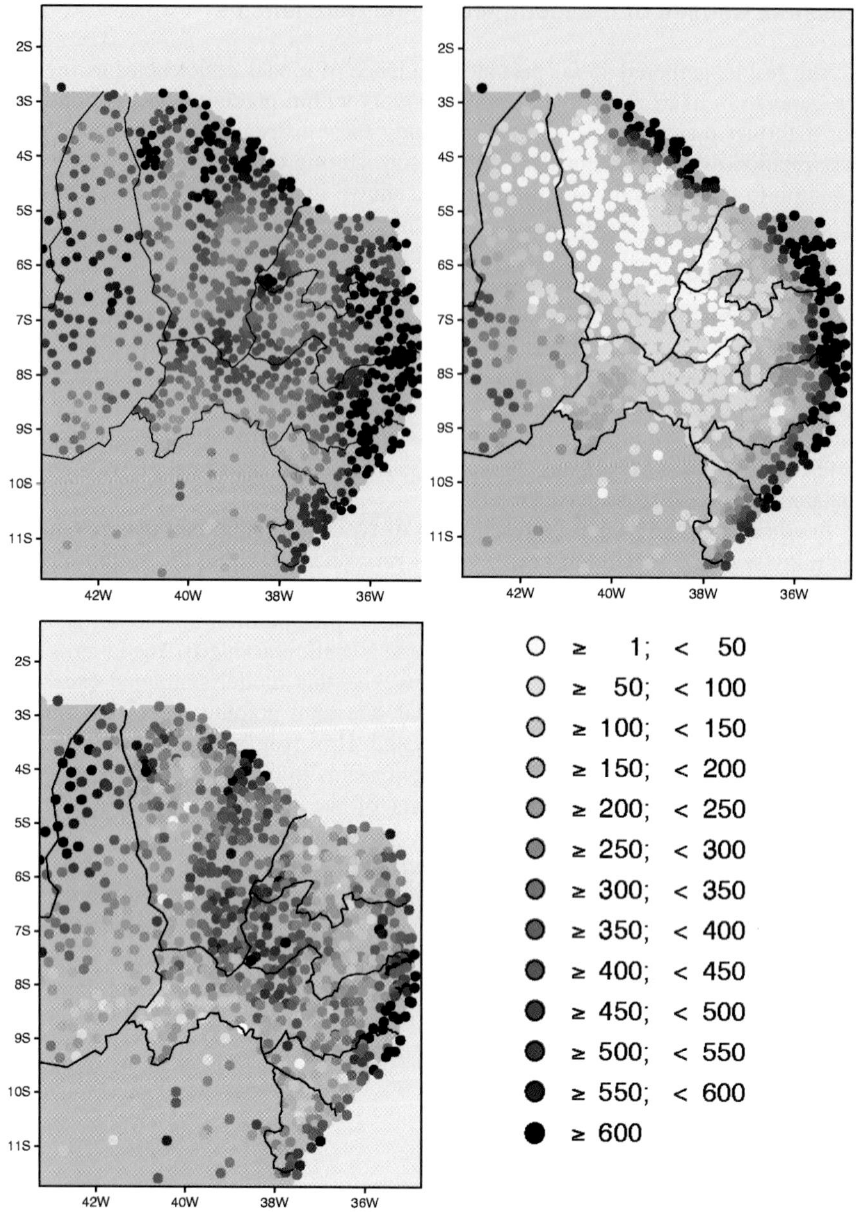

Fig. 5. Total precipitation during the rainy season (February - May) 1983 [mm]; LM (upper left) and REMO (upper right) model results when driven by ECMWF re-analyses and observations (lower left)

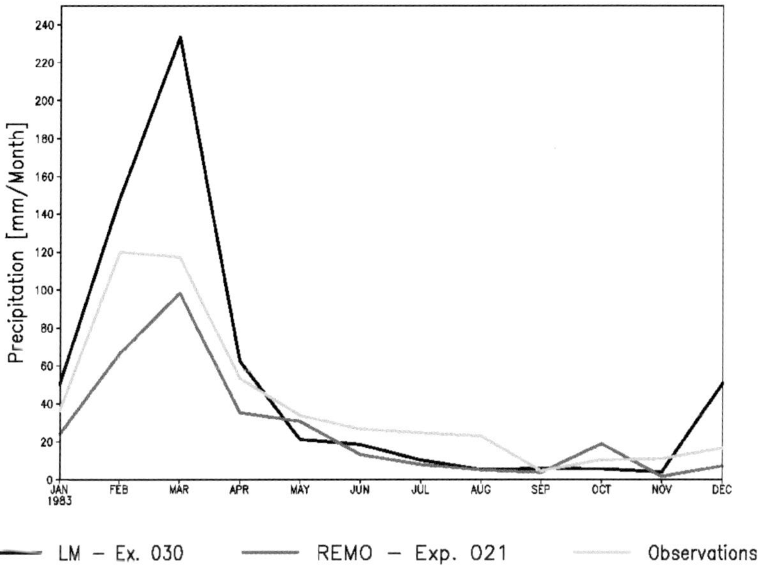

Fig. 6. Spatial mean of monthly total precipitation in the northeast of Brazil; LM and REMO results when driven by ECMWF re-analyses, compared to station observations for the year 1983

In the next step, we made snapshots and examined the month-by-month evolution of the spatial precipitation patterns for the entire model region. Since there is no dense station network over such a large area, we used the CRU gridded monthly observations here as reference. As shown in Fig. 7, this kind of diagnosis supports our previous findings. Even for March, the month with the strongest deviations between the LM results and observations, LM results over land are much closer to the reference data than those of REMO, especially in the Amazon region and in the vicinity of the Andes. The improvements in the latter region may hint at a better representation of vertical advection processes in the case of steep orographic gradients by LM as a representative of a non-hydrostatic model. Although both models underestimated the total rainfall over land, LM results showed a much better agreement with the observations for the area mean for all land grid points than the REMO results. Where air masses approach the coast of South America, however, LM generates excessive precipitation for the coast and in the land regions close to the coastline. Although the reason for this phenomenon is still unknown, it could be speculated that in this case over-strong vertical mass fluxes were simulated. As far as REMO results are concerned, a region with a rainfall amount of less than 1 mm is visible in Fig. 7 already in March 1983 in the north of South America. This feature extended later to the northeast of Brazil. By the end of the dry season, it covered the entire northern part of South America except for coastal regions. In the observations, however, no such dry conditions were identified for this region.

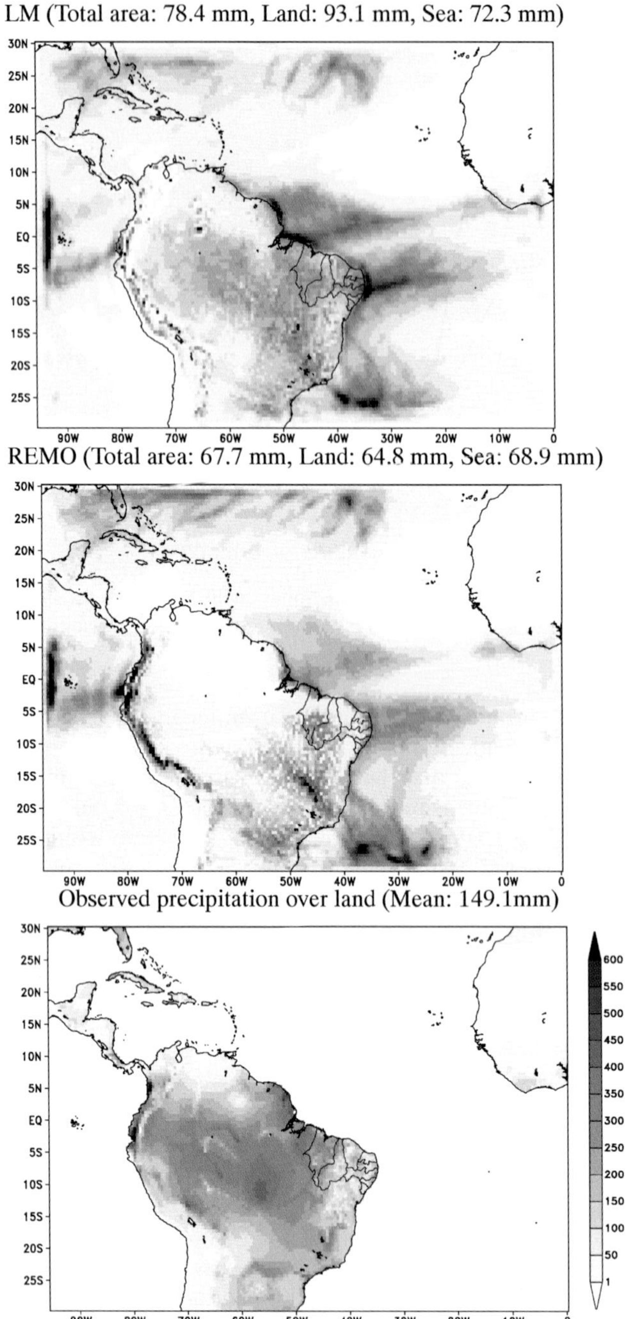

Fig. 7. Monthly total precipitation in March 1983; LM results (upper panel) and REMO results (middle panel), compared to gridded observations from CRU (lower panel)

Because no gridded observations from CRU are available over the oceans, we have compared the simulation results with ERA data there. No tendency to underestimate precipitation was found (not shown). Both models compare well with these reference data that provide an area mean over sea of 73.1 mm. This can be interpreted as a first hint that the reasons for the rainfall deficiencies detected over land have to be sought in processes linked to the soil.

In the remaining parts of this section, we focus mainly on the interpretation of the model deficiencies detected in REMO and the sensitivity experiments we performed to identify the reasons for these deficiencies and to test modified representations for the processes of concern.

From the derived findings, the hypothesis was postulated that a substantial amount of moisture, which was advected in REMO by easterly winds towards northern South America, could not be transformed into precipitation by ascending motions within the region of the intertropical convergence zone over land. Instead, moisture is advected further westward and leaves the continental regions. This is supported by markedly over-high near-surface temperatures in the REMO results over land (not shown), promoting evaporation in the air above and counteracting condensation processes. Such behavior could be triggered by soil that is too dry and warm and would be a further indication that the representation of the related soil processes plays an important role in generating the model deficiencies.

To assess the impact of different initial soil characteristics on the simulation results, an experiment was carried out in which REMO was driven by initial and boundary values that were derived from the global circulation model ECHAM4 instead of those from ECMWF re-analyses. In this experiment, we achieved different results for precipitation mainly during the first simulation months up to April 1983. For this period, more rainfall was generated for the northeast of Brazil compared with the experiment in which REMO was forced by ECMWF re-analyses. It can be shown that these discrepancies are closely linked to the different soil moisture initializations in both runs, since there is a correlation of 0.82 between precipitation and soil moisture for the ECHAM4-driven run and of 0.65 for the reference experiment. Furthermore, the ratio between total precipitation and soil moisture during the first month of integration is quite similar in both sensitivity experiments. There is also evidence of a good match between the spatial patterns of these two quantities, as shown in Fig. 8, e.g., for the ERA-driven REMO experiment.

After the above-mentioned initial period, however, both REMO experiments produced very similar deviations compared with the observed rainfall data sets. Therefore, the problem of over-low precipitation in the model results can probably not be attributed to the differences in terms of the soil initialization.

In a next step, we were able to eliminate the different representation of the large-scale circulation features by the driving model ECHAM4 and the ECMWF re-analyses as one of the possible sources of the detected model deficiencies. Remarkable differences exist between the two coarse data sets, e.g., regarding the location and the extent of such structures even after the initial phase of about 6 months, for the period from June to September. These differences, however, influenced the spatial means and also the monthly precipitation patterns in the regional model results only marginally. This may at least partially be a consequence of the large model region and the high degree of freedom the driven model has in the inner region.

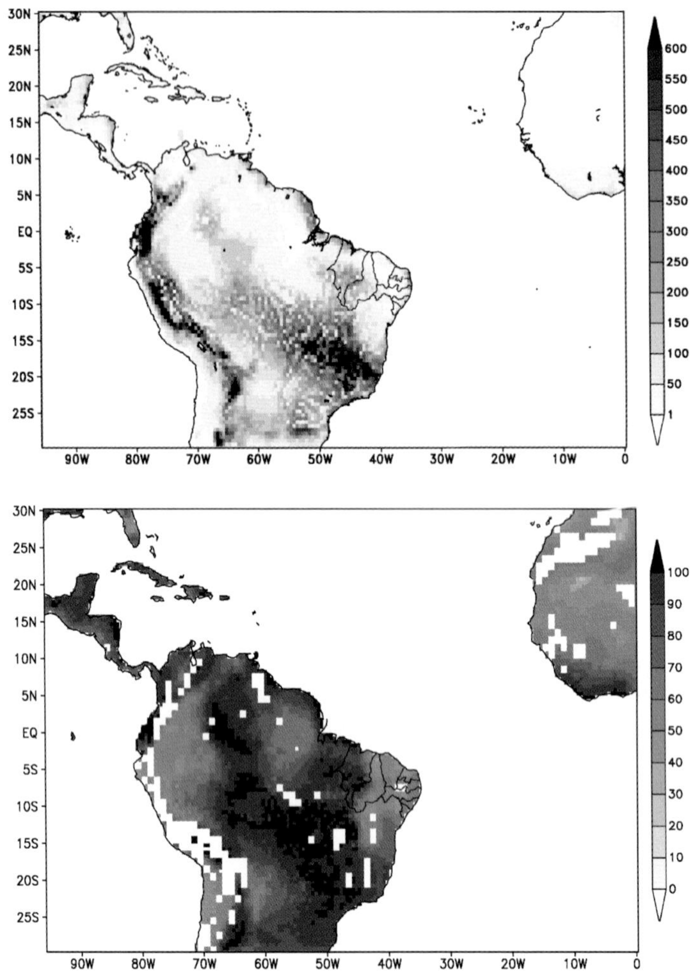

Fig. 8. Correspondence between monthly total precipitation (upper) and soil moisture (lower) patterns for December 1982, simulated by REMO when driven by ECMWF reanalyses

As a consequence of these findings, the hypothesis that soil processes play an important role in the identified model deficiencies was reinforced. To gain more confidence in this assumption, we further investigated some hydrological variables that contribute to the soil-water balance. First, we focused on the ratio between runoff and precipitation as a commonly used hydrological indicator. For the REMO experiment, using re-analyses as the driving force, a ratio of about 57% was estimated, whereas about 40% is given as a long-term climatological mean in well-known hydrological standard works. Since 1983, however, was a rather dry year, this ratio should be similar to the climatological value, or even lower.

Therefore, it seemed to be highly probable that too much runoff was generated by REMO. Because runoff is removed after each time step from the water balance and does not contribute to any infiltration or evaporation processes in neighboring grid cells, too little moisture would remain in the soil. As a consequence, too little water would be available for evapotranspiration and over-high surface temperatures would occur. As already mentioned, over-high surface temperatures really were identified in the REMO results. Furthermore, too little evapotranspiration as compared with ECMWF re-analyses was detected, as shown in Fig. 9.

We have implemented some modifications to the soil submodel in REMO to reduce the runoff formation and to increase both the maximum infiltration rate and the vertical moisture distribution in the soil. In all these sensitivity experiments, however, we were not able to achieve significant improvements of an order higher than 5–10%.

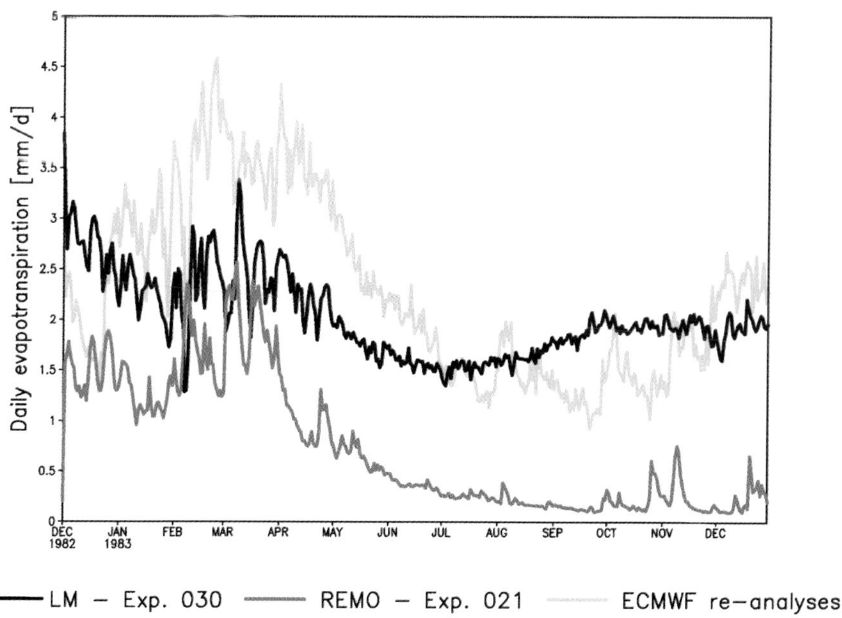

Fig. 9. Spatial mean of daily evapotranspiration in the northeast of Brazil; LM and REMO results when driven by ECMWF re-analyses, compared to the ERA data for December 1982 – December 1983

However, if we compare evapotranspiration in the REMO results with that for LM, a clear improvement is visible in Fig. 9, especially after the rainy season. These results altogether seem to give a strong indication that it is the formulation of moisture exchange at the surface, rather than the runoff generation, that is the reason for the model inaccuracies identified in the case of REMO. Using the approach of Dickinson (1984), more moisture is transported back to the atmosphere and is available for local precipitation recycling processes. Nevertheless, the excessive rainfall in LM close to the coast of South America is still an unsolved problem. Further investigations of near-surface energy and moisture fluxes and of the vertical structure of the lower atmosphere are required to achieve further improvements. Since we have implemented LM on a massive parallel computing architecture, such sensitivity experiments can now be performed much faster than for REMO.

Conclusions

As delineated in the goals of this WAVES research topic, we generated a four-dimensional spatially and temporally consistent data set involving atmospheric and soil variables for the drought period of the year 1983 with a horizontal resolution of 0.5° (about 50 km) and a recording interval of 6 h for use by the other WAVES working groups. As compared with known global data sets, regional structures are represented in more detail and more realistically in our model results. Compared with gridded observations, we additionally supply three-dimensional fields of atmospheric variables and their dynamic evolution.

To produce this data set, we applied a regional climate model and upgraded a new weather forecast model of a different type for long-term simulations in the northeast of Brazil. After a comprehensive validation, it can be stated that the latter model performs reasonably well and can be used for future climate scenario simulations in further steps.

To assess the model performance as extensively as possible in terms of derived climate parameters that can be used from other project groups, a new validation algorithm was developed and applied here. Using this technique, we examined, as an example, the representation in the model results of drought conditions that are important for agriculture. The drought severity was overestimated and the spatial patterns for drought were shifted northwesterly by the hydrostatic climate model REMO. These deficiencies could be attributed with a high degree of probability to the representation of soil processes, especially evapotranspiration, in the model. In this respect, the algorithm of Dickinson in the non-hydrostatic model LM gave better results than the Budyko method and about three times as much evapotranspiration for the year 1983. This model improves the representation of precipitation in the northeast of Brazil as compared with REMO by about 100%, which can be explained at least partially by these different soil model formulations. These findings are in agreement with the results of Ducharne et al. (1998), who pointed out that at least seasonal precipitation is sensitive to hydrological parameterizations within the ITCZ areas. Also, LM remarkably improves the spatial representation of the drought of 1983 in the northeast of Brazil as compared with REMO. The representation of the magnitude of drought severity by this model, however, has still to be enhanced to provide better results than those given by global models and to reduce the range of errors to an order of magnitude similar to that from gridded observations. To achieve this, further investigations of near-surface energy and moisture fluxes and of the vertical structure of the lower atmosphere are required.

References

Böhm U (1999) Eine Methode zur Validierung von Klimamodellen für die Klimawirkungsforschung hinsichtlich der Wiedergabe extremer Ereignisse. PhD thesis, Freie Universität Berlin, Fachbereich Geowissenschaften

Christensen JH, Machenhauer B, Jones RG, Schär C, Ruti PM, Castro M, Visconti G (1997) Validation of present-day regional climate simulations over Europe. LAM simulations with observed boundary conditions. Climate Dynam 13:489–506

Dickinson R (1984) Modeling evapotranspiration for three-dimensional global climate models. In: Climate processes and climate sensitivity. Geophysical monograph 29, Maurice Ewing vol 5, pp 58–72

Doms G, Schättler U (1999) The Nonhydrostatic Limited Area Model LM (Local Model) of DWD, part I. Scientific Documentation. Deutscher Wetterdienst, Offenbach

Ducharne A, Laval K, Polcher J (1998) Sensitivity of the hydrological cycle to the parametrization of soil hydrology in a GCM. Climate Dynam 14:307–327

Gerstengarbe F-W, Werner P-C (1999) The complete non-hierarchical cluster analysis. PIK-reports no 50. Potsdam Institute for Climate Impact Research, Potsdam, Germany

Gibson JK, Kallberg P, Uppala S, Hernandez A, Nomura A, Serrano E (1997) ECMWF reanalysis project report series. 1. ERA description. European Centre for Medium-Range Weather Forecasts (ECMWF), Reading, UK

IPCC (2001) Climate change 2000: contribution of Working Group I to the Third Assessment Report of the IPCC. Cambridge University Press, Cambridge

Jacob D, Claussen M, Majewski D, Podzun R, Rockel B (1995) REMO - a model for climate research and weather forecast. Research activities in atmospheric and oceanic modelling, WMO/JCSU/IOC Report, World Meteorological Organization, Geneva, Switzerland

Machenhauer B, Windelband M, Botzet M, Christensen JH, Déqué M, Jones RG, Ruti PM, Visconti G (1998) Validation and analysis of regional present-day climate and climate change simulations over Europe. MPIfM-report no 275. Max-Planck-Institute for Meteorology, Hamburg, Germany

New M, Hulme M, Jones P (2000) Representing twentieth-century space-time climate variability, part II. Development of 1901-96 monthly grids of terrestrial surface climate. J Climate 13:2217–2238

Roeckner E, Arpe K, Bengtsson L, Christoph M, Claussen M, Dümenil L, Esch M, Giorgetta M, Schleese U, Schulzweida U (1996) The atmospheric general circulation model ECHAM-4. MPIfM report no 218. Max-Planck-Institute for Meteorology, Hamburg, Germany

Schättler U, Doms G (2000) The Nonhydrostatic Limited Area Model LM (Local Model) of DWD, part III. User guide. Deutscher Wetterdienst, Offenbach

Schrodin R (1995) Documentation of the EM/DM-system. Deutscher Wetterdienst, Offenbach

Part 4: Water Availability, Water Management and Water Use

Investigations on Water Management and Water Quality in Picos/PI and Tauá/CE

Susanne Voerkelius, Christoph Külls, Maria Marlúcia F. Santiago, Horst Frischkorn, Ligia Alves dos Santos Semrau, Gerold Heinrichs, Margarita M. López Gil

Hydroisotop GmbH, Woelkestr. 9, 85301 Schweitenkirchen, Germany, hydroisotop@t-online.de

Introduction

The WAVES program (Water Availability, Vulnerability of Ecosystems and Society in Northeastern Brazil) aims at creating a knowledge base and strategies for sustainable development for the drought-prone northeast of Brazil. The perspective of climate change in a semiarid area with marginal water supplies requires strategic measures in order to avoid or at least mitigate severe water scarcity problems. Water scarcity results from the imbalance in a tripartite system between (1) the availability of natural resources with specific hydrochemical characteristics, (2) the state of development of water supply structures, and (3) water demand for specific uses. Seen from such a perspective, it is obvious that water quality is an important and integral aspect of water availability. Hence, the analysis of water availability needs to integrate quantitative and qualitative studies.

The Study Areas

The investigations were carried out in Picos (Piauí) and Tauá (Ceará), the two focus regions of the WAVES program. In these areas, water pollution is mainly caused by sewage and return flows from agriculture. Industrial wastewater plays a minor role. Only a few data on water quality were available, requiring own measurements. Water quality was assessed based on major anions and cations, N-components and physicochemical parameters (pH, conductivity). In selected cases pesticides were analyzed. Disinfected waters were analyzed on the content of chlorinated hydrocarbons. The main problem of microbiological investigations in the study area was the lack of transport and cool storage facilities. Consequently, microbiological testing on coliform bacteria and *Escherichia coli* was performed in situ using the sealing/indicator system Colilert, an EPA-approved test of IDEXX Laboratories (Edberg and Smith 1994). Finally, special investigations on stable isotopes, including groundwater ages, were made in order to shed light on processes linked to the genesis of groundwater. Finally, an analysis of the correlation between water supply structure and indicators of water quality was performed in order to allow a generalization of data.

The Natural Hydrochemical Background – Piauí

In Piauí, most of the area is covered by sediments providing storage and conveyance of groundwater. Groundwater use is an important backbone of water supply in the state of Piauí. The crystalline formations in Ceará, however, limit the effectiveness of groundwater recharge and groundwater storage. Water supply depends heavily on the use of surface water. The two focus regions Picos and Tauá reflect these geological differences, which consequently result in a different water supply structure. The samples taken from representative locations in Piauí show a low-to-medium degree of mineralization (Fig. 1).

In the region of Picos in an area of about 3000 km^2 samples from about 50 sites have been investigated. The existing hydrochemical quality does not impose any restriction for use following international drinking water standards. Groundwater samples taken in Picos were mostly of Na-HCO_3 type showing strong cation exchange but had low chloride concentrations, mostly below 75 mg/L. Only in a few samples from shallow springs and reservoirs were elevated concentrations of nitrate found, indicating pollution from private households and animal waste. In some deep wells, low, but detectable, concentrations of nitrate were found. This may be an early indicator for a starting contamination of groundwater resources. Pesticides were not detected in any groundwater samples. The quality of irrigation water was defined by the sodium adsorption ratio (SAR, Rehm 1989). Conductivity/SAR gives a direct indication of salinity and alkalinity hazards. Only one water sample in the municipality of Sussuapara shows a SAR value greater than 20. The continued use of this water could lead to serious sodium hazards. Where used on sensitive crops this water may be detrimental because of sodium phytotoxicity. In general, both the mineralization and the SAR index of ground- and surface water in Picos are low and water sources are suitable for irrigation.

The Natural Hydrochemical Background – Tauá

In the region of Tauá, due to natural conditions, many groundwater wells as well as some dams cannot be used directly because of high mineralization. In periods of drought, salty water is used after cost-intensive desalinization. Most water samples from dams have lower salinity; their major ion composition resembles that of floodwater with a balanced cation composition. The anion composition varies strongly and exhibits an increase in chloride due to evaporative enrichment (Fig. 2).

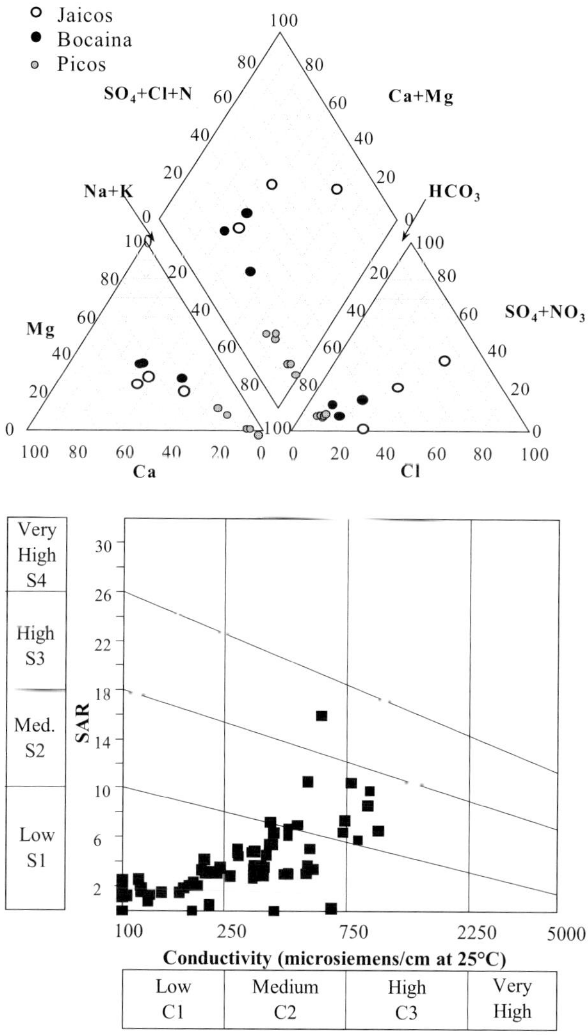

Fig. 1. Bulk water quality in Piauí (samples from Jaicos, Bocaina and Picos)

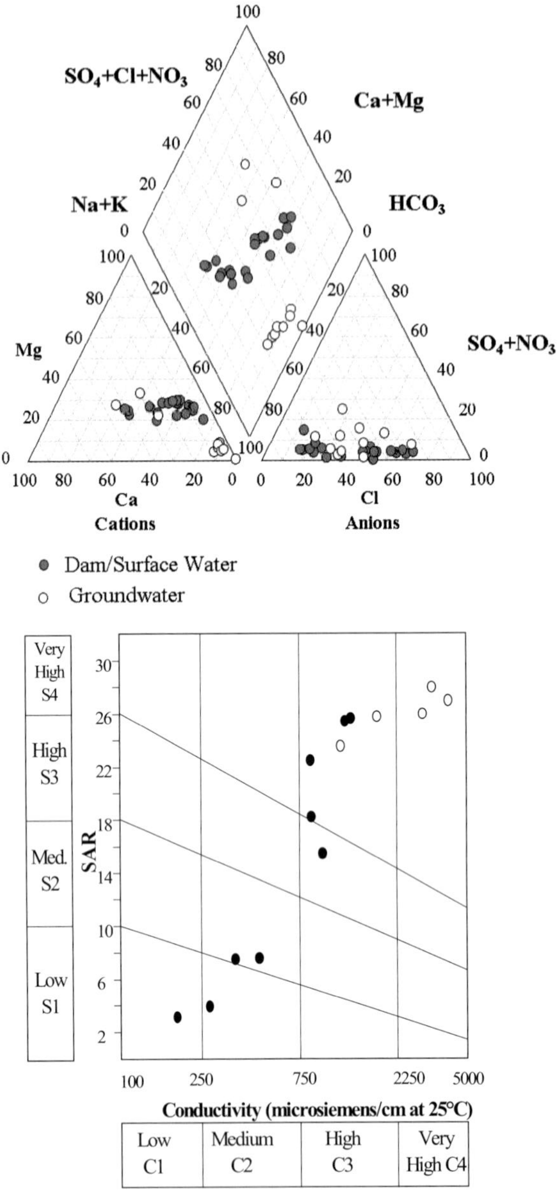

Fig. **2.** Bulk water quality in Tauá (ground and surface water)

The groundwater samples can be divided into two groups: one group has a distinct cation composition with more than 80% sodium; the chloride content varies between 25 and 75%. The other group is quite similar to surface water in its cation and anion composition. This group is probably influenced by recent recharge and can be classified as shallow groundwater, mostly found in the alluvium of rivers. The salinity found in many groundwater samples is too high to be used as drinking water. The SAR indices show that only some surface waters that exhibit little evaporative enrichment can be used for irrigation. Most groundwater samples and some samples from surface water have high SAR indices and are classified as not suitable for irrigation. However, the main constraining factor for irrigation is water scarcity.

Microbiological Contamination and Secondary Contaminants

The hygienic quality can be characterized by the content of coliform bacteria and *E. coli* (Kolsky and Blumenthal 1995; Moe et al. 1991). Surface water and groundwater as well as water from supply systems were analyzed on these indicators. In the Picos area, surface water is only used by a minority of people in detached locations, where microbiological water quality is relatively poor. In contrast, none of the samples taken from groundwater, the most common source for public or private consumption in Picos, showed coliforms or *E. coli*. In urban areas with a centralized water supply by companies, e.g., AGESPISA (PI) or CAGECE (CE), water is usually chlorinated on a routine basis and is microbiologically safe (e.g., Tauá). However, in remote parts of the water supply net, a slight but persistent microbiological contamination and no free chlorine were found. Central water services in rural areas had a slight but persistent contamination with coliform bacteria.

The system is disinfected only periodically in these smaller networks. Finally, in rural areas (Angical, south of Picos), groundwater samples contained a high number of coliform bacteria and *E. coli* on delivery to the consumers by water supply stations (chafariz). Hence, due to low raw water quality, disinfection is still a precondition for a safe drinking water supply. A severe disadvantage of disinfection by chlorine is the secondary formation of halogenated hydrocarbons if the raw water has high concentrations of organic substances. Such a condition can be expected in surface waters. Samples from Jaicos and Tauá, from the capitals Teresina (PI) and Fortaleza (CE), were analyzed for chlorinated hydrocarbons (CHC; Fig. 3). All samples show elevated concentrations of CHC. The measured contents reach up to 200 μg/L. In particular, trihalomethane is significantly above the international maximum standard of 30 μg/L.

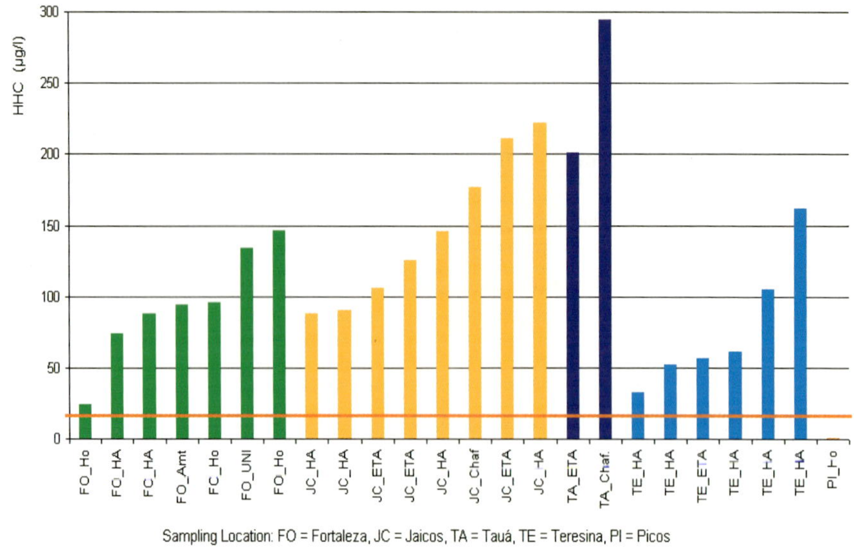

Fig. 3. Halogenated hydrocarbons in drinking water

Water Structure and Water Quality

In order to allow an up-scaling of water quality data, a structural analysis of existing water supply systems was made using a census evaluation based on interviews with local water users. The main items of the questionnaire were: characterization of the source of water, existing supply systems, water treatment, water demand, water costs, supply problems, domestic waste disposal and water-borne diseases.

The results of the census of Tauá were digitized in a form-based and user-friendly database system. This database system, linked to a GIS platform, allowed visualization of the results statistically and geographically (Fig. 4). As an open system it also provides a tool for extending, repeating and modifying the census. The source of water is mainly surface water and shallow groundwater from dug wells (top). The water structure in Tauá is characterized by a low degree of connection to supply network; individual transport from distribution points or from various water sources (dams, dug wells and boreholes) is dominant. Cisterns fed with rainwater are of increasing importance. The distance to the next surface water reservoir (middle-right) shows a regional cluster with long distances in the west. In drought situations, water supply cannot be sustained by water from boreholes: very few shallow dug wells are in use, the number of drilled boreholes supplying drinking water is still limited (lower-left). As in 1999, external supply still plays an important role („carro pipa“).

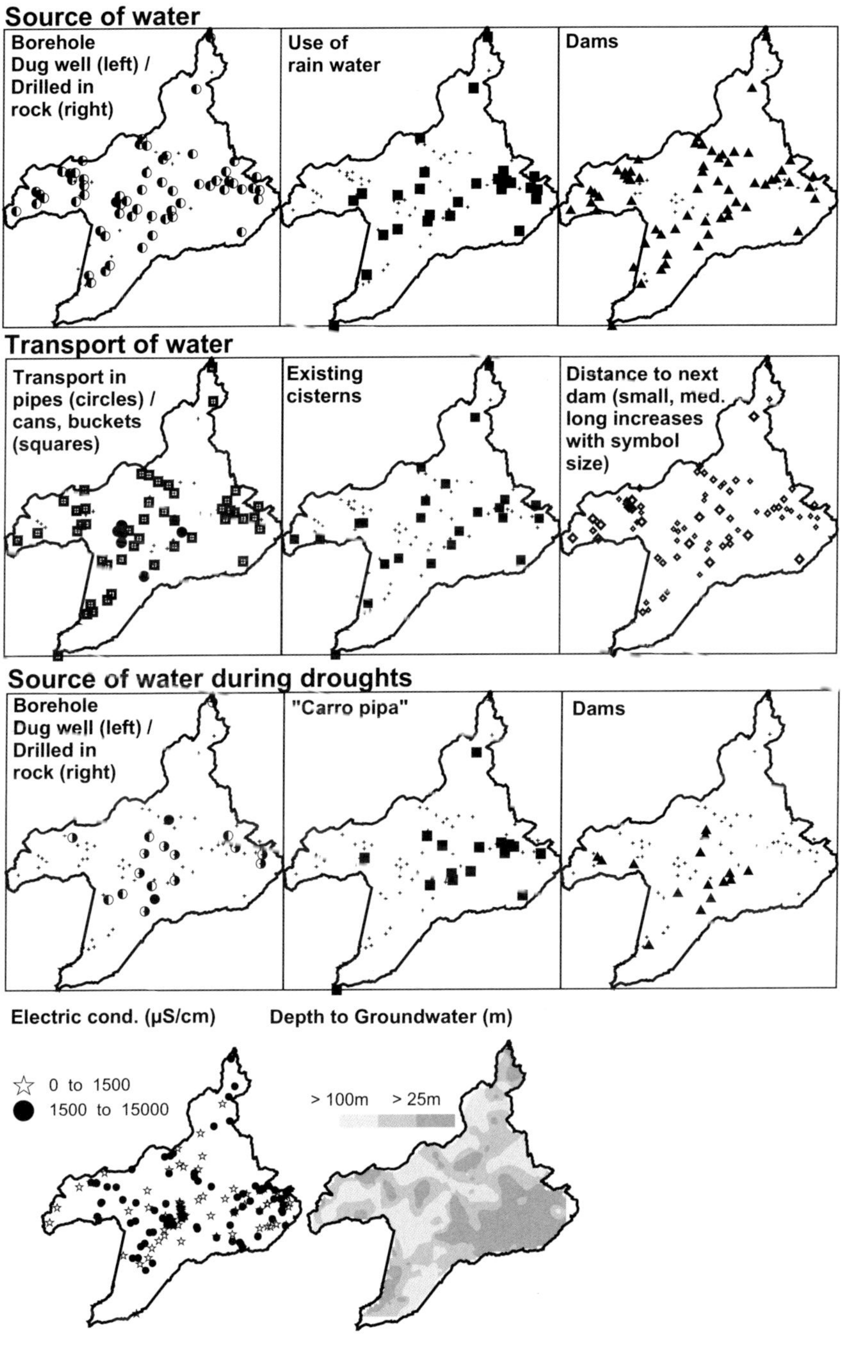

Fig. 4. Water structure and quality in Tauá

In the region of Picos, the water supply structure can be divided into three main groups, which differ in the extent of taking care of water quality. Central supplying systems in urban areas like Picos (>50,000 inhabitants) and Jaicos (>5000 inhabitants) are managed and controlled by specific companies. Water source is groundwater. Water quality and quantity are permanently controlled, and mostly the water is chlorinated. The majority of urban houses are connected to the supply net and consumers have to pay for the service (1998 in Picos: 10 \$/15m^3 /month ≈80 l/head/day).

In villages (500–5000 inhabitants) in rural areas like Bocaina (3000 inhabitants) or São João de Cana Brava (3600 inhabitants), central water supply systems have also been established and most of the houses are connected. The water source used most frequently is groundwater. Only sporadically are the quality and quantity of water controlled. A periodic disinfection by one-time treatment with calcium hypochloride prevents microbiological problems. For the residents, the water is often free of charge, but they have to pay for repairs.

A decentralized heterogeneous water supply exists in rural areas for small villages and settlements: public water-supply stations and private wells (deep and shallow), cisterns, rivers, lakes and tank trucks (carro pipa). We especially investigated public water-supply stations (Fig. 2) that provide for 40–200 families (360–1200 persons). In general, these people are living some 100 m from the water tap. Water is carried home in plastic and metal canisters on foot, bicycle, car or donkey. In the house, it is stored in clay vessels or refrigerators for cooling and often filtered through ceramic filters.

The investigations revealed a strong interdependence between water-supply structures and the hygienic quality of tap water, with quality deteriorating drastically in the order of the above presentation of the three supply types. As a result of the correlation of microbiological investigations with structural data, the following „valuation criteria“ were elaborated: ownership of the supplying system, number of inhabitants, technical installations within the supplying system, quality of water exploitation devices, water-treatment devices and prices for water.

Salinization Processes

There have been numerous studies discussing the possible causes of groundwater salinization in the crystalline rocks of northeastern Brazil (Reboucas 1973; Kreysing 1976; Salati et al. 1980; Frischkorn et al. 1992; Santiago et al. 2001); a review of pioneering works was given by Salati et al. (1974). The origin of salt is now widely recognized as being caused by the deposition of atmospheric salt from precipitation (1–5 ppm/L of rainfall) and additional input from dry deposition (Salati et al. 1980). A dominant contribution by water–rock interactions is unlikely, given the young ^{14}C age of many groundwater samples. The observed chloride-δ^{18}O relationship shows in fact a dual pattern: while surface water samples exhibit a strong enrichment in δ^{18}O with a slight parallel increase in chloride concentration, chloride increase is not correlated with isotopic enrichment in most groundwater samples. A correlation was only found locally, in larger alluvial aquifers with limited underflow or open discharge zones (i.e., Aldeota). The combination of the facts that for crystalline ground waters there is no dominant correlation between high salinity and $\delta^{18}O$ and that these waters, in general, are young (see „Water Resources of Ceará and Piauí", Frischkorn et al., this Vol.) and without any correlation between salinity and ^{14}C age, suggests that salinization is weakly related to aquifer dissolution, but more to the physical conditions of groundwater storage in the crystalline. A lack of connectivity of fractures may result in the selective loss of water (through evapotranspiration, evaporation through soil or complete evaporation in irrigation) without removal of salts from the aquifer or from the fracture system. In the flat areas and shallow subbasins, water is largely consumed by transpiration or totally lost by high evaporation during the dry season. As a result salts are trapped in the hydrogeological subsystems of the crystalline peneplain, accumulating in the highly variable microenvironment of fractures. Salinization is favored by the shallow depth of fracture systems: lateral subsurface drainage is low and limits the outflow of salty groundwater, surface runoff (about 8% of rainfall) does not contribute significantly to the flushing of salts from the system.

Initial recharge with only limited evaporation is assumed to take place only locally in elevated ridges (i.e., quartzite, migmatite and metamorphic limestones), where fresh groundwater has been targeted successfully. Shallow lateral subsurface drainage and underflow in alluvial bodies mainly of smaller tributaries may also provide fresh groundwater, at least on a seasonal basis (Santiago 1984; Santiago et al. 2001).

Conclusions

In the region of Picos the main problem with respect to water quality is poor hygiene. A clear dependence of the hygienic quality of tap water on the water supply structure was found. Hence, hygiene problems can be reduced by changes in water-supply structures: covers for open wells or open water tanks, installation of metal water taps, disinfection of taps, periodical cleaning of water tanks and pipes.

In Tauá, the main problem for an adequate water supply is the high salinity. Most of the enrichment of salts of atmospheric origin probably takes place in the saturated zone due to transpiration of groundwater and close to complete evaporation of soil water in the dry season. Only close to groundwater discharge zones can isotopic enrichment be found. In general, the low primary permeability of the basement and the resulting limited subsurface drainage induce a high degree of groundwater losses and chloride enrichment during recharge and redistribution processes (Fig. 5).

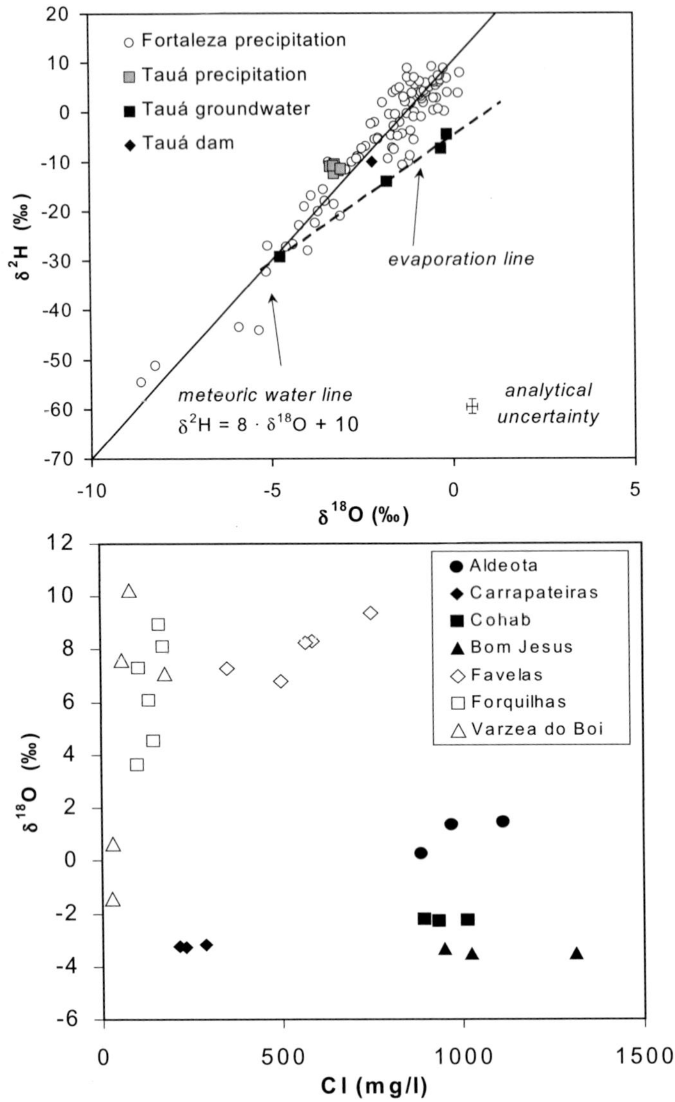

Fig. 5. Stable isotopes from selected locations in Tauá

In spite of the progress made in understanding the hydrogeological system of the crystalline basement in northeastern Brazil, little operational and technical use of this knowledge has been made so far. Common practices are generally focused on managing the problem of scarcity by damming surface runoff or on creating local technical solutions (desalinization without salt removal or disposal instructions). Future research should be directed at evaluating alternative methods such as the implementation of a decentralized water supply from existing fresh water lenses and the feasibility of measures to contain salinization and finally lower salt levels in some hydrogeologic subsystems under the given conditions of continuous salt input from the atmosphere. In addition, desalinization of hydrogeological subsystems by creating negative salt balance (such as proposed by Salati et al. 1980) could offer long-term perspectives for reducing water scarcity.

The disinfection of both surface and groundwater is still a primary requirement for a safe drinking water supply. Still the high concentrations of chlorinated hydrocarbons found in surface waters need to be reduced due to possible carcinogenic effects. A reduction in their level should be a second aim of water treatment. Where this is feasible, the use of groundwater should be developed.

Acknowledgements

Financial support by the German Federal Ministry for Education and Research (BMBF) and the Brazilian National Council for Scientific Development and Technology (CNPq) is gratefully acknowledged.

References

Companhia de pesquisa de recursos minerais (CPRM) (1993) Programa de água subterrânea para a região nordeste – Projeto hidrogeológico do Piauí, Picos – Perfil Hidrogeológico do Município. CPRM Report, Teresina, 28 pp

Edberg SC, Smith DB (1994) Comparison of the Colilert method and standard fecal coliform methods. AWWA Research Foundation, Denver, Colorado, pp 41–42

Forster M, Voerkelius S, Frischkorn H, Santiago da Silva M, Lantermann E, Brühl E (1996) Water availability within the semiarid northeast of Brazil investigating water quality, water resources, actual water utilization, and elaboration of first contributions for the development of regional water management concepts. IAEA-proceedings. IAEA, Vienna, pp 443–445

Frischkorn H, Santiago M.F. & Serejo A.N., 1992. Isotope study of wells in crystalline rock of the semi-arid northeast of Brazil. IAEA, Vienna, pp 73-89 (TECDOC-502)

Kolsky PJ, Blumenthal KJ (1995) Environmental health indicators and sanitation-related disease in developing countries: limitations to the use of routine data sources. Rapp. Trimest. statist. Samit. Mond. 48:132–139

Kreysing K (1976) Hydrogeologische Forschungsarbeiten im Kristallin NE-Brasiliens. Münster Forsch Geol Paläont 38/39:89–98

Moe C, Sobsey MD, Samsa GP, Mesolo V (1991) Bacterial indicators of risk of diarrhoeal disease from drinking water in the Philippines. Bull World Health Org 69:305–317

Reboucas AC (1973) Le problème de l'eau dans la zone semi-aride du Brésil. DSc. Thèse présentée à l'Université Louis Pasteur de Strasbourg, 291 pp

Rehm S. (1989) Handbuch der Landwirtschaft und Ernährung in den Entwicklungsländern, vol 3. Grundlagen des Pflanzenbaues in den Tropen und Subtropen, Ulmer Verlag, Stuttgart

Salati E, Leal JM, Campos MM (1974) Environmental isotopes used in a hydrogeological study of northeastern Brazil. Isotope techniques in groundwater hydrology 1974. Proc Symp Vienna 1:259–283

Salati E, Matrui E, Leal JM, Fritz P (1980) Utilization of natural isotopes in the study of salination of the waters in the Pajeú River Valley, northeast Brazil. IAEA-AG 158(10):133–151

Santiago MF (1984) Mecanismos de salinização em regiões semi-áridas. PhD Diss, Department of Geology, Univ São Paulo/Brasil

Santiago FM, Frischkorn H, Neto PS, Filho MF (2001) The recharge mechanisms in an alluvial aquifer zone in northeast Brazil. Ground Water 39:18–23

Water Quality in the Guaribas River Basin for Irrigation Use

Décio Frota, Margarita M. López Gil, Airton Ramalho, Francisco Dantas and Vladira Vieira

Departamento de Hídrometeorologia, Secretaria de Agricultura e Abastecimento, CEP 64002-150 Teresina- Piauí, Brazil, seab@seaab.pi.gov.br

Introduction

In order to achieve rational water resources management, knowledge about the available water in terms of quantity as well as quality is necessary as this forms the basis for planning and establishing an exact relationship between water availability and water demand for different uses.

In the Guaribas River basin, the focus area of the WAVES project, the main use of water is irrigation. It occurs at small scale along the main river, close to the margins and frequently in the riverbed. Irrigation represents a high potential for economic development in the area.

Since irrigation represents an alternative for the social and economic development of the semiarid area, where the basin is located, it should be used in a rational way, because the climatic conditions and the high salt contents in the irrigation water, together with restricted drainage, have provoked problems with soil salinity. Concerning the planning of irrigated agriculture it is important to know what kind of water is available in the basin to indicate adequate crops to be established in the area.

Together with the Bocaina Dam Project, it was planned to develop peripheral areas using the water from the reservoir according to the water availability and water demands. The area reserved for irrigation was located along the Guaribas River from the reservoir to the city of Picos (Fig. 1), in a strip of land 24 km in length, comprising about 2000 ha. Nowadays, the effective area using the water from the river is over 750 ha. Some farmers who live too far away from the Guaribas River use water from wells to irrigate. Flood irrigation is practiced along the river through pumping water from the Guaribas River in a diffuse way and without any control.

The irrigated areas are characterized by small properties with up to 3 ha, where rice, bean, corn, vegetables, banana, coconut, watermelon, mango, tomato and other crops are planted.

In connection with the dam construction, an analysis of soil salinity including the alluvial land along the Guaribas River was carried out. The results showed high contents of salt, especially near the city of Picos (EC up to 1000 μmhos/cm), that represents a serious risk of soil salinity in case irrigation increases in these areas, because intensive irrigation demands adequate drainage for salt leaching.

Besides salinity, there is a problem with sodium in soils, which in high concentrations can provoke clay dispersion and, consequently, can decrease the soil permeability. A few studies were accomplished in order to evaluate the water quality in the Guaribas River basin. In May 1995, the Irrigation Department of the Secretary of Agriculture, Water Supply and Irrigation of the state of Piauí collected samples from several rivers in the state in order to classify the water for irrigation use. The water from the Guaribas River was classified as „C_2S_1“ (water with medium salinity and lower concentration of sodium) and the Bocaina Dam water as „C_1S_1“ (water with low salinity and lower concentration of sodium).

As the campaign consisted of the analysis of just one sample for each point, this is not considered to be representative, because the rain pattern in the area (8 or 9 dry months and 3 or 4 months with rain) has a strong influence on the contents of the analyzed chemical elements.

Marwell (1997) calculated the available superficial yield to be 63.6 hm^3/year and estimated the total demand to be 52.4 hm^3/year. This author concluded that the superficial water resources in the basin are sufficient to supply the future demands (year 2020), being only necessary to know the availability in qualitative terms.

The aim of this research is to classify the water of the Guaribas River basin according to the preponderant use, irrigation, and to evaluate the available water in qualitative terms identifying potential problems and possible restrictions to its use.

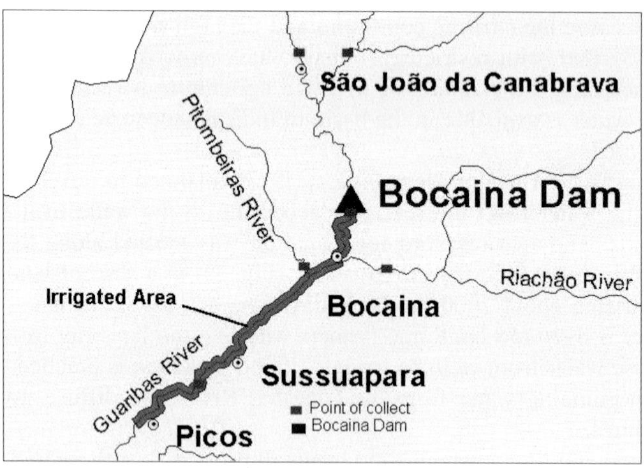

Fig. 1. Map with points of sample collection

Area Description and Methodology

Localization

The Guaribas River basin (Fig. 1) is localized in the semiarid area of Piauí. The rivers of the basin are intermittent, except for the Guaribas River, which is perennial along 30 km approximately, due to the Bocaina Dam whose maximum capacity to store water is 106,000,000 m^3.

Material Collection

Through the monitoring net implanted in the Guaribas River basin by DHME (Hydrometeorology Department) of the Agricultural Secretary of the state of Piauí, points of water sample collection were established. The samples were collected at eight points of the hydrographic net, three of them being in the main tributaries of the Guaribas River and five in the Guaribas River, including the Bocaina reservoir (Fig. 1). These points coincide with the ones with water discharge measurements. The sampling points are listed in Table 1.

The collections were carried out, monthly, during the period September/1998 until June/2000. At Guar-005, Guar-007, Guar-008 and Guar-010, monthly collections were not possible due to the intermittent character of the river.

Table 1. Sampling points for water quality analysis

Code	Point	River	Localization
Guar-001	Angical-Vigia	Guaribas	Downstream Picos city
Guar-003	Lagoa Grande	Guaribas	Upstream Picos city
Guar-004	Barragem	Guaribas	Downstream the reservoir
Guar-005	Guaribas Velho	Guaribas	Upstream the reservoir
Guar-007	Pitombeiras	Pitombeira	Downstream Bocaina city
Guar-008	Canabrava	Canabrava	Upstream the reservoir
Guar-010	Riachão	Riachão	Between Bocaina and Picos
Reservoir	Bocaina dam	Reservoir	

Generally, the characteristics that determine the water quality for irrigation and its restrictions are set in five basic parameters:

1. total salt concentration – salinity;
2. relative proportion of sodium in relation to the other cations – risk of reducing soil permeability;
3. concentration of toxic elements – toxicity;
4. bicarbonate concentration; and
5. considerations of sanitary aspects.

In this work, the following variables were measured: electrical conductivity (EC), pH, temperature, sodium, potassium, magnesium, calcium, carbon dioxide, chloride, sulfate, evaporation residue at 105 °C, and alkalinity (hydroxides, carbonates and bicarbonates). The analyses were accomplished in the DNOCS (DNOCS: National Department Against Drought) laboratory in Teresina-PI.

The results were compared with the „threshold values" given by Bernardo (1982). To classify the water, the approach of the Salinity Laboratory of the United States was used, based on the „electric conductivity (EC)", as a risk indicator of soil salinity, and on the „sodium adsorption ratio (SAR)", as a risk indicator of soil alkalinization (Bernardo 1982).

Results

According to the analyzed parameters, the classification of the water of the Guaribas River for irrigation use in its upper and medium section was C_1S_1 (water with low contents of salt and sodium). In its lower section it was C_2S_1 (water with moderate tendency to salinity and low contents of sodium; Table 2).

The tributaries of the Guaribas River located in the upper section, including the Bocaina reservoir, were also classified as C_1S_1. The tributaries located downstream of the reservoir were classified as C_2S_1. Concerning the quality of the water and the problems and restrictions presented by the analyzed water in the Guaribas River basin, the following observations were made:

Salinity

Only in its most downstream point (Guar-001) did the Guaribas River present moderate problems with salinity, with an EC of about 1.000 µmhos/cm in the dry months, the quality improving in the rainy months. The problem with salinity appears when salts from the irrigation water accumulate in the root zone with such concentrations that they cause yield losses due to water scarcity over a significant period of time.

The samples collected in Guar-001, the lowest section of the Guaribas River (after Picos city), showed a calculated SAR >3. This represents a moderate restriction for superficial irrigation, the most widespread irrigation method in this region.

Sodium and Chloride Toxicity

The concentrations of chloride and sodium and the medium values of SAR (SAR <3) showed that the water of the Guaribas River basin does not present restrictions due to sodium or chloride toxicity.

Risk of Reducing Soil Permeability

In the case of soil permeability, the water in the Guaribas River basin presented from moderate to severe restrictions. Irrigation water can cause reduction of the normal infiltration velocity, leading to water logging, oxygen deficiency in the rooting zone and increasing risk of salt accumulation. This problem also depends on other factors like soil structure, degree of compaction, content of organic matter and clay mineralogy.

Bicarbonate Concentration

The magnitude of the damage provoked by bicarbonates depends on the amount of absorbed ions and on the sensibility of the plants. Perennial crops are more sensitive. The damage is expressed as brownish borders of the leaves and chlorosis in the internerval area. When the accumulation of ions is sufficiently high, significant yield reduction is produced. Annual crops are more tolerant and, consequently, they are not affected by low concentrations of these elements.

The rivers of the basin did not present restrictions related to the bicarbonate concentration ($CaCO_3$), except the lowest section of the Guaribas River, which presented moderate restrictions. This problem is more evident when sprinkler irrigation is used, provoking spots of salt deposits on the leaves and fruits due to high contents of bicarbonate, gypsum or iron in the irrigation water. However, this kind of irrigation is not used in the area where the water presented such restrictions.

Table 2. Classification and restrictions of the water of the Guaribas River basin for irrigation use (EC in μmhos/cm and HCO_3^- in ppm)

Point	Parameter	Measured Values	Classification	WATER RESTRICTIONS			
				Salinity	Permeability	Toxicity (Na, Cl)	Bicarbonate conc.
Guar-001	EC	875.0	C_2S_1	Moderate	Moderate	None	Moderate
	SAR	3.5					
	HCO_3^-	193.0					
	PH	8.30					
Guar-003	EC	240.0	C_2S_1	None	Moderate	None	None
	SAR	1.1					
	HCO_3^-	78.5					
	PH	7.2					
Guar-004	EC	195.0	C_1S_1	None	Severe	None	None
	SAR	0.8					
	HCO_3^-	64.7					
	PH	7.2					
Guar-005	EC	130.0	C_1S_1	None	Severe	None	None
	SAR	0.7					
	HCO_3^-	38.5					
	PH	6.8					
Guar-007	EC	175.0	C_1S_1	None	Severe	None	None
	SAR	0.5					
	HCO_3^-	57.8					
	PH	7.1					
Guar-008	EC	205.0	C_1S_1	None	Moderate	None	None
	SAR	0.5					
	HCO_3^-	70.0					
	PH	7.2					
Guar-010	EC	375.0	C_2S_1	None	Moderate	None	None
	SAR	0.6					
	HCO_3^-	87.5					
	PH	7.3					
Reservoir	EC	200.0	C_1S_1	None	Severe	None	None
	SAR	0.9					
	HCO_3^-	75.0					
	PH	7.2					

Conclusions

The results of this research showed that the water in the Guaribas River basin does not present restrictions to irrigate any kind of crop based on their classification, which was either: C_1S_1 (low content of sodium and salt) or C_2S_1 (moderate tendency to salinity and low content of sodium). The only recommendation is to cultivate in soils with medium or high permeability or to assure drainage and leaching of salts when the cultivated soil has low permeability. Plants sensitive to salinity should be avoided. If the drainage is high enough, the control of salinity will simply demand that enough water be applied to the crops for salt leaching, maintaining the salt accumulation within the tolerance limits of the plants.

Recommended crops, according to the classification of the water, are: beans, vegetables, fruits, cotton, grass and soybeans. The water quality of the basin for irrigation uses is good in a general way, presenting only one sampling point with moderate restrictions due to salinity and bicarbonate concentration. All other sampling points had moderate to severe restrictions due to the risk of reducing soil permeability.

References

ABRH (1991) Hydrologia ambiental, vol 3. Associação Brasileira de Recursos Hídricos (Coleção ABRH), São Paulo

Ayers RS (1991) A qualidade da água na agricultura. Tradução de HR Gheaji, JF de Medeiros, FV Damasceno. Campina Grande, UFPB

Bernardo S (1982) Manual de irrigação. Viçosa, UFV, 463 pp

Marwell FP (1997) Modelo de Gestão dos Recursos Hídricos da Bacia do Rio Guaribas – Projeto BRA/IICA – 95/004, Teresina

PNUD/PRONI (1990) Revisão da legislação federal de meio ambiente pertinente a agricultura irrigada, BRA/004 B7/008, Brasília-DF

Porfírio EA, de Souza J, López MM (1997) Qualidade da Água na Bacia do rio Guaribas – Projeto BRA/IICA – 95/004 - Modelo de Gestão de Recursos Hídricos da Bacia do rio Guaribas. Teresina

Seixas BLS (1992) Usos e Características das Águas – Secretaria dos Recursos Hídricos, Saneamento e Habitação da Bahia – Coordenação de Recursos Hídricos – Publicação para uso interno, Salvador

Groundwater as Indicator for Paleoclimatic Change in the Northeast of Brazil

Horst Frischkorn, Maria Marlúcia F. Santiago and Markus Forster

Universidade Federal do Ceará, Departamento de Hidráulica e Ambiental, CEP 60451-970 Fortaleza-Ceará, Brazil, cariri@ufc.br

Introduction

The Earth's paleoclimate is not only a subject of human curiosity about the past, but has also become a matter of „understanding the past for predicting the future". With the Earth facing the threat of a man-made global change in climate, caused by the „greenhouse effect", the history of the world's climate has found new interest. Science needs large scale global as well as specific regional values for the calibration of models designed to foresee and quantify future changes in climate. (Such as CLIMAP – Climate: Long Range Investigation and Prediction, CLIMAP 1981). As instrumental meteorology reaches back to the seventeenth century only (based on the invention of the thermometer by Galileo, around 1590, and of the barometer by Torricelli, in 1643), paleoclimatic evidence for remoter periods has to be based on indirect measurements („proxy data"), analyzing the advance and retreat of glaciers, the stratification of loamy soils (varves), the width of annual growth-rings of trees, pollen as indicator for vegetation, historic records (for taxing normally) of crop yields, or the isotope composition of marine deposits, etc. Within this last category, the $^{18}O/^{16}O$ ratio of these stable oxygen isotopes is of special importance.

In natural oxygen, this ratio is roughly 2:1000; however, subject to variations in the order of some per mil due to various influences. During evaporation from a water surface, the vapor produced is depleted in ^{18}O whereas the remaining water shows an enrichment. This way, rain derived from that vapor shows a lower $^{18}O/^{16}O$ ratio than the original water, the actual value depending, principally, on the atmospheric temperature during the formation of the precipitation („temperature effect") and on the intensity of the rainfall („amount effect"). During glacial times, with a large part (up to one-third) of the earth's surface covered with ice, about one-fifth of the global water was retained, inhibiting its return to the oceans, which became, therefore, enriched in ^{18}O. Crustaceans that lived in equilibrium within that environment produced deposits with elevated ^{18}O concentration. In this way, the $^{18}O/^{16}O$ ratio of sediments depends on the extension of the world's ice sheet and, therefore, its temperature at the time of their formation.

It was the analysis of marine sediments that proved that the occurrence of glacial periods is cyclic with about 100,000 years, as postulated by Milutin Milankovitch already in the first half of the twentieth century. This astronomer attributed climatic changes to slight anomalies in the parameters for the earth's orbit, the eccentricity in the case, without any possibility to test his theory at that time. Later, Fourier analyses of the $^{18}O/^{16}O$ ratio versus age of sediments from the last half million years revealed principal cycle times of 100, 43, 24 and 19 thousand years.

In this paper we interpret ^{18}O results for groundwater from deep wells in sedimentary basins of the northeast (Fig. 1) through the temperature effect and the amount effect, together with analyses of the noble gases dissolved in those waters and carbon-14 dating of the dissolved carbon (for establishing a time scale) in order to reconstruct the paleoclimate of the northeast at the end of the Pleistocene and gain some knowledge on the climatic conditions that made possible the formation of the rich water resources in the deep aquifers of the sedimentary basins of the region.

Methods

Oxygen-18

Oxygen-18 was measured by mass spectrometry applying the classic method of water/carbon dioxide equilibration at KFA/Jülich – Germany, Hydroisotop/Schweitenkirchen – Germany, and CENA/Piracicaba – Brazil. Results are given in $\delta^{18}O$ ‰ (SMOW). In the following we shall explore the dependence of this parameter in rainfall on atmospheric temperature and rainfall amount. For a more detailed description of isotope methods in hydrology and hydrogeology, we refer the reader to Clark and Fritz (1997) and Moser and Rauert (1980).

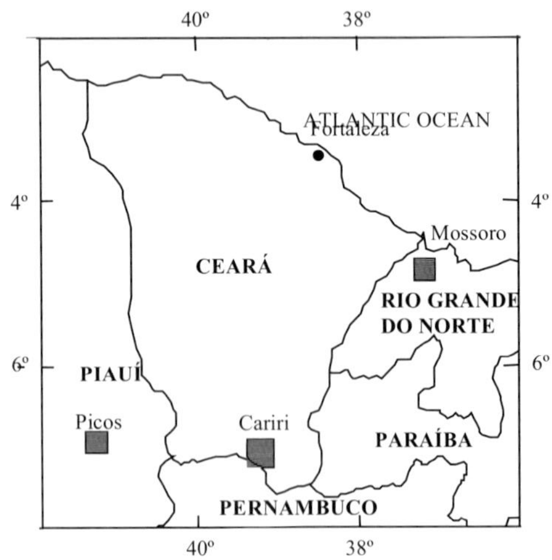

Fig. 1. Location of research areas

Carbon-14

As a measure of the „age" of groundwater, defined as the time elapsed since its infiltration through the unsaturated zone of soil, carbon-14 ages of the total dissolved inorganic carbon compounds (TDIC) are used. TDIC is extracted from the water in the field by precipitation with barium hydroxide. The method covers a range of ≈1 ka to about ≈40 ka. The datings used in this paper are of different origin and were obtained using various methods. The values given are „conventional ages" based on a carbon-14 half-life of 5568 years.

M. A. Geyh in Hannover/Germany and the Isotope Physics Laboratory (installed by Geyh in the 1970s) of the U.F.C./Fortaleza used gas counting in proportional tubes for dating early 1970s and 1980s samples from Piaui and Rio Grande do Norte. Hydroisotop/Schweitenkirchen/ Germany applied liquid scintillation counting to Cariri samples. Accelerator mass spectrometry was used at the Eidgenössische Technische Hochschule/Zürich – Switzerland for 1993 samples from Piaui.

Combined with the isotope methods described above, mass-spectrometric analysis of the noble gases dissolved in the paleowaters was used in order to determine the recharge temperature. This method is known as „noble-gas thermometry" (Mazor 1972; Rudolph 1981).

Noble Gases

During its infiltration through the unsaturated zone, water is in equilibrium with the atmospheric noble gases. The quantity dissolved is determined by the partial pressure of the gas and its physical solubility, which decreases with increasing temperature. The influence is stronger for the heavy gases, with solubility changing by a factor of 2.5 for xenon between 0 and 30 °C, compared to 1.1 for helium. After entering the aquifer, the gas concentration is maintained due to the action of hydrostatic pressure (even at elevated temperature preserving the temperature information). That way, the concentrations of noble gases dissolved and their relative abundance can be applied to calculate the soil temperature during the recharge period of the aquifer („noble-gas temperature"). The overall precision of the method is about ±1 °C (2σ).

Contamination of the water by mixing with air can be detected through the relative increase in light gases, which are more abundant in the air, whereas physical dissolution prefers the heavy ones. Helium is not evaluated because of possible radiogenic contamination.

Noble-gas sampling was performed, during sampling campaigns in Piaui in 1985 and 1993, by connecting malleable copper tubes of about 50 cm^3 volume directly to the well's discharge and filling them under pressure in order to prevent degassing. The tubes were sealed with wedge-shaped pliers. The gas determinations of the first sampling were made by J. Rudolph and M. Stute at the Environmental Physics Institute of the University of Heidelberg/Germany (Geyh et al. 1991) and those of the second campaign by Stute at the Lamont-Doherty/Palisades – USA (Stute et al. 1995).

Paleowaters in Sedimentary Basins of the Northeast

In more than 70% of the area of the Brazilian northeast, outcropping basement rock with zero primary hydraulic conductivity is the predominant hydrogeologic feature. However, some sedimentary basins with rich groundwater resources stand out against that background of groundwater scarcity. Aquifers in the basins Maranhão-Piauí, Potiguar, in Rio Grande do Norte, and Araripe/Cariri, in the very south of Ceará, at the border to states of the Pernambuco and Piauí exhibited artesianism with flowing wells.

The Serra Grande and Cabeças Aquifers/Piauí

In the mid-1970s, SUDENE, in cooperation with a German hydrogeological mission, initiated a systematic survey of the deep sandstone aquifers around Picos in the southeast of Piauí. Here, formations Serra Grande, Pimenteiras (aquitard), and Cabeças slope gently from south-east towards north-west (see Fig. 1 in „Water resources in Ceará and Piauí", Frischkorn et al., this Vol.). Fourteen groundwater samples from the Serra Grande aquifer, the locations of which are found in the geological map of Fig. 2, were analyzed for ^{14}C and ^{18}O. When plotting δ^{18}O vs ^{14}C ages, a well-defined step function was obtained (Fig. 3; Frischkorn et al. 1984, 1988; Geyh et al. 1991) with δ^{18}O changing, around 10 ka b. p., from about -6.2‰, for old waters, to about –3‰ for young ones. The latter value characterizes present-day rainfall. (We stress that this orderly behavior was found for the virgin aquifer only; ^{18}O values from the same aquifer under exploitation exhibit less clear correlation with age, but still show the same basic features.) It became clear that this change in ^{18}O had to be attributed to climate change at the Pleistocene/Holocene transition.

However, δ^{18}O is difficult to interpret quantitatively as it is basically influenced by both temperature and rainfall amount, which cannot be separated in ^{18}O measurements alone. Variations in the isotopic composition of oceans with paleoclimate represent a further complication as it determines the composition of the water vapor that produces the precipitation.

Application of the Noble-Gas Thermometer

Noble-gas temperatures calculated from the dissolved gas concentrations are shown in Fig. 4 as a function of the uncorrected conventional ^{14}C ages. As Ne values are considered to be of higher precision than the rest, we plotted them separately in the lower part and joined that for argon, crypton and xenon in the upper part. The step-like behavior at about 10 ka b. p. is the same as for δ^{18}O (Fig. 3). Modern and paleotemperatures are in surprisingly good agreement in the two graphs. However, the gas temperature exceeds the annual meteorologic mean of the area, which is 26.9 °C for Picos, by about 0.8 °C. This effect is well known as a superheating of the soil relative to the atmosphere in tropical areas with little vegetation. Whether this phenomenon also occurred in the Pleistocene or not is subject to discussion. A temperature step of $\Delta T = 4.7$ °C is extracted as the mean value from the two curves.

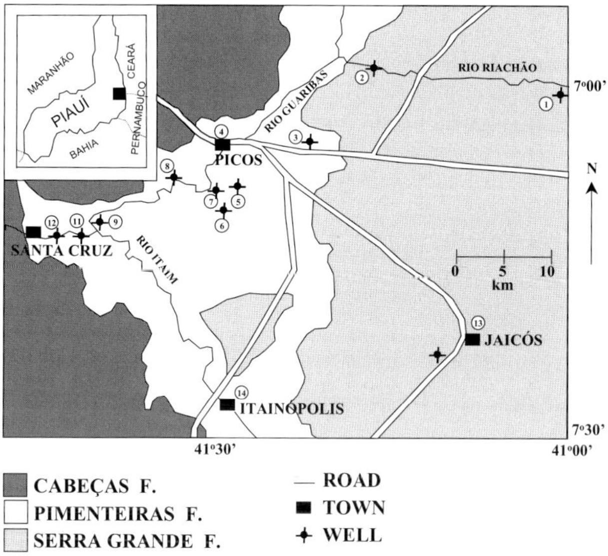

Fig. 2. Location of Serra Grande wells in the Picos region in Piauí (Geyh et al. 1991)

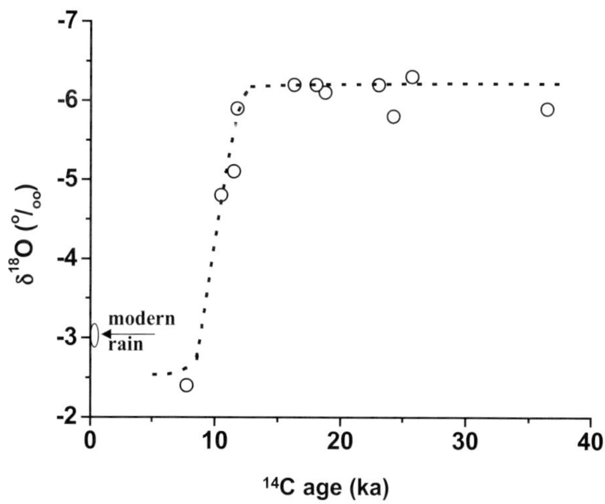

Fig. 3. Oxygen-18 as a function of carbon-14 age for drilled wells in the Serra Grande aquifer of Piauí

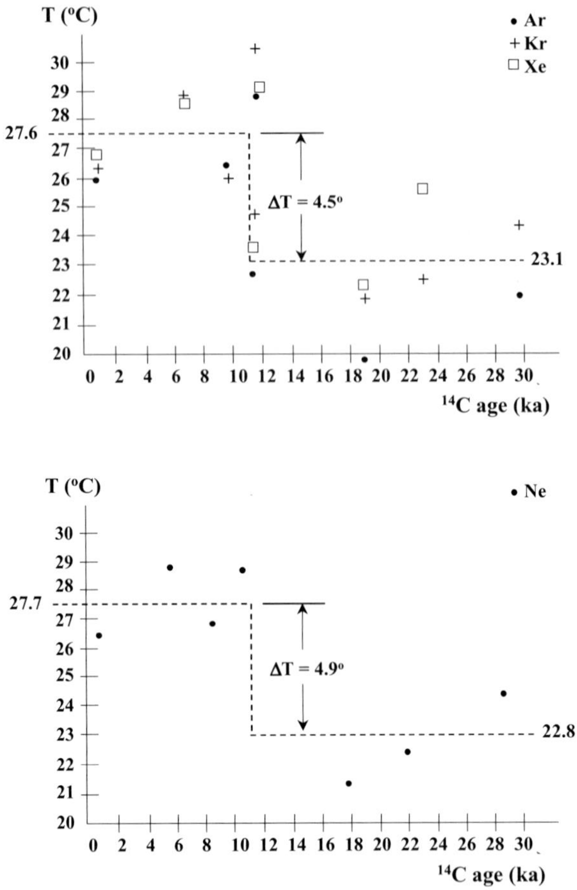

Fig. 4. Noble-gas temperature as a function of carbon-14 ages for drilled wells in the Serra Grande aquifer in Piauí (Geyh et al. 1991)

However, these results inspired disbelief as the expectation of paleoclimatologists was different. In early 1993, a new sampling campaign was performed, this time collecting 18 samples from both aquifers, Serra Grande and Cabeças. Careful evaluation of noble gases and ^{14}C ages suggests that the mean annual ground temperature from ≈35 ka to ≈10 ka was 5.4±0.6 °C lower than during the Holocene. Discussion on this result and consequences for climate modeling may be found in Stute et al. (1995). Here we restrict ourselves to stating that a climate change in tropical Brazil at ≈10 ka b. p. can be quantified through noble gases and that ^{18}O may be used as indicator for that event.

Thus, we denote as „paleowaters“ those that reveal their pleistocenic origins through ^{14}C ages higher than 10 ka and/or through $\delta^{18}O$ values significantly lower than that of present-day rainfall.

Paleobotanical and sedimentological observations suggest that the climatic change affected not only temperature, but also resulted in a reduction in rainfall that induced profound transformations of flora and fauna. The extinction of large herbivores happened in this period due to environmental changes.

The Açu Aquifer/Rio Grande do Norte

In the northeastern state of Rio Grande do Norte, the Açu sandstone aquifer crops out in a strip of about 10 km in width around the parallel 5°30’, about 60 km from the coast. It merges towards the north being superposed by the limestone of the Jandaira formation. On its way to the coast, waters gain ages higher than the limit of ^{14}C dating (≈40,000 years). In large areas of its confinement, wells were artesian flowing. However, pressure decayed rapidly under excessive exploitation and flowing conditions disappeared.

As is seen from Fig. 5, a $\delta^{18}O$ vs ^{14}C age plot exhibits qualitatively the same behavior as that for the Serra Grande aquifer of Piauí (Fig. 3). However, the step is smeared out and less pronounced; lowest ^{18}O values are –5‰. Considering that modern rainfall has $\delta^{18}O$ ≈–3.0‰, the presence of paleowaters is evident, however, without a clear indication in time for the climatic change. These features are certainly due to a strong interaction of the Açu aquifer with the Jandaira limestone aquifer producing mixes.

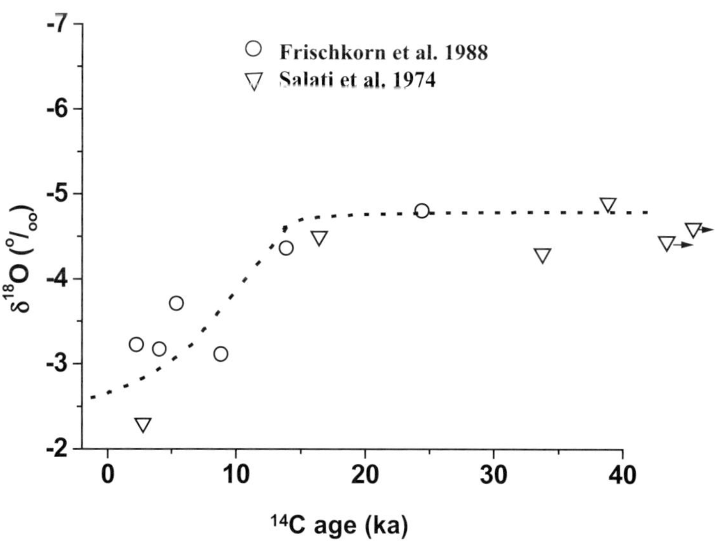

Fig. 5. Oxygen-18 as a function of carbon-14 age for the Açu aquifer in Rio Grande do Norte (two wells show ages beyond the limits of the method)

For example, $\delta^{18}O \approx -3.5‰$ for a sample with an (uncorrected) age of more than 10 ka can be explained by mixing 17% of very old water with $\delta^{18}O \approx -6.0‰$, (originating from the Açu aquifer) with 83% of water younger than 10 ka, with $\delta^{18}O \approx -3.0‰$, however originating from Jandaira limestone (which contains an elevated percentage of inorganic carbon, reducing the specific activity of ^{14}C and increasing, erroneously, the age by up to 5730 years).

In addition, heavy exploitation certainly caused strong perturbation and even inversion of groundwater flow in large areas, leading that way to intense mixing within the aquifer.

Cariri Aquifers/Ceará

The Araripe Basin constitutes a sedimentary „oasis“ in an environment of outcropping crystalline basement rock. It consists of a plateau, the Chapada do Araripe, of 7500 km^2 in area, at about 800 m of altitude, and a valley, the Vale do Cariri, north of it. The east/west axis of the valley marks the principal discharge area of regional groundwater flow systems from the plateau and from the northern border of the valley. However, strong tectonic features dissect these flow patterns and give rise to aquifer mixing. Despite these complications, the presence of paleowaters that „remember“ the past climatic regimen is easily detected through low ^{18}O values. Figure 6 joins groundwaters from the Araripe Plateau (the two samples with $\delta^{18}O \approx -5‰$) and from the Cariri Valley (those which form the slope of the curve). The latter stem from wells in Juazeiro do Norte and, as we shall see below, represent mixes from distinct aquifers.

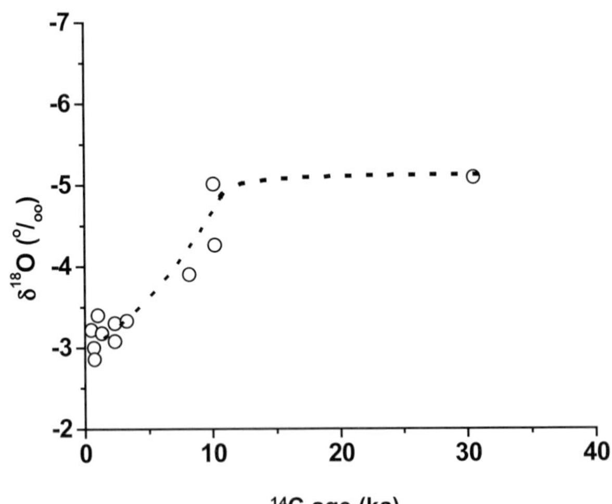

Fig. 6. Oxygen-18 as a function of carbon-14 ages for deep wells in aquifers of the sedimentary Araripe basin (Cariri)

Public water supplies of that town are based on two batteries of wells that exploit the Rio da Batateira aquifer, the upper one in the Cariri Valley. Seventeen wells are pumped 24 h/day producing 6500 m^3/day. In years of drought, the water table suffers a reduction of tens of meters; irreversible environmental changes have been induced (Mendonça 1996). We analyzed ^{18}O, ^{14}C and electric conductivity (CE) in samples from wells of the Riacho do Macacos battery. All of these parameters are satisfactorily explained as resulting from mixing of two components (Santiago et al. 1997): a young one, with pmC ≈90 (percentage of modern carbon), derived from modern rainfall ($\delta^{18}O \approx -3.0‰$) and low salinity (CE ≈150 μS/cm), and an old component (pmC ≈10) with ^{18}O more negative ($\delta^{18}O \approx -4.3‰$) and of higher salinity (CE ≈650 μS/cm).

The mixing lines are shown in Fig. 7. Computational mudflow simulations of observed water levels reveal that the old contribution is a result of aquifer leakage from deep aquifers caused by reduced pressure in the Rio da Batateira aquifer. That way, in years of drought, uprising paleowaters ease the deficit in the overexploited aquifer.

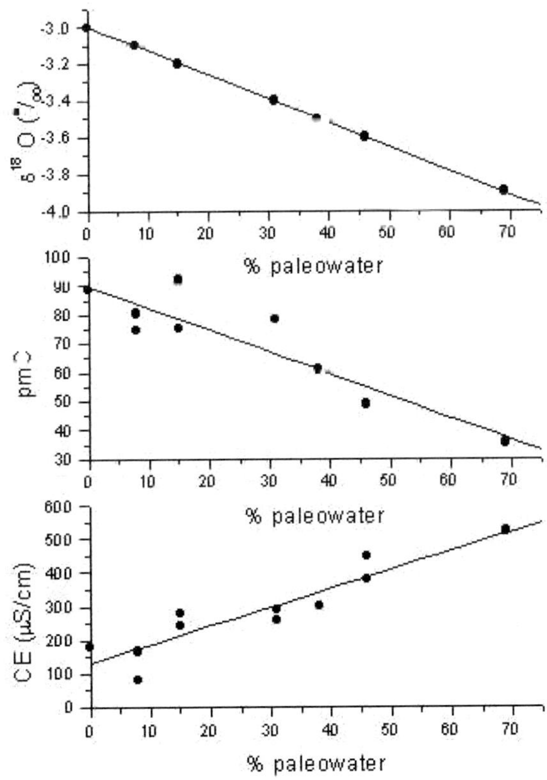

Fig. 7. Oxygen-18, carbon-14 and electric conductivity as a function of the percentage of paleowater in mixtures from wells of the Cariri valley of the sedimentary Araripe basin

Conclusions

A climate change in tropical Brazil is inferred from paleowaters in sedimentary basins of the northeast through measurements of dissolved noble gases and $\delta^{18}O$ values significantly lower than that of modern rainfall.

These paleowaters originate from the Pleistocene, when climatic conditions were different from actual ones, favoring the accumulation of these waters. The exploitation of paleowaters is to be compared with mining and must be made with maximum social and economic benefit, imposing a control of the volume withdrawn and of use. This is of special importance for aquifers with flowing wells, which constitute a seductive invitation to misuse and waste.

The Piauí and Rio Grande do Norte aquifers exhibited artesianism with flowing wells. Today, these conditions persist only in Piauí and at a much lower pressure than about 30 years ago when exploitation began. As pressure in artesian aquifers is related to aquifer matrix and water compressibility, it declines much more rapidly than in phreatic aquifers. This means that the generosity of nature to produce groundwater at zero costs, inviting in that way a squandering of the resource, does not last long under exploitation. Unfortunately, authorities are not always aware of that fact. We have witnessed the drilling of artesian wells just for the illusion of exuberance and the pleasure of extravagance.

References

Clark I, Fritz P (1997) Environmental isotopes in hydrogeology. Lewis Publ, New York

CLIMAP (1981) Project members. Map chart series 36. Geological Society of America, Boulder, Colorado

Frischkorn H, Torquato JR, Santiago MF (1984) Medidas isotópicas em aqüíferos profundos na região centro-leste do Piauí. In: 3° Congresso Brasileiro de Águas Subterrâneas, Fortaleza/BR, pp 42–51

Frischkorn H, Santiago MF, Torquato JR (1988) Dados isotópicos e hidroquímicos da porção oriental da Bacia Potiguar. In: 5° Congresso Brasileiro de Águas Subterrâneas, São Paulo/BR, pp 144–153

Geyh MA, Stute M, Frischkorn H, Santiago MMF (1991) Contribuição para a história climática do nordeste do Brasil. 20 anos de Cooperação Científica Tecnológica Brasil – República Federal da Alemanha. KFA, pp 159–165

Mazor E (1972) Paleotemperatures and other hydrological parameters deduced from noble gases dissolved in groundwaters. Geochim Cosmochim Acta 36:1321–1336

Mendonça LAR (1996) Modelagem matemática, química e isotópica de uma bateria de poços na cidade de Juazeiro do Norte/ Ceará. MSc thesis, Universidade Federal do Ceará, Centro de Tecnologia, Fortaleza/Brazil

Moser H, Rauert W (1980) Isotopenmethoden in der Hydrologic. Borntraeger, Berlin

Rudolph J (1981) Edelgastemperaturen und Heliumalter ^{14}C datierter Paläowässer. PhD thesis, Ruprecht-Karls-Universität, Heidelberg

Salati E, Leal JM, Campos MM (1974) Environmental isotopes used in a hydrogeological study of northeastern Brazil. Hydrology – SM-182/13. IAEA, Vienna, pp 259–283

Santiago MF, Silva CMS, Mendes Filho J, Frischkorn H (1997) Characterization of groundwater in the Cariri (Ceará/Brazil) by environmental isotopes and electric conductivity. Radiocarbon 39:49–60

Stute M, Forster M, Frischkorn H, Serejo A, Clark JF, Schlosser P, Broecker WS, Bonano G (1995) Cooling of tropical Brazil (5 °C) during the last glacial maximum. Science 269:379–383

Sedimentation of Reservoirs in Semiarid Brazil

José Carlos de Araújo, Lia Fernandes, José Christiano Machado Júnior, Márcia Regina Lima Oliveira and Tácito Cunha Sousa

Universidade Federal do Ceará, Departamento de Hidráulica e Ambiental, CEP 60451-970 Fortaleza-Ceará, Brazil, jcaraujo@ufc.br

Introduction

The water supply in the Brazilian semiarid region relies mostly on surface water stored in reservoirs called *açudes*. For example, 93% of the water delivered with minimum confidence in the state of Ceará comes from these reservoirs. Therefore, surface water conservation, both from quality and quantity aspects, is strategic for the sustainable development of the region. The transport and deposition of sediments are definitely among the most significant processes that contribute to the decrease in the water availability in this region. In order to intervene in this process, it is necessary to understand its causes and effects in semiarid regions, which is the main objective of this research. Pursuing this objective, a model called HidroSed was developed to evaluate the sedimentation process in reservoirs of this region. This model was based on the investigation of several watersheds in the state of Ceará and its main goals are to quantify the storage capacity of the *açudes* throughout the time and to assess the impact of future scenarios, such as land use and global changes, on surface water.

Methodology

Study area. Seven watersheds controlled by reservoirs were selected as experimental areas for the model validation (see Table 1). In addition, a small watershed (São Jorge watershed – Fortaleza) was adopted for the evaluation of the urban per capita sediment contribution. It is important to note that these experimental watersheds include all geographical regions of the state and they are indeed representative of the whole state of Ceará. Furthermore, catchment areas range from 3–1220 km^2, initial storage capacities range from 1–126 hm^3, and simulation periods range from 45–95 years. Each watershed was divided into cells considered homogeneous. The watershed divisions were based on 1:100,000 scale maps developed by SUDENE, except for the Santo Anastácio watershed division that was based on a 1:10,000 scale map. The size of the cells vary within each watershed and from one watershed to the other, but average cell areas were defined as follows: 4 km^2 for Acarape do Meio, Canabrava, Cedro, São Mateus and Várzea da Volta; 25 km^2 for Várzea do Boi; and 1 km^2 for Santo Anastácio.

Table 1. Characteristics of experimental watersheds

Watershed	Area (km^2)	Simulation period (years)	Initial volume (hm^3)	Characteristics
Acarape do Meio	208.19	74	34.1	Mountain humid region, NE of the state
Canabrava	2.85	57	1.2	Cariri tropical semihumid region, south of the state
Cedro	219.96	95	125.7	Very dry region of Sertão, center of the state
São Mateus	226.94	46	10.3	Dry (transition) region of Sertão, center east of the state
Várzea do Boi	1,221.20	45	12.5	Very dry region of Inhamuns, SE of the state
Várzea da Volta	155.24	82	51.9	Sertão–coast transition region, NW of the state
Santo Anastácio	9.34	75	1.0	Coast region, urban area, NE of the state
Average	292.0	68	33.8	(-)

Sedimentation model HidroSed. The model, summarized in Fig. 1, evaluates the sedimentation in the reservoirs based on rainfall, topography, soil and land use. It works under watershed and planning scale, i.e., there is no specific concern on the event scale.

The deposition mass rate in a reservoir, dM_S/dt (ton $year^{-1}$), can be computed using Eq. (1), where dM_E/dt is the rainfall sediment yield rate (ton $year^{-1}$), SDR is the sediment delivery ratio, Pop is the urban population in the watershed, ω is the per capita sediment urban contribution, and η is the reservoir retention coefficient.

$$\frac{dM_S}{dt} = \left(\frac{dM_E}{dt} \cdot SDR + Pop \cdot \omega \right) \cdot \eta \quad (1)$$

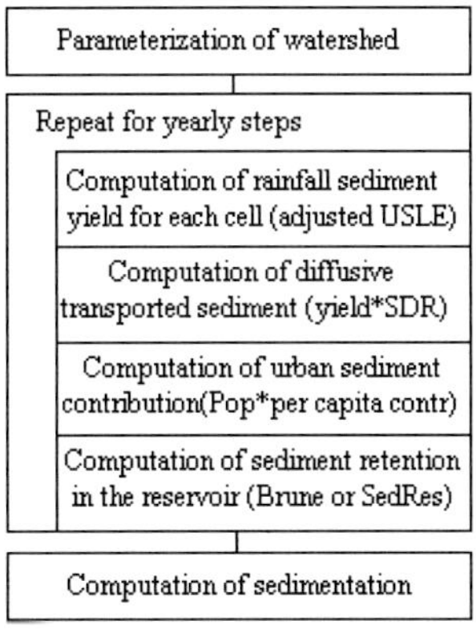

Fig. 1. Basic concept of the HidroSed sedimentation model

Rainfall sediment yield computation. The calculation of the rainfall sediment yield rate is based on Eq. (2), which is the universal soil loss equation, USLE (Wischmeier and Smith, 1978), considering the adjustments to the investigation region conditions.

$$\frac{dM_E}{dt} = A \cdot \left(R \cdot K \cdot LS \cdot C \cdot P\right) \tag{2}$$

A is the area (ha), R is the erosivity factor for rainfall and runoff (MJ. mm ha^{-1} h^{-1} year^{-1}), K is the soil erodibility factor (ton h MJ^{-1} mm^{-1}), LS is the slope–length–steepness factor, C is the cover and management factor, and P is the support practice factor. In order to validate the model, the rainfall sediment yield was calculated for each cell and each year from the reservoir construction date to the control measurement date. The simulation periods range from 45–95 years. Erosivity factor is calculated using Eq. (3), presented by Lombardi Neto and Moldenhauer (Bertoni and Lombardi Neto 1990).

$$R_m = 67.355 \cdot \left(\frac{P_m{}^2}{P_a}\right)^{0.85} \quad ; \quad R = \sum_{m=1}^{12} R_m \tag{3}$$

where P_m is the monthly total rainfall (mm), P_a is the yearly average rainfall (mm), R_m is the monthly erosivity factor (MJ mm ha^{-1} h^{-1} month^{-1}) and R is the yearly erosivity factor. This equation was developed for southeast Brazil, but it is applicable to northeast Brazil as well. The soil erodibility factor (ton h MJ^{-1} mm^{-1}) is determined according to Wischmeier and Smith (1978). The main soil types were identified for each watershed and samples were collected and analyzed in the laboratory. The K factor mean value for each cell is computed based on the percentages of each soil type within the cell area. The slope–length–steepness factor (%) is determined using Eqs. (4) and (5) (Carvalho 1994; Fernandes 2000), where

D is the slope steepness (%), Lr is the slope length (m), A_c is the cell area (m^2), and L_{dr} is the drainage length (m).

$$LS = 0{,}00984 L_r^{0{,}63} D^{1{,}18} \tag{4}$$

$$L_r = \frac{A_c}{4\sum L_{dr}} \tag{5}$$

In this calculation, rivers, brooks, lake contours and roads account for drainage length. Due to survey scale problems, most information is available only in 1:100,000 maps, and the lengths measured must be multiplied by a correction factor (Table 2) obtained by comparison between drainage length in map scale and in photo scale (1:10,000) for several small watersheds in the Cedro area.

Table 2. Drainage length correction factor for 1:100,000 scale maps

Steep watersheds	Average-to-steep watersheds	Flat-to-average watersheds
3.50	3.00	2.50

The cover and management factor and support practice factor (%) are computed together for each cell, where each type of land use is identified. Then, the mean cell CP factor is computed by weighted average on land use area. The CP values were estimated for reference years using reports, maps, satellite images and field survey. For non-reference years, the CP factor is estimated by linear interpolation. Due to the lack of specific research on cover and management factor (%) in the Brazilian semiarid region, comparative results from West Africa and South America (Reining 1992) were used (Table 3).

Table 3. Range of cover–management–support practice factor (CP) in the Brazilian semiarid

Land use system	C factor	P factor	CP factor
Natural vegetation	0.001–0.005	1.0	0.001–0.005
Agricultural use	0.1–0.4	0.8–1.0	0.080–0.400
Urban use	0.0	(-)	0.0

Diffusive Transport of Sediment Computation. Only a small fraction (25% average) of the rainfall sediment yield reaches the drains, leaving the origin watershed. The rest is retained diffusely. According to Khanbilvardi and Rogowski

(1984), there are three main equations to compute sediment delivery ratio (SDR): the Maner equation (Eq. 6), the Roehl equation (Eq. 7), and the Williams and Brendt equation (Eq. 8).

$$\mathrm{Log}\left[\mathrm{SDR}(\%)\right] = 2.943 - 0.824 \times \mathrm{Log}\left(\frac{\mathrm{L_m}}{\mathrm{F_r}}\right) \tag{6}$$

$$Log\left[SDR(\%)\right] = 4.27 - 0.23 \times \mathrm{Log(A)} - 0.51 \times \mathrm{Log}\left(\frac{\mathrm{L_m}}{\mathrm{F_r}}\right) - 2.79 \times \mathrm{Log(B)} \tag{7}$$

$$\mathrm{SDR}(\%) = 62.7 \times \left(\mathrm{S_m}(\%)\right)^{0.403} \tag{8}$$

where L_m is the maximum length in the watershed parallel to the main river (m), F_r is the height between the average watershed divide and the outlet point (m); A is the watershed area (square miles), B is the bifurcation rate and S_m is the main river slope (%). According to the Maner equation, the terrain slope is the main parameter that influences delivery ratio. The Roehl equation considers not only terrain slope, but also watershed area and bifurcation rate. In addition, the Williams and Brendt equation assumes that the main river slope is the parameter that defines SDR.

Per Capita Urban Sediment Contribution. In order to determine the urban per capita sediment contribution (%), a small urban watershed (São Jorge watershed – Fortaleza) was monitored for 3 months. The watershed population, the rainfall, the discharge and the sediment concentration were measured and analyzed. The results revealed that the average per capita contribution is 0.6 kg/year. This value was used to compute the Santo Anastácio sedimentation.

Reservoir Retention Computation. Sediment retention within a reservoir is estimated using the Brune equation (Carvalho 1994) as a function of the volume to annual inflow ratio. Brune's retention ratio is usually very high, on average above 97% (Table 4), to be applied to semiarid reservoirs. Observed ratios in the region have shown lower values (Fernandes 2000). This can be explained by the high correlation coefficient between sediment income (heavy rain seasons) and sediment outlet. In order to improve the assessment of this ratio, a physical approach to the problem is under development (model SedRes).

Model Field Validation. The validation procedure for each watershed consists of five steps: (1) plotting height–volume curve before the dam was built; (2) generating updated height–volume curves by means of topographic and bathymetric surveys; (3) assessing sedimented soil density in the laboratory; (4) simulating sedimentation by HidroSed from construction to control month; and (5) comparing sedimented mass measured and simulated.

Results

The main results obtained in this research are summarized in Tables 4 (results of simulation using HidroSed) and 5 (results of validation of model HidroSed). A brief description of the results and some important features of each experimental watershed are presented in the following subitems.

Acarape do Meio Watershed. The erosivity factor computation was based on a data set from three gauge stations: Redenção, Acarape and Palmácia. The K factor was determined by analysis of six samples. Although the terrain slope is steep, drainage in the Acarape watershed is rather dense, so that the average LS factor equals 2.77. In terms of vegetation, the area has two main natural covers: humid and dry forests. The humid forest is composed of trees that can be more than 30 m in height and can be found in areas 600 m above sea level. The dry forest is the transition between the humid forest and the „caatinga“, typical vegetation of the Brazilian semiarid. The land use evaluation took into account four reference years: 1924 (DNOCS report), 1958 (DNOCS aerial photographs), 1970 (SUDENE maps), and 1997 (COGERH satellite images), as well as field survey. According to Sales (1997), factor P=1 before 1970, when no soil conservation practice was observed. After 1970, joint rice/corn crops were used in order to reduce soil loss, therefore P=0.8. According to the Maner equation (Eq. 6), the SDR=33.4%. A value close to that, SDR=34.5%, was obtained experimentally by Fernandes (2000). The field work verified a volume reduction of 2.67 hm^3 (7.8%) in the Acarape do Meio reservoir within 74 years.

Canabrava Watershed. The R factor was calculated using the Missão Velha gauge station and the K factor was determined by analysis of two soil samples. The LS factor is intermediate (steep terrain but dense drainage) and the average LS=1.80. Personal reports, SUDENE maps (1970) and a field survey were used to estimate CP values. In 1944 this watershed was still untouched. In the 1970s, there was an agricultural boom in the area and soil losses were high (P=1.0), but from 1985 on there was a continuous recuperation of natural conditions, especially the riparian vegetation and support practice (P=0.8). SDR determined was high, close to that observed in the Acarape watershed (35.3%). According to the field survey, the Canabrava reservoir lost 0.084 hm^3, which is 6.9% storage capacity, in 57 years.

Cedro Watershed. Cedro is the oldest dam in Ceará (Araújo 1990); its construction began in 1885 and finished in 1906. The gauge stations Cedro Dam and Quixadá were used to compute the R factor. Eight soil samples were collected and analyzed to estimate K factor distribution in the watershed. The factor LS is relatively high (1.97), due to steep watershed divide and weak natural drainage. The Cedro experimental area is located in the dry „sertão“ and its natural vegetation is the caatinga (high C factor). Agricultural land use began in 1906 after the dam construction. Cattle, corn and beans crops became the most important economic resource in the area after the cotton boom was over (1980). The land use was estimated using IFOCS/DNOCS reports, DNOCS images (1958), SUDENE maps (1970), aerial images (1985) and satellite images (1999), as well as field survey.

According to the Maner equation, only 18% of rainfall yield reaches the drains. Topographic and geotechnical surveys show that the Cedro Reservoir lost 16.7% (20.95 hm^3) of its initial storage capacity within 95 years.

São Mateus Watershed. This dam was built in 1954 by DNOCS to supply water to the town of Canindé. The R factor was calculated using two gauge stations: Salão (1954–2000) and Canindé (1989–2000). The K factor distribution was estimated using six soil samples. The terrain slope is mild and well drained, so that average LS=1.30. The historical land use determination was based on four sources: DNOCS report (1954), aerial images (1958), maps (1970), satellite images and field survey (1999). The average sediment delivery ratio (Maner) is low (12%) but compatible with field observation and topographic conditions. The model validation showed that São Mateus lost 1.46 hm^3, that is 14.1% of its volumetric capacity, within 46 years.

Várzea do Boi Watershed. The Várzea do Boi Reservoir was built in 1954 (Araújo, 1990) by DNOCS. The erosivity factor determined was low (6119 MJ mm ha^{-1} h^{-1} yr^{-1}) compared with the average value in Ceará. This can be explained by the location of this watershed being a drought area. Due to the size of the watershed, 12 soil samples were collected to determine the K factor. The CP computation was based on reports, aerial images, maps and satellite images and five reference years were adopted (1906, 1954, 1958, 1965 and 1997). The Maner SDR value is low, 15.6%, but compatible with this flat region. The storage capacity was decreased to 5.82 hm^3 (11%) within 54 years.

Várzea da Volta Watershed. The gauge stations of Meruoca and Várzea da Volta were used to perform R factor calculation. Four soil samples were collected to assess soil erodibility factor distribution. This experimental watershed presents three distinct regions: the high region, relatively steep and well drained; the very steep region, where no soil yield is observed (almost vertical rock formation); and the very flat region, where SDR tends to zero. Due to this formation, the slope factor LS=1.62 is close to the state average (1.67). Degradation of the natural conditions can be easily observed in the high region, but the natural conditions are well preserved in the rest of the catchment area. The land use reference years were based on reports (1919), aerial images (1958), maps (1970), satellite images and field survey (1999). The Maner SDR=33.7% is a very high value compared with observed data, because it is a function of average terrain steepness. The steep–very steep–very flat terrain distribution cannot be considered by the Maner equation (Eq. 6). A modified value was computed separately for each region, and a weighted SDR average estimated: 15.8%, which reduced the model error from 181% to 32%. The storage capacity of the Várzea da Volta Reservoir was reduced to 1.53 hm^3 (12,2%) between 1919 and 2000.

Santo Anastácio Watershed. The Santo Anastácio urban watershed is controlled by a dam built in 1918. The lake is now under an advanced eutrophication process, once it receives intense solid waste and wastewater contributions. When it was built, the area was completely rural (80% natural vegetation), but its urbanization, which started in the late 1930s, has turned the area to completely urban since 1980. In order to estimate the erosivity factor R, two gauge stations, Fortaleza (1918 to 1981) and Campus do Pici (1982–1992), were used. The erodibility factor was calculated by means of three soil samples. The average slope factor, LS=2.04, is relatively high, taking into account the low steepness, but it is reasonable considering the poor natural drainage in the area. The CP factor was computed using reports (1918), aerial images (1965), maps (1970) and satellite images (1995). The SDR values according to the Maner equation (7.2%) and the Williams and Brendt equation (7.3%) are practically the same, but the SDR according to the Roehl equation (82.5%) is a very improbable result. Urban contribution is very important in this watershed so the estimation of population is also important. In 1918, less than 2000 people lived in this area whereas, in 1992, there were more than 78,000 inhabitants (over 2,062,000 inhabitants/year altogether). This shows a population growth rate of 5% per year, very high but consistent with the rate observed for Fortaleza in the century. Considering a per capita contribution of 0.56 kg/year, a global contribution of less than 1500 tons would have been disposed by urban contribution in this reservoir. The observed value is much higher considering exactly USLE and Maner SDR. The calibration would show a per capita contribution of 0.2 kg/day. This high difference can be explained by the presence, inside the catchment area, of industries and the campus of the Federal University of Ceará facilities, which were not considered as „population“ in the computation. Bathymetric survey in the lake is not possible today due to eutrophication, so 1992 data (Dept Fishing Engineering, UFC; Fig. 2) had to be used for validation purposes. The reservoir showed a volumetric reduction of 136,000 m^3 (26.7%) in 75 years, that represents 3.6% reduction per decade, twice as much as in the rural areas.

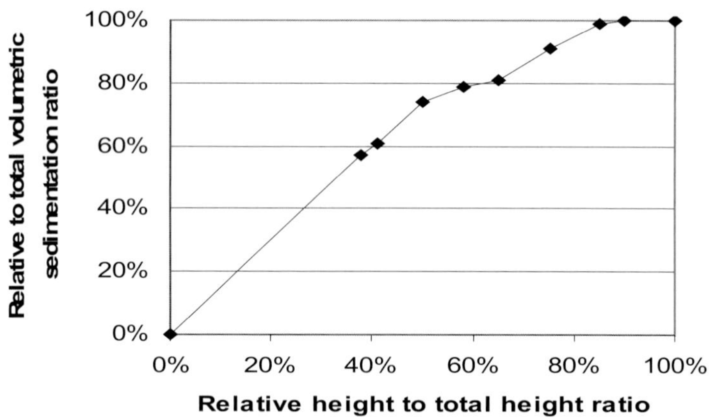

Fig. 2. Sedimentation distribution in the Santo Anastácio Reservoir (1918–1992)

Conclusions

The sedimentation process was assessed by means of the HidroSed model for seven representative watersheds in Ceará and their respective field validations were performed. The research allowed the determination of hydrosedimentological parameters for seven different areas in the Brazilian semiarid (Table 4). The HidroSed model presented encouraging results (Table 5, Fig. 3), having an average error of 29% for a 68-year average simulation period. The Maner formulation (error 29%) proved to be the best for sediment delivery ratio computation since the Roehl equation presented 76% error and the Williams and Brendt 306%. The average reservoir sedimentation rate in Ceará is 1.85% per decade, i.e., the state diminishes its storage capacity by 23 million m^3 every year.

Acknowledgements

The authors would like to acknowledge the support of CNPq and BMBF to the WAVES Program, as well as the University of Kassel for the financial aid for field work. The authors are also grateful to FUNCEME – Ceará Meteorological and Water Resources Foundation, COGERH – Water Resources Management Company and Department of Fishing Engineering of UFC, for sharing important data with this research group.

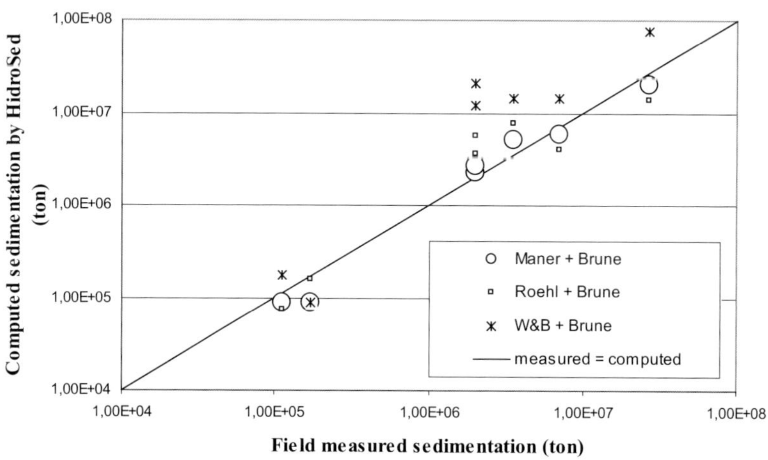

Fig. 3. Comparative results of computed and measured sedimentation for seven watersheds

References

Araújo JA de A (1990) Dams in the northeast of Brazil, 2nd edn. DNOCS, Fortaleza

Bertoni J, Lombardi Neto F (1990) Conservação do Solo. Ícone, São Paulo

Carvalho NO (1994) Hidrossedimentologia Prática. CPRM, Rio de Janeiro

Fernandes L (2000) Processo Hidrosedimentológico na Bacia do Açude Acarape do Meio, Ceará. Master's Diss, Department Hydraulic and Environmental Eng. UFC, Fortaleza

Khanbilvardi RM, Rogowski AA (1984) Quantitative evaluation of sediment delivery ratios. Water Res Bull 20:865–874

Reining L (1992) Erosion in Andean hillside farming: characterization and reduction of soil erosion by water in small scale cassava cropping systems in the southern central cordillera of Colombia. Margraf, Hohenhein tropical agricultural series no 1

Sales MTBF (1997) Caracterização de processos erosivos em áreas florestadas e antropizadas na serra de Baturité-CE. Master's Diss, Geology Department, UFC, Fortaleza

Wischmeier WH, Smith DD (1978) Predicting rainfall erosion losses. A guide to conservation planning. Agriculture handbook, no 537. US Department of Agriculture, Washington, DC

Table 4. Results of simulation using the HidroSed model for seven watersheds in Ceará: average (standard deviation)

Variable	R	K	LS	CP	SDR			Brune coeffi-cient	Global erosion	Simula-tion time (dt)
					Maner	Roehl	W&B			
Unit	MJ.mm/(ha.h)	Ton.h/(MJ.mm)	(-)	(-)	(-)	(-)	(-)	(-)	mm/year	Years
Acarape do Meio	8135 (3257)	0.037 (0.005)	2.77 (2.09)	0.0[illegible]4 (0.013)	0.334	0.493	0.912	0.98	0.246	74
Canabrava	7347 (1659)	0.032 (-)	1.80 (1.59)	0.058 (0.034)	0.353	0.292	0.684	0.98	0.401	57
Cedro	6828 (4198)	0.020 (0.005)	1.97 (2.45)	0.192 (0.160)	0.182	0.[illegible]22	0.676	0.98	0.716	95
São Mateus	6106 (4839)	0.019 (0.004)	1.30 (1.15)	0.130 (0.095)	0.121	0.[illegible]96	0.647	0.97	0.159	46
Várzea do Boi	6119 (4267)	0.032 (0.008)	0.22 (0.22)	0.139 (0.094)	0.156	0.105	0.374	0.98	0.076	45
Várzea da Volta	9683 (5921)	0.032 (0.011)	1.62 (1.55)	0.028 (0.027)	0.158	0.349	1,248	0.98	0.155	82
Santo Anastácio	9923 (4072)	0.013 (0.004)	2.04 (0.84)	0.039 (0.023)	0.072	0.325	0.073	0.95	0.007	75
Average	7734	0.026	1.67	0.093	0.197	0.340	0.659	0.974	0.252	68
Coefficient of variation	54%	25%	75%	73%	54%	74%	57%	1.2%	96%	28%
Maximum average	9923	0.037	2.77	0.192	0.353	0.825	1.248	0.98	0.716	95
Minimum average	6106	0.013	0.22	0.014	0.072	0.105	0.073	0.95	0.007	45

Table 5. Validation of the HidroSed model for seven watersheds in Ceará

Variable	Sedimentation volume (dV)	Sedimentation rate (dV/V/dt)	Soil density	Sedimentation (dM1)	Sedimentation rate (dM1/A/dt)	Simulated sedimentation (dM2)			Simulation error (dM2/dM1-1)		
						Maner	Roehl	W&B	Maner	Roehl	W&B
Unit	hm^3	% per decade	ton/m^3	ton	$ton/km^2/year$	ton	ton	ton	(-)	(-)	(-)
Acarape do Meio	2.67	1.06	1.32	3.52E6	228.5	5.21E6	7.69E6	1.42E7	+47.9%	+118.1%	+303.5%
Canabrava	0.08	1.21	1.38	1.12E5	689.9	8.94E4	7.40E4	1.73E5	–20.4%	–34.1%	+54.4%
Cedro	20.95	1.77	1.28	2.68E7	1282.5	2.03E7	1.36E7	7.53E7	–24.2%	–49.2%	+181.2%
São Mateus	1.46	3.06	1.36	1.99E6	188.6	2.26E6	3.66E6	1.21E7	+13.7%	+84.4%	+508.6%
Várzea do Boi	5.82	2.39	1.20	6.98E6	127.0	5.96E6	4.01E6	1.43E7	–14.7%	–42.5%	+104.3%
Várzea da Volta	1.53	1.51	1.32	2.02E6	158.7	2.67E6	5.78E6	2.10E7	+32.4%	+191.4%	+942.8%
Santo Anastácio	0.14	3.57	1.26	1.71E5	244.1	8.79E4	1.57E5	8.97E4	–47.7%	+8.5%	–47.7%
Average	4.66	1.83	1.30	5.94E6	360.4	5.22E6	4.99E6	1.96E7	28.7%	75.5%	306.1%
Coefficient of variation	160%	46%	46%	160%	102%	134%	94%	131%	50%	83%	106%
Maximum average	20.95	3.57	1.38	2.68E7	1282.5	2.03E7	1.36E7	7.53E7	+47.9%	+191.4%	+942.8%
Minimum average	0.08	1.06	1.20	1.12E5	127.0	8.79E4	7.40E4	8.97E4	+13.7%	+8.5%	–47.7%

Remarks: (i) for sedimentation rates, weighted average were used; (ii) for simulation errors, absolute values were used.

Large-Scale Hydrological Modeling of a Semiarid Environment: Model Development, Validation and Application

Andreas Güntner and Axel Bronstert

Potsdam Institute for Climate Impact Research (PIK), P.O. Box 601203, 14412 Potsdam, Germany, guentner@pik-potsdam.de

Introduction

Semiarid environments are areas of water scarcity. The assessment of water availability is an essential prerequisite for understanding the natural and socioeconomic aspects and their interactions within the system. In particular, the quantification of water resources is required to mitigate the impacts of droughts and to plan future sustainable development in the light of changing climatic, environmental or socioeconomic conditions. An improved and adapted water management of semiarid regions in the context of a changing climate is of particular importance, due to the fact that unmanaged water systems and/or systems under stress are likely to be most vulnerable to climatic change. Hydrological models are indispensable tools for this purpose. However, the number of models that have been specifically developed for dry-land hydrology is small and models are often restricted to small-scale applications (for an overview, see, e.g., Lange et al. 1999). Large-scale models developed for humid areas often perform less well when applied to semiarid environments (e.g., Abdulla and Lettenmaier 1997). Thus, there is a need for further research in this field of hydrology. Taking additionally into account the specific requirements within the interdisciplinary framework of the WAVES program, the main tasks for the development of the hydrological model WASA (Model of *W*ater *A*vailability in *S*emi-*A*rid environments) in this study are (1) to quantify water availability in terms of river runoff, reservoir storage and soil moisture (soil moisture is not only required for an adequate representation of runoff generation, but also as input for the agricultural production model; Krol et al., this Vol.); (2) to provide spatially distributed results for a large study area (400,000 km²); (3) to capture the influence of climate and land-use change on the water balance, which calls for a physical representation of the hydrological processes involved; (4) to be applicable for semiarid hydroclimatological and physiographic conditions, e.g., high temporal and spatial variability of rainfall, high rainfall intensities, intermittent river runoff, sparse vegetation cover, large number of reservoirs; and (5) to provide adequate interfaces to adjacent modules within the integrated model SIM (Krol et al., this Vol.) in terms of input/output variables and their spatial and temporal scale.

Model Concept

The hydrological model WASA is a deterministic model, working with a daily time-step and in a spatially distributed mode. The largest spatial units in the model are subbasins or municipalities (level 1 in Fig. 1). The latter, administrative, units were required within the integrated modeling approach applied in the WAVES program to provide an interface with adjacent modules, e.g., the water-use sector, for which data are available only at the level of administrative units.

In order to capture the influence of spatially variable land-surface properties on soil moisture patterns and runoff generation, subbasins or municipalities are subdivided into smaller modeling units. A hierarchical top-down disaggregation scheme of the landscape, based on the SOTER concept (Soil and Terrain Digital Database; FAO 1993) was applied and parameterized with terrain and soil characteristics of the study area as given in the SPICE database of Gaiser et al. (this Vol.) and in DNPM (1973a,b, 1981). Within this structure, the focus in WASA was placed on taking into account lateral surface and subsurface flow processes at the hillslope scale, which is usually not perceived in large-scale hydrological models. Thus, modeling units were defined as terrain patches with similar characteristics referring to lateral flow processes (landscape units, level 2 in Fig. 1). Terrain components (level 3) represent areas of similar slope gradient, topographic position and soil associations within the toposequence of a landscape unit. In order to describe heterogeneity of soils and land use within each terrain component, terrain components are subdivided into soil-vegetation components (SVC, level 4). Finally, at the lowest level of this hierarchy (at the scale of representative soil profiles), vertical hydrological processes are represented by conceptual, physically based approaches. For example, infiltration is based on the Green-Ampt approach (Green and Ampt 1911), evapotranspiration is described by a modified Penman-Monteith approach, particularly taking into account soil evaporation (Shuttleworth and Wallace 1985), and the soil water balance is calculated by a multi-layer storage approach. Lateral surface and subsurface runoff is redistributed among SVCs and between terrain components; e.g., surface and subsurface runoff generated in one terrain component (e.g., slopes) is transferred to the next terrain component with lower elevation (e.g., lowlands), allowing for reinfiltration of surface runoff or return flow of subsurface runoff there (see Fig. 1).

Level of hierarchy	Type and criteria of delimitation	Function

1 Sub-basin / Municipality / Grid cell

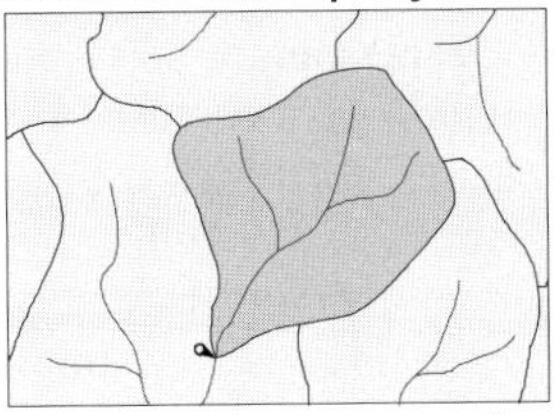

- Polygons with geographic reference
- Source of basins: terrain analysis of 30"-USGS-DEM and digitized topographic maps
- Municipalities: administrative boundaries (municipios)

➤ Runoff routing in river network between sub-basins, including retention in reservoirs and water use

2 Landscape unit

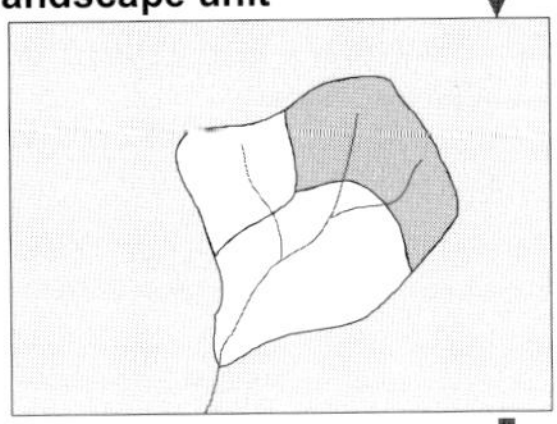

Polygons with geographic reference

Similarity of
- major landform
- general lithology
- soil associations
- toposequences

➢ Modelling unit with similar characteristics referring to lateral processes
➢ Composed of 1 - 3 terrain components

3 Terrain component

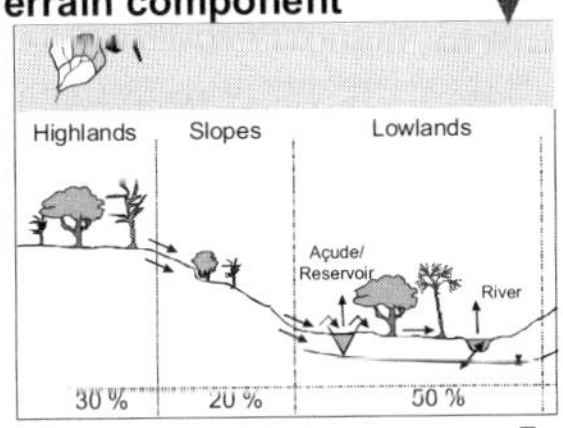

Fraction of area of landscape unit (no geographic reference)

Similarity of
- slope gradients
- position within toposequence
- soil associations

➢ Lateral transfer of surface and subsurface runoff between terrain components of different topographic position by upland-lowland relationships
➢ Reinfiltration and exfiltration in component with lower topographic position

4 Soil-Vegetation component

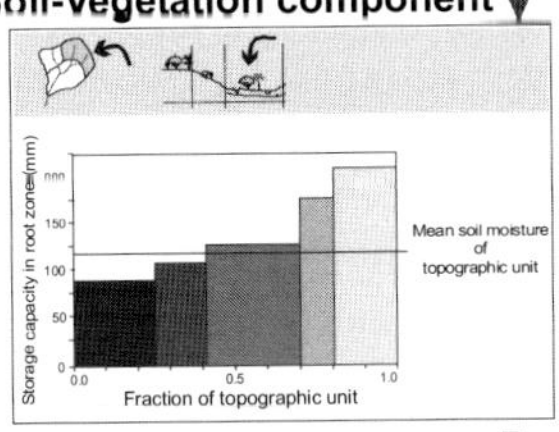

Fraction of area of terrain component

Characterized by specific combination of
- Soil (sub-)type
- Vegetation / landuse class

➢ Variability of soil moisture within terrain component
➢ Lateral redistribution of surface runoff among soil-vegetation components

5 Profile

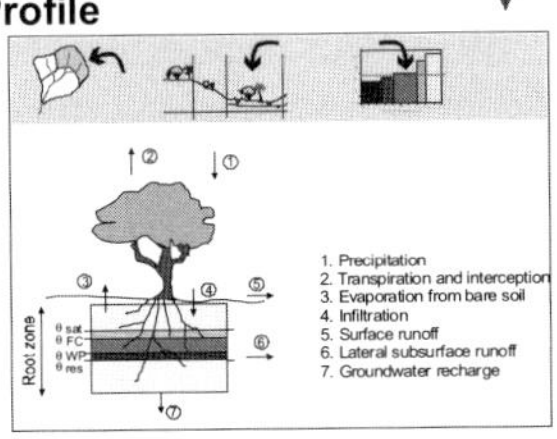

- Several soil horizons of variable depth
- Lower limit by depth of root zone or bedrock

➢ Calculation of water balance in the root zone
➢ Determination of vertical and lateral fluxes for each soil-vegetation component

Fig. 1. Hierarchical approach in WASA for disaggregation of river basins/municipalities into smaller modeling units at different scales

Another important component of hydrological systems in semiarid areas are surface reservoirs. For each large reservoir (storage capacity $>50\times10^6$ m^3) in the study area, the water balance is calculated explicitly in WASA. The large number of small reservoirs (about 8000 dams in the state of Ceará) cannot be taken into account with geographic reference. Instead, small reservoirs are captured by means of their distribution among five storage volume classes, for which the number of reservoirs in each class is known for all municipalities (SRH 1992). In each class, the water balance is calculated for one reservoir of average characteristics, which is assumed to be representative for all reservoirs in this class. A simple cascade scheme is applied, which routes runoff exceeding the storage capacity of one class to the reservoirs of the next larger storage class. The water-use model is coupled by withdrawal and return flow directly to the reservoirs and rivers within WASA (Döll et al., Krol et al., both this Vol.).

Model Validation and Sensitivity

Validating the model with respect to its capability of adequately representing the governing hydrological processes is an important prerequisite when subsequently applying the model for impact studies. At a small spatial scale, daily data of nested catchments in the municipio of Tauá (Ceará), being representative for large areas on crystalline bedrock, were used (Cavalcante et al. 1990). The smallest catchment (Caldeirão, 0.77 km^2) was not influenced by any water retention in reservoirs, so model performance of WASA with respect to runoff generation could be directly evaluated. Figure 2a–c shows good model results on a daily basis. Mean annual observed and simulated runoffs are 88 and 98 mm, respectively. This discrepancy of about 10% is mainly attributed to an overestimation of runoff in the first simulation year (1981). Nevertheless, the dynamics of basin response between dry and wet years were well represented. A multiplicative scaling factor, which reduces the hydraulic conductivity of the soil in the infiltration routine of WASA, was used. This is to compensate for the underestimated real rainfall intensities when using daily rainfall input. The scaling factor was derived as the ratio of rainfall intensities of daily and hourly wet intervals. By this means, simulations resulted in a fraction of infiltration-excess runoff on total runoff (f_h) of 50%. This is in good correspondence to f_h=41% which arises from model runs with an hourly time-step, using disaggregated rainfall volumes (Güntner et al. 2001). The importance of this simple scaling approach for adequately representing an important hydrological process of the area (infiltration-excess runoff) was revealed when running the model without the scaling factor. This resulted in f_h=12% and underestimated runoff events in dry years or at the beginning of the rainy season (Fig. 2d).

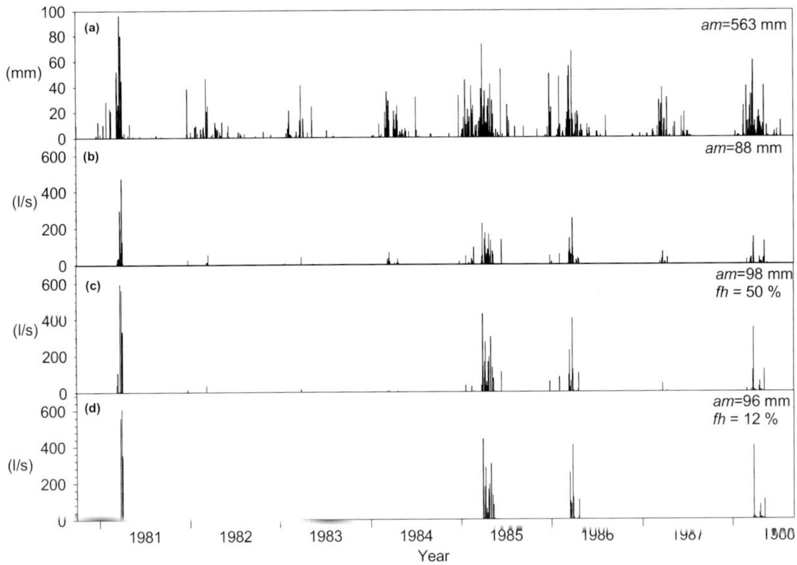

Fig. 2. a Precipitation, **b** observed runoff, **c** simulated runoff with scaling factor, and **d** simulated runoff without scaling factor (see text). Basin Caldeirão (upper Jaguaribe River, Ceará), daily values, period 10/1980 09/1988. *am* Annual mean, *fh* fraction of infiltration excess runoff on total runoff

Validation results of WASA for a larger spatial scale (state of Ceará) are presented in Fig. 3. When interpreting these results, one has to take into account the overall poor quality of observed validation data in accuracy and temporal extent of available time series. Daily rainfall input data for the simulations where interpolated with Kriging methods to a 10×10 km^2 grid, based on time series of about 230 stations. No model calibration was carried out, as the parameters were derived from available physiographical data or from the literature. In general, simulated discharge is in reasonable correspondence to observed values. Model performance on average improves with increasing basin area, as the effect of coarse and uncertain estimates of model input and parameters on simulation results is smoothed out at larger scales. This applies also for differences between the basin area in the model and in reality. The deviation of the simulated from the observed mean annual runoff is on average about ±20% (Fig. 3a). WASA is well capable of representing the variability of runoff dynamics in time. This can be seen for interannual variability (Fig. 3b) as well as for the seasonal dynamics of river runoff in terms of its mean monthly distribution (Fig. 3c). Even when comparing the continuous monthly observed and simulated time series, the model efficiency (according to Nash and Sutcliffe 1970, with optimum value of 1.0) is above 0.8 for most basins.

Another way to perform model validation is by comparing measured and simulated storage volumes of large reservoirs in the study area. Reasonable results are obtained (Fig. 4). Deviations between observations and simulations, however, are larger as in the case of discharge, because additional factors of uncertainty are comprised in the model, i.e., area–volume relationships of the reservoirs or operation rules of the reservoir outlet.

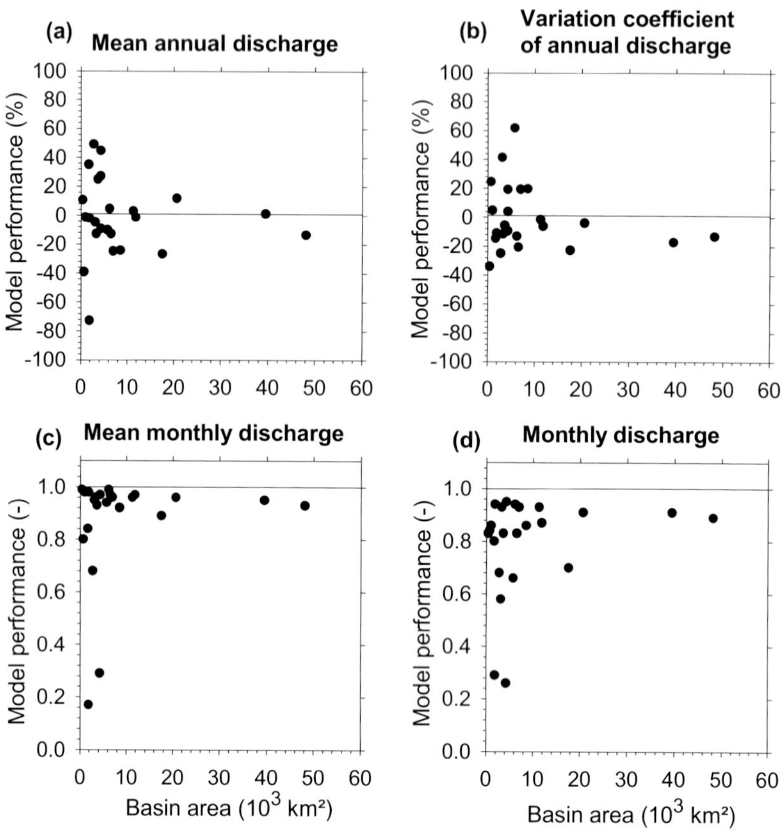

Fig. 3. Validation of WASA for 24 gauging stations in Ceará, simulation period 1960–1998, validation period for individual stations is between 6 and 30 years. Deviation of simulations from observations with regard to the **a** mean annual discharge, **b** coefficient of variation of annual runoff. Nash-Sutcliffe criteria for **c** mean monthly discharge and **d** continuous monthly discharge time series

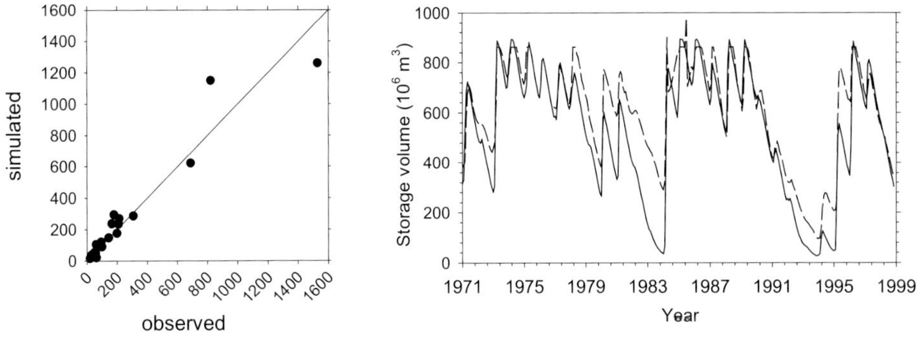

Fig. 4. Validation of WASA for storage volumes of 22 large reservoirs in Ceará, simulation period 1960–1998, validation period for individual reservoirs is between 6 and 34 years. **a** Observed vs simulated mean storage volumes at the end of the rainy period (June) and **b** example of observed (*dashed line*) and simulated (*solid line*) monthly reservoir storage for the reservoir Açude Paulo Sarasate (Ceará), period 1971–1998, in 10^6 m^3

As an example for the sensitivity of simulation results on input data, the effect of variation in precipitation on runoff is presented in Fig. 5. At the scale of sub-basins in Ceará, two rainfall data sets were used, the first based on around 230 stations (see above) and the second on only those 29 stations with time series long enough to be used for construction of regional climate scenarios (see Gerstengarbe and Werner, this Vol.). It can be concluded from this experiment that the effect of changes in rainfall input on runoff is highly non-linear. Changes in runoff are larger than the underlying changes in precipitation by a factor of about 3 (Fig. 5); similar results for semiarid environments are given, e.g., by Arnell (2000). Thus, limited availability of rainfall data may considerably constrain the quality of the model results.

Retention in reservoirs and extraction by water use may have another important effect on discharge. For instance, at the most downstream gauging station of the Jaguaribe River, upstream reservoirs and water use add up to a reduction in annual discharge of on average 20% (Fig. 6). This example demonstrates that, firstly, water retention and water use has clearly to be taken into account when validating the model against runoff data. Secondly, it shows the potential of WASA to be used for assessing dynamically the effects of water use on downstream availability.

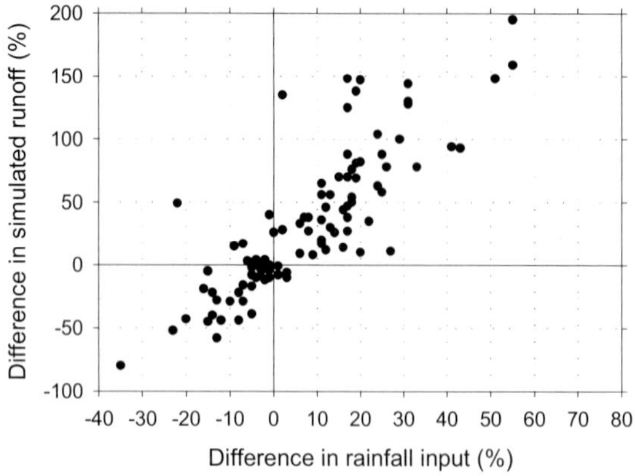

Fig. 5. Effect of variation in rainfall input on simulated runoff for 108 subbasins in Ceará, mean annual values period 1960–1998. Differences in rainfall are caused by different interpolation results due to differences in method and number of stations

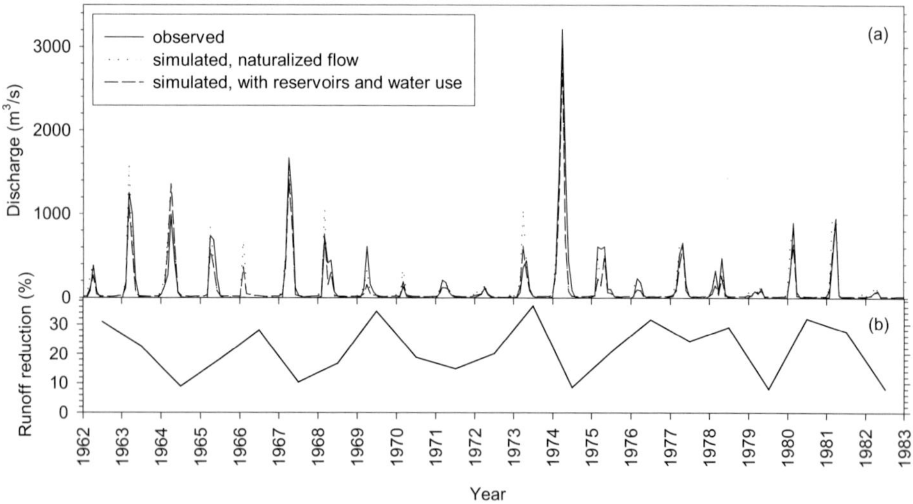

Fig. 6. Monthly discharge at gauging station Peixe Gordo, Jaguaribe River, Ceará, catchment area 47500 km^2, period 1962–1982. **b** Discharge reduction by upstream retention in reservoirs and water use as compared with natural discharge, annual values

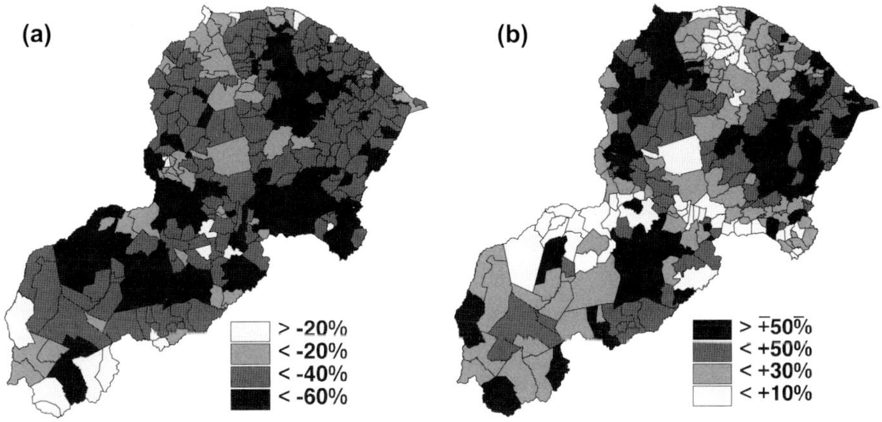

Fig. 7. Trends of runoff generation at the scale of municipios over the period 2001–2050, calculated with WASA, based on climate scenarios derived from **a** ECHAM4 and **b** HADCM2

Model Application for Climate Change Scenarios

Scenario calculations with WASA for the period 2001–2050 were performed for two climate scenarios, one with an expected decrease in rainfall (based on the GCM ECHAM4), and one with increasing rainfall in the study area (based on HADCM2; see Krol et al., this Vol.). Other factors like land use were not changed for these simulations. The results show that, for both scenarios, the expected change in runoff is more pronounced than the corresponding change in rainfall (Table 1). For the ECHAM4-based scenario, a sharp decrease in runoff can be expected for the period until 2050 (Table 1, Fig. 7). The expected change in other climate elements than rainfall, like temperature and radiation, are small, causing only slight changes in potential evaporation. However, due to the lower availability of soil moisture by decreasing rainfall, actual evapotranspiration and groundwater recharge are expected to decrease. For the HADCM2-based scenario, the simulations result in a significant increase in runoff. Actual evapotranspiration, however, increases only slightly, which is, on the one hand, due to a small decrease in potential evaporation. On the other hand, the length of the rainy period does not increase with increasing annual rainfall volumes for the given scenario. Thus, more rainfall is quickly transformed into runoff (and groundwater recharge) instead of being evaporated or transpired.

Table 1. Impact of climate change scenarios on components of the hydrological cycle in the federal states of Piauí and Ceará, simulations with WASA, area average trends (%) for the period 2001–2050; climate scenarios based on ECHAM4 and HADCM2 global circulation model results

	ECHAM4		HADCM2	
	Piauí	Ceará	Piauí	Ceará
Precipitation	–22	–26	+9	+11
Potential evaporation	–4	–3	–4	–3
Actual evapotranspiration	–17	–15	+3	+4
Groundwater recharge	–36	–47	+40	+24
Runoff	–47	–55	+42	+34

Changes in water availability are even more pronounced for municipalities at downstream locations in river basins, as they are exposed to the aggregate effect of all changes in upstream areas. This can be seen, for instance, when looking at expected storage volumes in reservoirs, assuming no changes in storage capacity during the scenario period. Large reservoirs, which usually receive a considerable amount of inflow from a large upstream catchment area, are relatively more affected by the decreasing runoff volumes in the case of the ECHAM4 scenario than smaller reservoirs with rather local catchment areas (Fig. 8).

Comparing both climate change scenarios, there is indication that future changes are more likely in line with the ECHAM4 scenario. This is based, firstly, on the fact that ECHAM4 represents present-day precipitation in the area of northeastern Brazil better than HADCM2. Secondly, for a more recent version of the Hadley Center model (HADCM3), precipitation is expected to decrease in northeastern Brazil, comparable with ECHAM4 (see Arnell 1999).

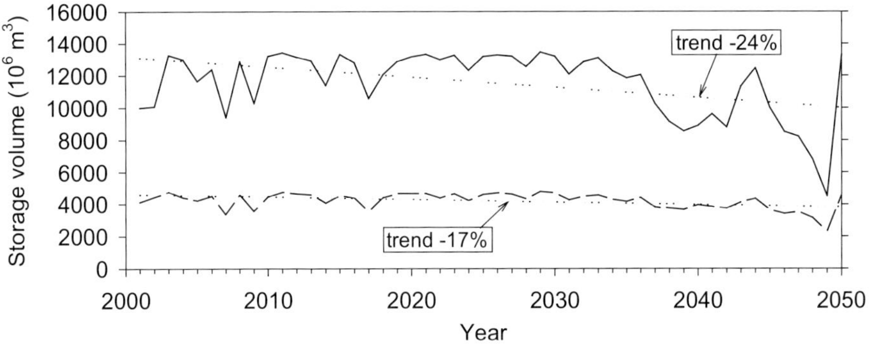

Fig. 8. Total storage volume at the end of the rainy season (June) in Ceará. Large reservoirs (storage capacity >50 × 10^6 m³; *solid line*), small and medium-sized reservoirs (storage capacity <50 × 10^6 m³; *dashed line*), and trend lines (*dotted*), simulations with WASA based on ECHAM4 climate scenario, period 2001–2050

Concluding Remarks

In view of the significant heterogeneity of land surface properties, hydrological models should focus on dominant processes and terrain characteristics in order to keep a model structure simple enough to be applicable at a large scale. The multiscale top-down approach used in WASA allows delineation of modeling units with respect to the key land surface properties, to the objectives of model application and to the resolution of available data. Hydrological processes can be represented at their relevant scale, while smaller-scale features are represented as subscale variabilities. Within this structure, focus is placed on capturing lateral redistribution of runoff and soil moisture at the hillslope scale, using a simple but efficiently applicable scheme for large areas. Model applications at different scales gives reasonable results when compared with measured discharge and reservoir storage volumes. However, due to limited data availability and resolution, uncertainty of results is high. Precipitation is one of the most important factors controlling model performance, due to its strong non-linear effect on basin response, particularly in semiarid environments. No calibration of WASA was carried out for the model application to the large geographic domain in this study. For applications of WASA to individual smaller-scale basins, the model should be adjusted to their specific characteristics if additional information is available. For the assessment of global climate change scenarios on regional water availability, the additional uncertainty induced by diverging results of global circulation models for future climate in the study area has to be taken into account.

References

Abdulla FA, Lettenmaier DP (1997) Development of regional parameter estimation equations for a macroscale hydrological model. J Hydrol 197:230–237

Arnell NW (1999) Climate Change and global water resources. Global Environ Change 9:31–49

Arnell NW (2000) Thresholds and responses to climate change forcing: the water sector. Climatic Change 46:305–316

Cavalcante NMC, Doherty FR, Cadier E (1990) Bacia hidrográfica representativa de Tauá-CE, Relatório Final. Ministério do Interior, SUDENE-DPG-PRN, Recife, Brasil

DNPM (1973a) Projeto Radam, Levantamento de recursos naturais, vol 1, Folhas 23/24 Rio São Fransisco/Aracaju. Ministério das Minas e Energia, Rio de Janeiro, Brasil

DNPM (1973b) Projeto Radam, Levantamento de recursos naturais, vol 3, Folhas 23/24 São Luis/Fortaleza. Ministério das Minas e Energia, Rio de Janeiro, Brasil

DNPM (1981) Projeto Radam, Levantamento de recursos naturais, vol 23, Folhas 24/25 Jaguaribe/Natal. Ministério das Minas e Energia, Rio de Janeiro, Brasil

FAO (1993) Global and national soils and terrain digital databases (SOTER). Procedures manual. World Soil Res Rep no 74. Roma, Italy

Green WH, Ampt GA (1911) Studies on soil physics I. The flow of air and water through soils. J Agric Sci 4:1-24

Güntner A, Olsson J, Calver A, Gannon B (2001) Cascade-based disaggregation of continuous rainfall time series: the influence of climate. Hydrol Earth Syst Sci 5(2): 145–164

Lange J, Leibundgut C, Greenbaum N, Schick AP (1999) A noncalibrated rainfall-runoff model for large, arid catchments. Water Res Res 35:2161–2172

Nash JE, Sutcliffe JV (1970) River flow forcasting through conceptual models, part I. A discussion of principles. J Hydrol 10:282–290

Shuttleworth J, Wallace JS (1985) Evaporation from sparse crops – an energy combination theory. Quart J R Met Soc 111:839–855

SRH (1992) Plano estadual dos recursos hídricos, vol 1. Diagnóstico, Secretaria dos Recursos Hídiricos, Governo do Estado do Ceará, Fortaleza, Brasil

Reservoir Yield Vulnerability Under Global Change

José Nilson Beserra Campos, Ticiana Marinho de Carvalho Studart and Patrícia Freire Chagas

Universida*de* Federal do Ceará, Departamento de Hidráulica e Ambiental, CEP 60451-970 Fortaleza-Ceará, Brazil, nilson@ufc.br

Introduction

There is no scientific consensus on how climate changes will modify the hydrological cycle. Climate models still need to be improved in order to better evaluate accurately these changes. Predictions of regional effects on the hydrological cycle vary significantly among different models. Nevertheless, there is a consensus on global precipitation and evaporation increase. Despite all these uncertainties, some likely trends in the hydrological cycle can be stated:

- An increase in air temperature leads to an increase in evapotranspiration and, as a result, an intensification of the hydrological cycle. For a mean global temperature increase of 2 °C, potential evapotranspiration increases by up to 40% (Hoff 2001).
- There is a tendency for an increase in the total precipitation over the world. Besides, for some regions, a decrease in precipitation is expected, despite the glo bal trend to a higher precipitation, due to changes in global circulation and atmospheric conditions (Hoff 2001).

The northeast of Brazil (NEB) has a semiarid climate subject to recurrent droughts, and one of the main concerns of researchers is evaluating the impact of global changes on water availability in the region. In NEB the trend has been towards a culture of storing water in surface reservoirs, as much as possible. It is common to read in the historiography of droughts that, in order to solve the drought problem, it would be necessary to store in reservoirs down to the last drop of water from the intermittent rivers of the region. Therefore, the estimation of the impacts of climate on water availability in NEB is strongly related to evaluation of the impacts on yields of reservoirs.

The present study intends to evaluate the impacts of climate change on the availability of water in NEB, taking as a case study the Várzea do Boi dam basin. The main focus is on the water yielded by surface reservoirs, which represents the main source for most regions of the state of Ceará.

The Study Area

The study area comprises Várzea do Boi Reservoir's drainage basin, in the municipality of Tauá, situated in the microregion of „Sertões dos Inhamuns“ (Fig. 1). Várzea do Boi Reservoir is in the upper part of the Jaguaribe River basin, which is the largest intermittent river in the world.

The climatic and hydrological conditions in the Várzea do Boi Reservoir correspond to typical semiarid conditions in the state of Ceará: highly variable precipitation and runoff patterns, both within-the-year as well as in yearly scales, associated with high negative water balance. The mean annual precipitation is around 520 mm, mean annual inflow (μ) approximately equal to 42.6 hm^3, standard deviation of annual inflows (σ) equal to 66.8 hm^3, coefficient of variation of annual inflows (CV) equal to 1.57, and mean annual evaporation during the dry season (E_v) equal to 1.438 mm. The drainage area (A) is 1220 km^2 and the reservoir's capacity (K) is 51.8 hm^2 (Campos et al. 2000).

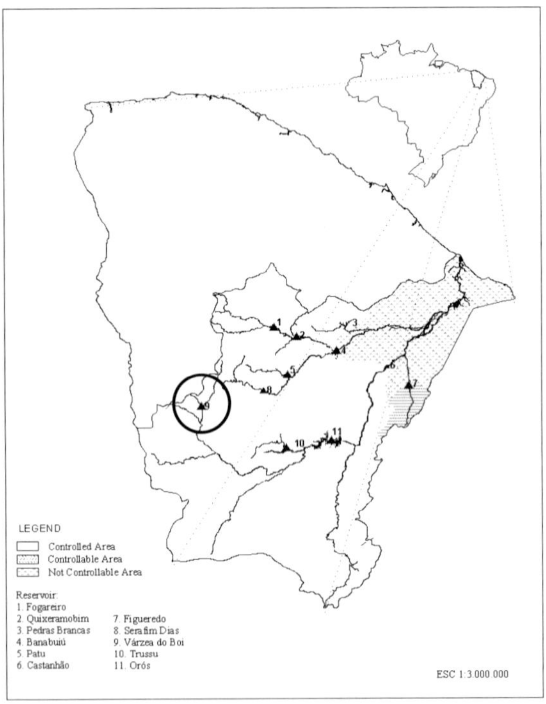

Fig. 1. Location of the Várzea do Boi Reservoir in the Jaguaribe Basin (state of Ceará)

Methodology

A reservoir works as a hydrologic transformation system. The water provided in a very irregular manner by nature is stored and released according to demand. In this process, the inflows are partitioned in three parts: (1) evaporation from the lake; (2) reservoir's spills to downstream, and (3) controlled volume, which is termed „reservoir yield", and represents the system's availability. This third part represents the reservoir yield, or water availability, which is the focus of the present study.

The reservoir yield was estimated using two approaches. In the first approach, the yield was estimated solving the reservoir budget equation, in the parametric formulation of Campos (1987, 1996), using Monte Carlo simulation. In the second, the yield was computed for an infinite reservoir according to an equation, also obtained using Monte Carlo simulation and the parametric formulation mentioned previously (Campos and Ibiapina 1997; Campos et al. 2001).

To introduce climate change into the study, we analyzed two standards for hydrological modifications in the Várzea do Boi dam basin. For each standard, scenarios were built for different values of precipitation, lake evaporation and reservoir inflows. The yield estimations were performed using both procedures: Monte Carlo simulation and the infinite reservoir yield equation. In addition, we also estimated the global change impact on reservoir yield for the climate scenario estimated in the WAVES project.

Standards Assumed for Hydrological Changes

Due to the uncertainties in the expected hydrological modifications in the Várzea do Boi dam basin, the impacts of climate change on water availability were studied under two different standards of hydrological modifications in the basin. In each standard, 10 scenarios were built and grouped in ensembles, named ensemble 1 and ensemble 2, respectively. In *ensemble 1*, precipitation, evaporation and inflows increase at the same ratio (K_X). In *ensemble 2*, precipitation and evaporation increase at the same ratio (K_X) and the coefficient of surface runoff increases according to the Aguiar (1978) polynomial rule (K_R).

Ensemble 2 is a kind of optimistic assumption, once Aguiar's empirical rule is valid for a constant value for evapotranspiration. So, once an increase in evapotranspiration is expected as a consequence of global change, it is also expected that the runoff increases a little less than that obtained by Aguiar's formulation. On the other hand, the first ensemble is a kind of pessimistic scenario regarding reservoir inflow, because some increase in runoff is expected due to the increase in precipitation.

The computations to estimate the global change impacts on water availability were performed in the following steps:

1. The initial condition is that in the year 2000, with data from the Jaguaribe River Basin Water Management Plan (COGERH 2001)

2. For each ensemble, 10 hydrological scenarios were built (K_X varying from 1.02 to 1.20) and the values of precipitation, evaporation and reservoir inflows computed. The coefficient of variation was assumed constant and equal to the conditions in the scenario for the year 2000
3. For each scenario, estimates were made of the reservoir yield and efficiency, according to two methodologies: (1) using Monte Carlo simulation and (2) using an empirical formulation for an infinite reservoir (Campos and Ibiapina 1997; Campos et al. 2001)

In addition, an estimate was made of the impact on the reservoir yield for the global change in precipitation, evaporation and runoff estimated in the WAVES project. This estimation was carried out only by the infinite reservoir procedure.

Yield Estimation Using Monte Carlo Simulation

The reservoir water budget was solved using the Campos (1987, 1996) formulation. The procedure consists of making a parameterization of the budget equation and computing the yield using Monte Carlo simulation. The water budget can be represented by Eqs. (1) and (2), as follows:

$$Z_{t+1}=Z_t+I_t-M-(1/2).(A_{t+1}+A_t).E-S_t \tag{1}$$

with

$$S_t=\max\,(Z_t+I_t-M.\ (1/2).(A_{t+1}+A_t).E_v-K;\ 0) \tag{2}$$

where Z_{t+1} and Z_t are storage at the beginning of the $(t+1)^{th}$ and t^{th} years, respectively; I_t is inflow into the reservoir during the tth year; M is release from the reservoir during the tth year (M is assumed constant from year to year); A_{t+1} and A_t are lake areas at the beginning of the $(t+1)^{th}$ and t^{th} years; E_v is mean evaporation depth during the dry season (E_v is assumed constant from year to year); K is reservoir capacity; and S_t is spill from the reservoir during the tth year.

The reservoir and lake morphologies are described, respectively, by $Z(h)=\alpha.h^3$ and $A(h)= 3.\alpha.h^2$. The α value can be estimated from a regression equation between the lake level (h) and the stored volume (Z). Considering Z(h) and A(h), Eq. (1) can be rewritten as Eq. (3):

$$Z_{t+1} = Z_t + I_t - 3\alpha^{1/3} - \left(\frac{Z_{t+1}^{2/3} + Z_t^{2/3}}{2}\right) E - S_t \tag{3}$$

The reservoir budget equation is reduced to the following parametric presentation:

$$f_M = \Phi\ (CV, G, f_K, f_E) \tag{4}$$

where G is the reservoir reliability (assumed 90%) in this study, f_K is the dimensionless capacity (K/μ), f_E is the dimensionless evaporation factor computed by $(3\alpha^{1/3}E_V)/\mu^{1/3}$ and f_M is the dimensionless release in steady-state conditions estimated by $f_M=M/\mu$.

Equation (4) can be solved for synthetic inflows, for a gamma distribution function, using a coefficient of variation of annual inflows equal to 1.57.

Yield from an Infinite Reservoir

To estimate global impacts on an infinite reservoir yield (theoretically, the maximum yield from a basin controlled by a single reservoir), we used the equation defined by Campos et al. (2001). Assuming reliability (G) of 90%, the authors obtained a general equation for the efficiency of a reservoir (η_M) on the steady state of the storage process (Eq. 5).

$$\eta_M=0.99 \exp [-f_E/(1.5031-1.7104CV+0.8555CV^2-0.1528CV^3) \quad (5)$$

Only two input parameters are necessary: the coefficient of variation of annual inflows (CV) and the dimensionless factor of evaporation (f_E). This equation is valid for CV ranging from 0.5 to 1.6 and f_E ranging from 0.05 to 2.0. In this study the efficiency of the reservoir will be estimated for CV=1–1.6 and for the various values of f_E defined in ensembles 1 and 2.

The infinite reservoir efficiency (η_M) and the dimensionless factor of release (f_M) are related by $\eta_M=0.95\ f_M$ and the reservoir annual release (M) is given by $M=1.05.\mu.\eta_M$.

Results

The results obtained were grouped and analyzed by hydrological modifications standards (ensembles) and by the methodology used in yield evaluation.

Ensemble 1 with Yield Estimated by Monte Carlo Simulations

The results (Table 1) show that there is a negative trend in water availability and in reservoir efficiency. For example, if the mean precipitation and evaporation increases by 4%, the reservoir yield drops from 5.48 to 5.39 hm^3/year (approximately 1.7%). Regarding efficiency of the reservoir, it remains constant for a significance of 0.01.

Table 1. Values of annual yield (M) and reservoir efficiency (η) for 90% reliability, computed by Monte Carlo simulations, in ensemble 1

K_X	μ (hm^3/year)	σ (hm^3/year)	f_E	M (hm^3/year)	f_M	$\eta = 0.95\ f_M$
1.00	42.58	66.85	0.36	5.48	0.13	0.12
1.02	43.43	68.18	0.36	5.47	0.13	0.12
1.04	44.28	69.52	0.37	5.39	0.12	0.12
1.06	45.13	70.86	0.37	5.39	0.12	0.11
1.08	45.98	72.20	0.38	5.29	0.12	0.11
1.10	46.84	73.53	0.38	5.23	0.11	0.11
1.12	47.69	74.87	0.39	5.17	0.11	0.10
1.14	48.54	76.21	0.39	5.15	0.11	0.10
1.16	49.39	77.54	0.40	5.06	0.10	0.10
1.18	50.24	78.88	0.40	5.05	0.10	0.10
1.20	51.09	80.22	0.41	4.99	0.10	0.09

Note: The reference scenario (year 2000) is shaded.

Ensemble 1 with Yield Estimated by the Infinite Reservoir Equation

The analysis (Table 2) shows that there is a slight positive trend in reservoir yield. Notice that this yield represents the maximum controllable yield in the basin. So, regarding maximum regulation capacity, in case of hydrologically balanced changes (ensemble 1), the impact of global change on the Várzea do Boi dam basin will be slightly positive.

Ensemble 2 with Yield Estimated by Monte Carlo Simulations

Ensemble 2 embraces the scenarios where precipitation and evaporation increase in the same ratio (K_X) and runoff increases at a rate K_R, according to Aguiar's rule (Eq. 6)

$$R(mm)=28.53H-112.9H^2+351.91H^3-118.74H^4 \tag{6}$$

where H is the mean annual precipitation over the basin and R(mm) is the runoff, in millimeters, for an unit basin. Using H=0.52 as the initial condition, one can determine the reference runoff for a unit basin (R°mm) to be equal to 25.02 mm. The other values of R(mm) (for each precipitation H) are evaluated by Eq. 6. The K_R values are obtained by dividing the correspondent R(mm) by R°mm. The mean annual inflow (μ) in each scenario K_X is estimated by multiplying the mean annual inflow of the initial conditions (42.58 hm³/year) by the correspondent K_R. The results are shown in Table 3.

Table 2. Values of annual yield (M) and reservoir efficiency (η) for 90% reliability for an infinite reservoir in ensemble 1

K_X	μ (hm³/year)	f_E	M (hm³/year)	$\eta=0.95\ f_M$
1.00	42.58	0.36	15.16	0.12
1.02	43.43	0.36	15.24	0.12
1.04	44.28	0.37	15.32	0.12
1.06	45.13	0.37	15.40	0.11
1.08	45.98	0.38	15.47	0.11
1.10	46.84	0.38	15.54	0.11
1.12	47.69	0.39	15.61	0.10
1.14	48.54	0.39	15.67	0.10
1.16	49.39	0.40	15.73	0.10
1.18	50.24	0.40	15.78	0.10
1.20	51.09	0.41	15.83	0.09

Note: The reference scenario (actual values) is shaded

Table 3. Values of annual yield (M) and the reservoir efficiency (η) for a 90% reliability, computed by Monte Carlo Simulations, in ensemble 2

K_X	H (m)	R (mm)	K_R	μ (hm³/year)	σ (hm³/year)	f_E	M (hm³/year)	f_M	$\eta=0.95\ f_M$
1.00	0.52	25.02	1.00	42.58	66.85	0.36	5.48	0.13	0.12
1.02	0.53	26.39	1.05	44.91	70.51	0.36	5.64	0.13	0.12
1.04	0.54	27.81	1.11	47.34	74.32	0.36	5.74	0.12	0.12
1.06	0.55	29.29	1.17	49.85	78.27	0.36	5.93	0.12	0.11
1.08	0.56	30.82	1.23	52.46	82.37	0.36	6.08	0.12	0.11
1.10	0.57	32.41	1.29	55.17	86.61	0.36	5.96	0.11	0.10
1.12	0.58	34.05	1.36	57.96	91.00	0.36	6.00	0.10	0.10
1.14	0.59	35.75	1.43	60.86	95.54	0.36	5.98	0.10	0.09
1.16	0.60	37.51	1.50	63.84	100.24	0.37	5.95	0.09	0.09
1.18	0.61	39.32	1.57	66.93	105.08	0.37	6.02	0.09	0.09
1.20	0.62	41.19	1.65	70.11	110.08	0.37	6.00	0.09	0.08

It can be seen from Table 3 that there is a positive trend in water availability for K_X ranging from 1.02 to 1.08. For instance, when precipitation increases 4%, the reservoir yield increases 4.7%. That is, in the global change assumptions of ensemble 2, a reasonable gain in water availability is expected. For values of K_X greater than 1.08, the value of reservoir yield decreases and increases in an irregular way.

Ensemble 2 with Yield Estimated by the Infinite Reservoir Equation

The analysis showed that there is positive trend in reservoir yield. Regarding maximum capacity of regulation, in the case of hydrologically balanced changes (ensemble 2), the impact of global change on the Várzea do Boi dam basin will be positive regarding water availability. Regarding the efficiency, however, there is a slightly negative trend (Table 4).

Table 4. Values of annual yield (M) and reservoir efficiency (η) for 90% reliability and infinite reservoir in ensemble 2

K_X	K_R	μ (hm³/year)	f_E	M (hm³/year)	$\eta=0.95\ f_M$
1.00	1.00	42.58	0.36	15.16	0.34
1.02	1.05	44.91	0.36	15.95	0.34
1.04	1.11	47.34	0.36	16.78	0.34
1.06	1.17	49.85	0.36	17.63	0.34
1.08	1.23	52.46	0.36	18.51	0.34
1.10	1.29	55.17	0.36	19.43	0.33
1.12	1.36	57.96	0.36	20.38	0.33
1.14	1.43	60.86	0.36	21.36	0.33
1.16	1.50	63.84	0.37	22.37	0.33
1.18	1.57	66.93	0.37	23.41	0.33
1.20	1.65	70.11	0.37	24.48	0.33

Scenario Waves: According to the WAVES Program Climate Group and running the hydrological model named WASA, there will be a decrease in precipitation (–12%), runoff (–32%) and evaporation (–0.1%). The coefficient of variation of annual inflows is almost the same (present: 1.21 and future: 1.22). In that case, there will be a decrease both in annual yield and reservoir efficiency (Table 5).

Table 5. Annual yield (M) and efficiency (η) for a infinite reservoir considering WAVES global change (it is assumed G=90%)

	f_E	CV	M	η
Reference scenario	0.27	1.21	23.25	0.52
Waves scenario	0.38	1.22	11.87	0.39

Concluding Comments

The results of this research regarding impacts of global change on water availability in the Várzea do Boi dam basin lead to the following conclusions:

- for the WAVES global change scenario, the water availability in assumption of an infinite reservoir (maximum yield in the basin), the reservoir yield drops from 23.25 to 11.87 hm^3/year (49%), and the reservoir efficiency drops from 52 to 39%;
- for the hydrologically balanced changes „ensemble 1" (precipitation, evaporation and inflows increasing at the same ratio), there is a negative trend in the reservoir yield and efficiency in the actual topology of the Várzea do Boi dam Basin. On the other hand, in the assumption of an infinite reservoir (basin runoff totally controlled), there is a slightly positive trend in reservoir yield and a negative trend in reservoir efficiency;
- for the hydrologically balanced changes „ensemble 2" (precipitation and evaporation increasing at the same ratio, and inflows increasing according to Aguiar's polynomial rule), there is a positive trend in the reservoir yield for the actual topology of the Várzea do Boi dam basin as well as for the totally controlled basin. On the other hand, in both situations, a decrease in reservoir efficiency is predicted.

The general conclusion is that, in case of the WAVES scenario, there are substantial losses in water availability in the basin. On the other hand, if the global change is hydrologically balanced, as previously defined, most of the results point to an increase in water availability in the basin. Only from the pessimistic assumption of evaporation, precipitation and inflows increasing at the same rate, will there be small losses in the reservoir yield. In the case of totally controlled basin and hydrologically balanced changes, it is expected that there will be an increase in water availability in the basin.

Acknowledgements

The authors would like to thank CNPq for supporting the research.

References

Aguiar FG (1978) Estudo Hidrométrico do Nordeste Brasileiro. Departamento Nacional de Obras Contra as Secas. Boletim Técnico 36(2). Reimpressão

Campos JNB (1987) A Procedure for Reservoir Sizing on Intermittent Rivers Under High Evaporation Rate. Fort Collins, Colorado State University, PhD thesis

Campos JNB (1996) Dimensionamento de Reservatórios: O Método do Diagrama Triangular de Regularização. UFC, Fortaleza.

Campos JNB, Ibiapina NG (1997) Uma Equação para a Máxima Capacidade de Regularização em um Reservatório. In: XII Simpósio Brasileiro de Recursos Hídricos, Vitória, ES. Associação Brasileira de Recursos Hídricos

Campos JNB, Vieira Neto J, Queiroz EA (2000) Impacto Cumulativo da pequena açudagem: um estudo de caso do Açude Várzea do Boi, em Tauá, Ce. In: V Simpósio de Recursos Hídricos do Nordeste, Natal

Campos JNB, Studart TMC, Ibiapina NG (2001) Computing the yield from an infinite reservoir. In: American Geophysical Union 21st annual hydrology days, Colorado State University, Colorado, USA, April

COGERH (2001) Plano de uso racional da água para irrigação nos vales do Jaguaribe e Banabuiú. Companhia de Gestão dos Recursos Hídricos do Estado do Ceará, Fortaleza, Ceará.

Hoff H (2001) Climate change and water availability. In: Lozán JL, Graβl H, Hupter P (eds) Climate ot the 21st century: changes and risk. Wissenschaftliche Auswertungen, Hamburg

Modeling of Present and Future Water Use in Piauí and Ceará as a Basis for Water Resources Planning

Petra Döll, Maike Hauschild, E. Mario Mendiondo and Jose Carlos de Araújo

Center for Environmental Systems Research, University of Kassel, 34109 Kassel, Germany, doell@usf.uni-kassel.de

Introduction

Sustainable water use is an important prerequisite for the sustainable development of a region. The challenge of water resources planning is to provide sufficient water of appropriate quality for the most beneficial uses while safeguarding the environment. In order to match water supply and water demand, it has been common to satisfy the increasing demand for water by constructing a new physical water infrastructure (reservoirs, wells, pipelines, canals, ...). While in Piauí and Ceará it is at many locations and for certain purposes still necessary to improve water supply, it is equally important to manage water demand. So, on the one hand, more efforts should be made to fulfil the basic human needs for drinking water and sanitation (only about half of the population is connected to the public water supply), while, on the other, the water used for all but the basic human needs should be regarded as a scarce resource and an economic good which must be put to its best (i.e., economically most efficient) use (see, e.g., World Water Commission for the 21st Century 2000).

Sustainability-oriented water resources planning in general and water demand management in particular require an assessment of present and future water uses as well as of renewable water resources. Such an assessment is preferably based on models that are used to simulate the current situation and to derive scenarios of the future (for an overview of the scenario approach that was applied in WAVES, see Döll et al., this Vol.). Sustainability (with respect to water quantity) can then be evaluated, e.g., by comparing the volume of water used with the available water resources or, in a more sophisticated way, by applying a comprehensive water resources management tool for river basins, which needs water use and resources estimates as input, e.g., RiverWare (Zagona et al. 1998). There is a wide range of hydrological (rainfall-runoff) models to simulate the quantity of renewable water available in a drainage basin and how it is affected by climate change and certain water supply management measures, e.g., reservoir construction. However, very few models exist that estimate the water use in a region and how it is affected by global change and water demand management measures. There are some data-intensive models for municipal water use in urban areas (e.g., USBR 1991; Clarke et al. 1997) and many models to compute irrigation requirements (e.g., FAO 1992). IWR-MAIN (2000) is a commercial water use forecasting system that can cover all water use sectors that has been widely applied for US-American cities and river basins.

In order to support water resources planning in Piauí and Ceará, we developed the regional-scale water use model NoWUM (Nordeste Water Use Model), which computes municipality-specific water uses and is designed to assess the impact of global change and demand management measures on water use. In the following, we present the model as well as the simulated water uses under current conditions (around 1996/98) and in 2025, for the reference scenarios RS A „Coastal Boom and Cash Crops“ and RS B „Decentralization – Integrated Rural Development“ (cf. Döll et al., this Vol.) and for selected interventions.

Methods

Two methods are combined to generate model-based scenarios of water use, mathematical modeling and scenario development. The regional water use model NoWUM simulates water use as a function of many driving forces. Quantitative assumptions on the future development of these driving forces (and of the model parameters), which are based on storylines, are developed and then implemented in NoWUM to compute alternative water use scenarios. Döll and Hauschild (2001a) and Hauschild and Döll (2000) provide a detailed model description of NoWUM, and, in the first paper, model uncertainties are discussed. The storylines of the two reference scenarios and the quantitative assumptions about the development of many driving forces of water use can be found in Döll et al. (this Vol.). Döll and Hauschild (2001b) present the quantitative assumptions on the development of all the NoWUM driving forces (shown in Fig. 1), discuss the resulting water use scenarios and the impact of selected demand-management measures, and show the sensitivity of domestic and industrial water use to the assumed prices and price elasticities. Below is a description of NoWUM. Due to limited space, the quantified scenario assumptions cannot be presented here.

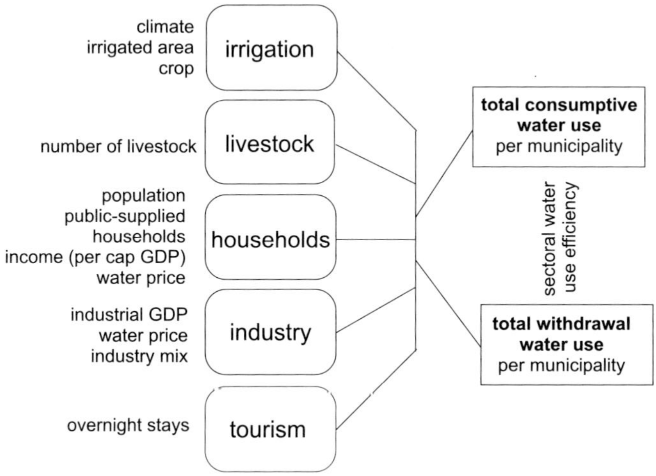

Fig. 1. Overview of the regional-scale water use model NoWUM; on the left-hand side, the driving forces of the water use in the five sectors are indicated

NoWUM computes withdrawal and consumptive water use in each of 332 municipalities in Piauí and Ceará. Withdrawal water use is the quantity of water taken from its natural location, while consumptive water use is the part of the withdrawn water that is lost by evapotranspiration. The ratio between consumptive and withdrawal water use is called water use efficiency. In situ water use is not taken into account in NoWUM. Model design has strongly been influenced by the availability of data. NoWUM distinguishes five water use sectors: irrigation, livestock, households, industry and tourism. Each sectoral water use is computed as a function of a water use intensity (e.g., per-capita domestic water use of the self-supplied population) and a driving force of water use (e.g., self-supplied population). Figure 1 shows the driving forces of each water use sector that are taken into account by NoWUM. In a scenario of future water use, both the water use intensities and the driving forces might differ from present-day conditions. Although all sectoral water uses are expected to vary at least to a certain degree with climate and the season of the year, only the strong seasonality and climate dependence of irrigation water use is simulated by the model.

Irrigation Water Use

Irrigation water use is calculated as a function of climate, irrigated area and type of irrigated crop. Consumptive water use, or net irrigation requirement, is defined as the amount of water that must be added to the soil such that the crop can evapotranspirate at the optimal (potential) rate. Crop-specific consumptive water use per unit of irrigated area during the growing period, I_c, is computed according to the CROPWAT methodology (FAO 1992) as:

$$I_c = \sum_{i=1}^{n} K_{c,i} E_{p,i} - P_{eff,i} \tag{1}$$

i: 10-day-period
n: number of 10-day periods within the growing period (crop-specific)
K_c: crop coefficient (crop-specific) (–)
E_p: potential evapotranspiration (mm/day)
P_{eff}: effective precipitation (mm/day)

Usage of effective precipitation P_{eff} instead of total precipitation P takes into account that part of the rain runs off and is therefore not available for crops. According to the USDA Soil Conservation Service Method (as given in FAO 1992):

$$\begin{aligned} P_{eff} &= P(4.17 - 0.2\,P)/4.17 \quad \text{for } P < 8.3 \text{ mm/day} \\ P_{eff} &= 4.17 + 0.1\,P \quad \text{for } P \geq 8.3 \text{ mm/day} \end{aligned} \tag{2}$$

The total consumptive irrigation water use of each municipality is computed by taking into account the crop-class-specific irrigated areas. Withdrawal use is calculated from consumptive use by applying an irrigation water use efficiency of 0.6 for 1996/98, based on information on irrigation technology and the respective efficiencies for northeastern Brazil (FAO 2000).

Irrigated Areas in 1996/98. The Brazilian agricultural census of 1995/96 (IBGE 1998a,b) provides crop-specific harvested areas of different irrigated crops and the total irrigated area for all municipalities in Ceará and Piauí. However, Brazilian experts agree that the total irrigated areas are strongly overestimated by the census. Besides, the harvested areas and the total irrigated area often seem to be inconsistent. Therefore, we derived a best estimate of the crop-specific irrigated areas in each municipality by taking into account, for Ceará (1) information on crop-specific irrigation licenses 1998 in the most important irrigation areas from the Water Resources Ministry in Ceará, (2) information from the agricultural census for 23 municipalities, (3) estimates of Brazilian experts for eight municipalities, and for Piauí, (1) information from the agricultural census and (2) estimates of Brazilian experts. The estimated total irrigated area is 43,000 ha in Ceará (compared with 77,000 ha according to the agricultural census), and 13,000 ha in Piauí (compared with 18,000 ha).

Crops. Based on the Brazilian agricultural census of 1995/96 (IBGE 1998a,b), nine irrigated crop classes are distinguished which cover almost all crops irrigated in Ceará and Piauí: banana, beans, cotton, fruit trees, grass, maize, rice, sugarcane and vegetables. Crop-coefficients and growing season length of each crop class were taken from FAO (1992) (for fruit trees and vegetables, respectively, the FAO values for citrus trees and tomato were used), while crop-class-specific planting dates were defined based on a variety of local sources. Due to the lack of information on multicropping and the uncertainty with respect to the irrigated areas, each crop class was assigned to have only one growing season per year.

Climate Data. Daily values of precipitation and (Penman-Monteith) potential evapotranspiration in each municipality were generated for the period 1921–1980 by spatial interpolation of daily data from 30 measurement stations (Gerstengarbe and Güntner 1999, pers. commun.). Average net irrigation requirements per unit irrigated area under present-day conditions were computed by first simulating 30 years of irrigation with the climate of 1951–1980 and the irrigated areas of 1996/98, and then averaging the resulting annual requirements. This procedure is more appropriate for the computation than to use long-term average climatic data as input, which, due to the nonlinearity of Eq. (1), would lead to an underestimation of the of long-term average irrigation water requirement.

A climate change scenario for the years 2000–2050 (daily values of precipitation and potential evapotranspiration) was derived (for each municipality) by a statistical downscaling method which takes into account precipitation change in northeast Brazil as computed by the ECHAM4 climate model of the Max-Plank-Institute, Hamburg, Germany, for a 1% yearly increase of greenhouse gas concentrations, and historic station data for 1921–1980 (Werner and Gerstengarbe 1997).

Livestock Water Use

Annual values of livestock water use are calculated from the number of animals per species in each municipality (IBGE 1998a,b) and a species-specific water use typical for northeast Brazil. Withdrawal water use is assumed to be equal to consumptive water use.

Domestic Water Use

Domestic water use includes not only residential water use, but also water use by commerce and by public institutions like schools. It is computed as the product of the per-capita water use and the population. The water use in self-supplied households is differentiated from that in households connected to the public water supply, i.e., households with tap water (Fig. 1). The municipality-specific per-capita water use in connected households for the current situation is calculated from municipality-specific data provided by the local water suppliers. No information is available on the per-capita water use in self-supplied households. As a first estimate, we assume a per-capita withdrawal water use of 50 l/day, a value defined by Gleick (1996) as the basic human water requirement, which is equal to approximately half of the per-capita water use of the supplied population. Water use efficiency in the domestic sector is assumed to be 0.2 (Solley et al. 1998). The same efficiency is also used for the industrial and the tourism sectors.

While per-capita water use in self-supplied households is assumed to remain unchanged in the future, per-capita domestic water use in connected households is assumed to depend on water price and income. The econometric concept of elasticity is used to compute per-capita annual withdrawal water use U as a function of water price P per unit volume of water (price elasticity) and income (here per-capita gross domestic product *GDP*; income elasticity). Price elasticity ε_p, which is generally negative, is defined as:

$$\varepsilon_p = \frac{\frac{dU}{U}}{\frac{dP}{P}} \tag{3}$$

And income elasticity ε_i, which is generally positive, as:

$$\varepsilon_i = \frac{\frac{dU}{U}}{\frac{dGDP}{GDP}} \tag{4}$$

Industrial Water Use

Industrial water use is computed as a function of the industrial GDP (IGDP) of 19 industrial branches in each municipality and the branch-specific water use per IGDP (industrial water use intensity). Branch-specific industrial water use intensities were assumed to be the same as in Germany because no Brazilian data were available. Future industrial water use per industrial output is modeled as a function of water price according to Eq. (3), where *U* is annual withdrawal water use per IGDP.

Tourism Water Use

Tourism water use is modeled as a function of the number of tourists in a municipality as well as the average length of stay and the water use per tourist and day, which are regionally differentiated (coast and interior).

Results and Discussion

Table 1 presents the computed sectoral water withdrawals for 1996/98 and for both 2025 reference scenarios (climate is assumed to remain constant). Total withdrawal water use in the study area Piauí and Ceará in 2025 is 71% higher than in 1996/98 for RS A, and 39% higher for RS B. The irrigation sector accounts for 75–80% of the change in total water withdrawals; irrigation withdrawals as a ratio of total withdrawals increase from 45% in 1996/98 to 59% (RS A) and 54% (RS B) in 2025. In northeast Brazil, rural development is considered to be strongly linked to irrigated agriculture. Therefore, large increases in irrigated area are foreseen in both reference scenarios, with an average over the study area of 3.8%/year in RS A and 2.5%/year in RS B (between 1990 and 1998, irrigated areas in Brazil only increased by 1.2%/year, FAO 2000). Consumptive water use in the study area is 500 million m^3/year in 1996/98, 49% of the withdrawals, and increases by 104% (RS A) and 67% (RS B) until 2025. It grows more strongly than withdrawal water use because the irrigation sector, where consumption is a larger fraction of withdrawal than in the domestic, industrial and tourism sectors, becomes more important, and because it is assumed that irrigation water use efficiency improves from 0.6 to 0.7. Irrigation consumptive use as a ratio of total consumption increases from 54% in 1996/98 to 70% (RS A) and 63% (RS B) in 2025. As consumptive use represents the amount of water that certainly cannot be reused (while the return flow part of the withdrawn water has the potential to be reused if water quality is sufficient), its future development is of particular interest. Water use in the second most important water use sector, the domestic sector, rises by 22% in the case of RS A and by 11% in the case of RS B. These values result from a combination of decreased per-capita water use in the households connected to the public water supply (due to an annual price increase of 6%, which is below the value observed in Ceará between 1989 and 1998), an increased fraction of public-supplied households and the population change. Per-capita water use in public-supplied households decreases, on average, by 17% in case of RS A and by 22% in case of RS B, from 139 l/day in 1996/98.

Table 1. Sectoral water withdrawals in Piauí and Ceará in 2025 for both reference scenarios as compared with withdrawals in 1996/98 (climate change is neglected)

Scenario region (Population 1996, in millions)		Withdrawals, in 10^6 m^3/yr					
		Irrigation (irrig. area, in 1000 ha)	Livestock	Domestic	Industry	Tourism	Total
Piauí	1996/98	127.4 (13)	65.1	123.6	4.1	2.1	322.4
(2.7)	2025 RS A	279.5 (45)	64.3	130.3	4.0	8.3	486.5
	2025 RS B	209.3 (29)	77.8	125,6	3.7	6.2	422.6
Ceará	1996/98	323.9 (43)	81.3	225.6	46.3	14.3	691.3
(6.7)	2025 RS A	745.0 (116)	81.7	196.9	55.2	64.6	1243.5
	2025 RS B	547.7 (83)	93.9	260.4	42.8	43.5	986.6

While the state values of water use in the case of RS A are modeled to be larger than the case of RS B (Table 1), the opposite is true for many scenario regions (compare Döll et al., this Vol.) and municipalities. For example, irrigation and domestic water use in the „Regions with low potential water resources“ in Ceará are higher in RS B than in RS A due to the higher population (less migration to the coastal region) and the stronger extension of private irrigated areas, where products for the regional markets are produced (for a more detailed representation of the spatial distribution of water use, see Döll and Hauschild 2001b). In the case of RS A, all public irrigation projects planned in 1998 according to Lopes Neto (1998) are assumed to be implemented, but only one-quarter of the project area in RS B; however, private irrigation based on local initiative, which is distributed more equally over the study area, expands more strongly in RS B than in RS A.

Based on the results for the two reference scenarios, the impact of water demand measures can be assessed. For the domestic sector, we calculated both the impact of an extension of public water supply and the impact of water pricing. In 1996/98, only about 50% of the population of the study area was connected to the public water supply, and in the reference scenarios, an increase to 66% (RS A) and 58% (RS B) of the population in 2025 is foreseen. A possible intervention is to improve the water supply situation in 2025 by extending public water supply more strongly than in the reference scenarios. However, average per-capita withdrawal for public-supplied households is more than double the value assumed for self-supplied households (50 l/day). Therefore, an extension of public water supply leads to a higher domestic water use, and the desirable supply of safe drinking water might put additional pressure on the scarce water resources of Piauí and Ceará. If 83% of the population had access to public-supplied tap water in the case of RS A, instead of only 66%, domestic withdrawals in the study area would increase by 12%, and total withdrawals by less than 3%. In the case of RS B, if the connection rate is 79% instead of 58%, domestic withdrawals would rise by 14% and total withdrawals by less than 4%. Thus we conclude that extended public water supply would not put a high additional stress on water resources.

There is wide recognition that an appropriate pricing of water helps (1) to achieve the economically most efficient use of water, (2) to avoid its wasteful use and thus to protect the environment, and (3) to generate monies to improve water supply and demand management. The current water prices only cover operational and maintenance costs, and not investment costs, e.g., for an extension of the water supply, or environmental costs (information from the water provider of Ceará CAGECE, pers. commun.). Currently, 1% of the average per-capita GDP in the study area is spent on a (public) water supply, similar to in Germany. Actual price increases in the last decade were larger than the yearly increase assumed for both reference scenarios (6%/year) from 1996/98 to 2025. A price increase of only 2.5%/year results in an approximately constant fraction of the total income that is spent on water, and it is unlikely that with such a price development investment costs could be covered. A price increase of 11%/year leads, on average, to a water price that, with the same per-capita water use, amounts to about 10% of the per-capita GDP. Such a pricing policy has the potential of full cost recovery. Besides, prices per volume of freshwater will effectively double when households are connected to the sewage system (in 1996/98, only about 10% of the urban withdrawals are discharged into a sewage system). With a 2.5%/year water price increase, total domestic water use (including self-supply) increases by about 40% as compared with the reference scenarios (where 6%/year was assumed), while with 11%/year, total domestic water use decreases by approximately 45%. Please note, however, that these results strongly depend on the assumed income elasticity (0.7 according to Gómez 1987, determined for a Brazilian city of 1.5 million inhabitants) and price elasticity (linear decrease from –0.55 in 1996 to –0.3 in 2025; -0.55 is a typical value for northeast Brazil according to BNB/PBLM 1997, –0.3 a typical value for Europe and North America).

The study area suffers from strong interannual climate variability. Figure 2 shows that the impact of climate variability on irrigation water requirements is much larger than the impact of climate change. The increase in irrigation requirements in 1970, a 1-in-10 dry year (in 90% of all years, the study area gets more precipitation), as compared with the climate-average value (Fig. 2a), is much higher than the differences due to the assumed climate change (Fig. 2b). Therefore, climate change was neglected in Table 1.

As a type of validation of the irrigation sector modeling of NoWUM, the crop-specific irrigation requirements per unit irrigated area are compared with recommendations given by agricultural advisors for northeast Brazil (Table 2). For banana, cotton, fruit trees and vegetables, the correspondence is satisfactory, while the lower values for maize, beans and rice might be due to (1) overestimation of the irrigation water use efficiency, which for surface irrigation, should be about 0.3 and not 0.6, and (2) neglecting multicropping.

Conclusions

The newly developed regional-scale water use model NoWUM is an instrument that is suitable for regional water resources planning. It provides the first consistent and comprehensive assessment of the current sectoral water uses in all municipalities of Ceará and Piauí. Previous studies on water use in Ceará and Piauí either did not achieve results that were consistent for all municipalities (SRH 1992, for Ceará), or only provided water use in a few river basins but not in municipalities (Marwell Filho 1995, for Piauí). Besides, in both studies, the domestic water use is estimated by assuming typical per-capita values that depend on the size of the town, while standard (neither climate-dependent nor crop-specific) withdrawal values are adopted to compute irrigation water use. In NoWUM, however, municipality-specific information on domestic water supply and the number of connected households is included, and the dependence of irrigation water use on irrigated crops and climate is taken into account. More importantly, NoWUM is an operational model that can be applied to generate scenarios of future water use in a flexible and consistent manner. In addition, it provides information on consumptive water use that is necessary to compute water availability with hydrological models, as the natural discharge is reduced by consumptive water use. NoWUM results can be used to assess water availability and water scarcity in river basins by aggregating the municipality values along the drainage directions in the basin.

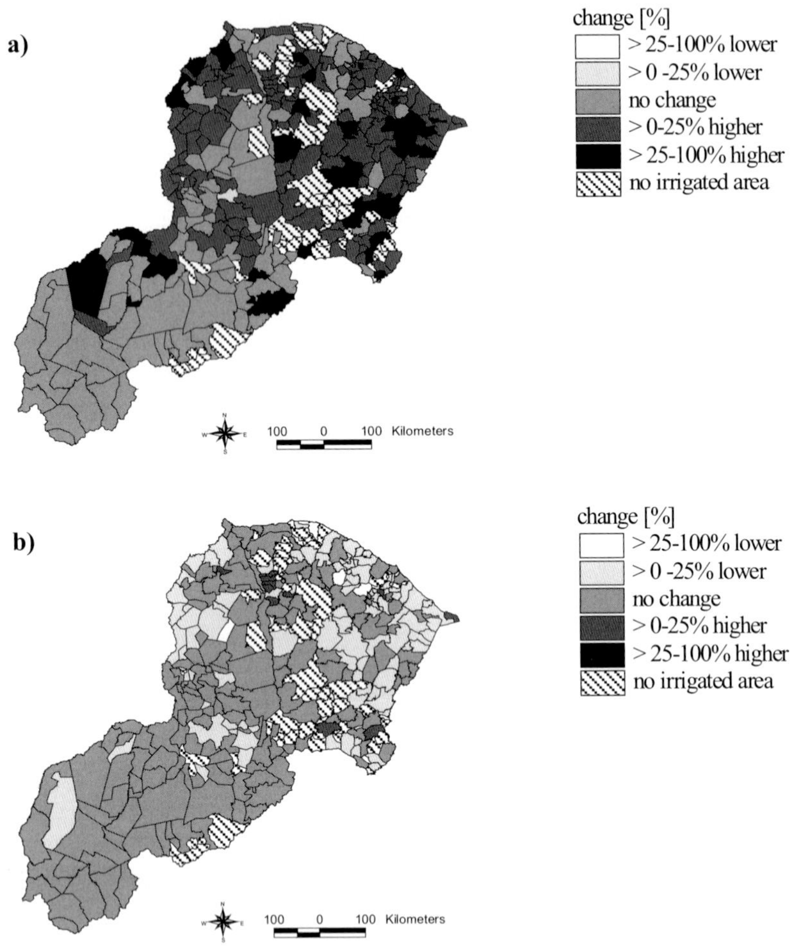

Fig. 2. a Irrigation requirement per unit irrigated area in the 1-in-10 dry year 1970 as compared with the requirement under long-term average climatic conditions (1951–1980); **b** irrigation requirement under a changed climate (approx. 2025, i.e., average 2011–2040) as compared with the requirement under long-term average climatic conditions (1951–1980), on average 4% lower

NoWUM integrates scarce and uncertain data; therefore, its output is uncertain, too. NoWUM possibly underestimates irrigation water use due to a conservative estimate of irrigated areas, the neglecting of multicropping, and a high value of the irrigation water use efficiency (only relevant for withdrawal water use). It probably overestimates industrial water due to assuming industrial water use intensities typical for humid Germany. The domestic water use estimate in municipalities with a large self-supplied population is uncertain due to the lack of quantitative information on water use in households that are not connected to the public water supply. Despite these uncertainties, NoWUM results can serve to assess the problem of water scarcity in Ceará and Piauí and to develop appropriate water management strategies.

Table 2. Comparison of crop-specific gross irrigation requirements per unit irrigated area: Withdrawals recommended by northeast Brazilian agricultural advisors compared with the climate-dependent and thus municipality-specific values computed by NoWUM (332 municipalities)

Crop	Recommended withdrawal per growing period[a] (mm)	Withdrawals per growing period computed with NoWUM[b] (mm)	
		Average ± standard deviation	Range of values in different municipalities
Banana	2000	1559±212	1069–2093
Beans	500	132±53	41–288
Cotton	800	513±153	292–911
Fruit trees	1000–1200	1267±166	950–1742
Grass	No information	1966±251	1481–2638
Maize	600	201±88	61–474
Rice	800	290±78	128–508
Sugarcane	No information	1846±237	1390–2487
Vegetables	600	835±147	444–1123

[a] BNB (1997). [b] Calculated for 1951–1980 climatic conditions and with a water use efficiency of 0.6.

With respect to the future situation, the pressure on scarce water resources will increase significantly in Piauí and Ceará if the irrigation is strongly extended either by the implementation of the planned public irrigation projects (RS A) or by more disperse private irrigation (RS B) – even if the irrigation water use efficiency is improved. Domestic and industrial water use will increase in regions with high immigration, but water use intensities can be controlled by an appropriate water pricing. Extending public water supply to more households than assumed in the reference scenarios will not increase considerably the pressure on the scarce resource water.

A significant improvement in the developed water use scenarios is only possible if more reliable data on water use and its driving forces become available. This could be achieved if sectoral water use data in each municipality were acquired regularly, e.g., by adapting the approach taken in the USA (Solley et al. 1998) and if studies were performed on water use behavior. Then, the water use model could be refined (e.g., with respect to domestic water use by the self-supplied population) and its reliability could be improved. Furthermore, scenario construction could be enhanced by integrated modeling; a coupled modeling of agricultural economics, water use and water availability would lead to improved scenarios of irrigated areas and crops. Finally, for the generation of regional-scale scenarios of water use, it is desirable to model water-use-related decision-making in smaller-scale units, e.g., river basin committees.

Acknowledgements

The research was funded by the German Federal Ministry of Education and Research (01 LK 9705).

References

BNB (Banco do Nordeste do Brasil) (1997) Agenda do produtor rural 1997. BNB, Recife, Brazil

BNB/PBLM (1997) Execução de servicios técnicos sobre a demanda de água no Nordeste do Brasil. Relatorio Final, Recife, Brazil

Clarke G, Kashti A, McDonald A, Williamson P (1997) Estimating small area demand for water: a new methodology. J Chart Inst Water Environ Manage 11:186–192

Döll P, Hauschild M (2002) Model-based regional assessment of water use: an example for semi-arid Northeastern Brazil. Water Int (submitted)

Döll P, Hauschild M (2002) Model-based scenarios of water use in two semi-arid Brazilian states. Reg Environ Change (submitted)

FAO (1992) CROPWAT – a computer program for irrigation planning and management. FAO irrigation and drainage paper 46. FAO, Rome

FAO (2000) Irrigation in Latin America and the Caribbean in figures. Water Rep 20. FAO, Rome

Gleick PH (1996) Basic water requirements for human activities: meeting basic needs. Water Int 21:83–92

Gómez C (1987) Experience in predicting willingness to pay on water projects in Latin America. In: Monatanari FW (ed) Resource mobilization for drinking water and sanitation in developing nations. American Society of Civil Engineers, New York, pp 242–254

Hauschild M, Döll P (2000) Water use in semi-arid northeastern Brazil. World water series 3. Center for Environmental Systems Research, University of Kassel, Germany, 30 pp + appendix, <http://www.usf.uni-kassel.de/usf/archiv/dokumente.en.htm>

IBGE (Instituto Brasileiro de Geografia e Estatística) (1998a) Censo agropecuário 1995–1996 – Piauí. IBGE report no 10. IBGE, Rio de Janeiro, Brazil

IBGE (Instituto Brasileiro de Geografia e Estatística) (1998b) Censo agropecuário 1995–1996 – Ceará. IBGE report no 11. IBGE, Rio de Janeiro, Brazil

IWR-MAIN (2000) <http://www.iwrmain.com/> (Nov. 20, 2000)

Lopes Neto A (1998) Possibilidades de Modernização Rural do Ceará através da Agricultura Irrigada e da Fruticultura. CNPq, SECITECE, Fortaleza, Brazil

Marwell Filho P (1995) Análise des Sustentabilidade do Estado do Piauí quanto aos Recursos Hídricos. Projeto ÁRIDAS: Uma Estratégia de Desenvolvimento Sustentável para o Nordeste, Tema 7, Governo do Estado do Piauí - Secretaria de Planejamento - Grupo Recursos Hídricos, Teresina, Brazil

SRH (Secretaria de Recursos Hídricos) (1992) Plano Estadual dos Recursos Hídricos – Estudos de Base I, II, vol 2. SRH, Fortaleza, Brazil

Solley WB, Pierce RS, Perlman HA (1998) Estimated use of water in the United States in 1995. USGS circular 1200. USGS, Reston, Virginia, USA

USBR (US Bureau of Reclamation) (1991) Utah Area Water Demand Model. <http://www.usbr.gov/hmi/91inv.html> (20 Nov 2000)

Werner PC, Gerstengarbe F-W (1997) Proposal for the development of climate scenarios. Clin Res 8:171-182

World Water Commission for the 21st Century (2000) A water secure world — Vision for water, life and the environment. Report to World Water Council, World Water Vision (http://www.worldwatcrvision org/reports.htm)

Zagona EA, Fulp TJ, Morgan H, Shane RM (1998) RiverWare: a general river and reservoir modeling environment. Proceedings of the 1st Federal Interagency Hydrologic Modeling Conference, Las Vegas, Nevada, 19-23 April 1998, 5:113-120, <http://cadswes.colorado.edu/riverware/vegas98.pdf> (20 Nov 2000)

Assessment of Water Costs in Semiarid Brazil

José Carlos de Araújo, Cláudia Beghini Rodrigues Abreu, Cláudio Pacheco Barbosa and Elano Lamartine Leão Joca

Universidade Federal do Ceará, Departamento de Hidráulica e Ambiental, CEP 60451-970 Fortaleza-Ceará, Brazil, jcaraujo@ufc.br

Introduction

In semiarid regions, water supply infrastructure costs must be computed as precisely as possible, once society and the vulnerability of the ecosystem strongly depend on water availability. Water management federal and state laws also consider bulk water tariff an important management instrument especially where conflicts already exist. Therefore, in order to foresee public and private costs for water infrastructure for the coming years, unit water costs must be assessed. This research has provided two different types of water costs: investment (capital) costs as well as operational, maintenance and administration (O&M) costs. Different water supply possibilities were investigated: surface water supply by means of dams; groundwater in both crystalline and sedimentary basins; water reuse from stabilization ponds; rain water tanks; and water delivered by trucks (a rather common water source not only in dry years in rural communities of the Brazilian Semiarid). An important part of the investigation is the assessment of water demand management costs by means of improvements in irrigation technology. Indeed, irrigation demands about 70% of the bulk water in the state, and its technological level is still not good enough, so water managers can definitely decide to subsidize technological improvement in irrigation systems and, therefore, save water.

Surface Water Costs

Surface water represent more than 90% of all water supply in Ceará due to artificial reservoirs, built from dams, once all rivers in the state are intermittent. However, because of hydrological constraints, surface water yield can only be assessed in terms of supply guarantee, computed by means of stochastic simulation. According to experience in the Brazilian Semiarid, 90% guarantee yield from a reservoir (Q_{90}) is usually 30–50% of annual average inflow (Qi; Campos 1996). For water planners, though, licensable yearly volume (Qo) is an important parameter: it can be evaluated as Q_{90} minus water losses (to atmosphere and aquifer) and negligible demand. According to federal and state water laws, negligible water demand (in Ceará, lower than 2 m^3/h individually) does not require official licensing. Licensable yearly volume can be as low as 60% of Q_{90} in Ceará (de Araújo 1996).

Investment Costs

The *cost recovery* approach is used to compute surface water costs, in which annualized costs are divided by yearly volume. In order to obtain infrastructure costs, Mota (1995) updated costs in dam budgets for 40 reservoirs in Ceará. De Araújo (1996) used these data to compute investment costs (50 years, 8% yearly interest rate) for 20 dams with storage capacity range from 5–450 hm^3, so that Q_{90} costs could be assessed. The results for 17 dams (statistical filter excluded three) are shown in Fig. 1 and Eqs. (1) and (2), in which C$ are investment costs US$/$m^3$; R(%) is hydrological efficiency (=Q_{90}/Qi); and f is the correction factor dependent on residence time T_R (=V/Qi, where V=reservoir volume). After applying this methodology for all strategic dams in planning regions of Ceará and dividing annual costs per average flows (Q_{90} or Qo, respectively), surface water costs were estimated (see Table 2). Minimum Q_{90} cost found was 0.022 US$/$m^3$ in Curu, maximum 0.080 US$/$m^3$ in Litoral and average 0.035 US$/$m^3$. Licensable water Qo costs 0.067 US$/$m^3$ in Ceará.

$$C\$ = f.\ 0.3505.\exp\left[-0.067.R(\%)\right] \tag{1}$$

$$f = (TR/1.5)^{-0,25} \tag{2}$$

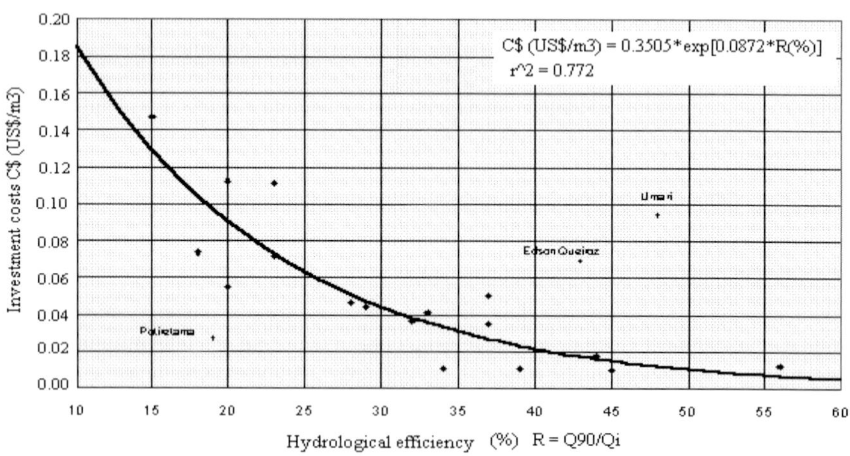

Fig. 1. Investment surface water costs (Q_{90}) for Ceará as a function of hydrological efficiency

O&M Costs

Operational, maintenance and management costs (O&M) of surface bulk water are computed considering annual costs of basin management offices, maintenance, personnel, administration and pumping. Annual costs are then divided by Q_{90} or Qo, depending on which water supply category cost is desired, plus groundwater flow, if not only surface water is to be managed. Results (see Table 3) show that O&M costs are less than 10% of investment costs.

Groundwater Costs

In order to assess groundwater costs, yearly expenses (both investment and O&M) are divided by exploited yearly volume, i.e., flow of the wells in use pumping 12 h/day, according to field observation. Information about almost 14,000 wells was used from the databases of CPRM, FUNCEME and PHD (*apud* Barbosa 2000). For each planning region in the state, after statistical computation, two *reference wells* were generated (one for the crystalline and one for the sedimentary aquifer) and average costs calculated. Analysis shows that about 190 hm^3 groundwater are exploited in Ceará every year: 77% from sediment and 23% from crystalline basins. High spatial concentration can be observed: 74% of groundwater flow is in the Salgado and Metropolitan regions. The Médio Jaguaribe region, for example, produces only 0.5% of state groundwater.

Investment Costs

For groundwater investment costs assessment, updated prices were taken and costs annualized for 30 years (Tancredi 1996) and 8% interest rate. Costs of the pumping station were annualized for 6 years (Tancredi 1996) and 8% interest rate as well.

O&M Costs

Groundwater O&M costs include the following annual expenses: pumping (energy), basin management offices, maintenance (admitted 5% of respective investment expenses; Barbosa 2000) and administration, similar to the surface water approach. Basin management offices and administration expenses are evaluated as COGERH respective expenses divided by summation of ground- and surface water flows.

Global Costs

Average global groundwater cost in crystalline is 0.073 U\$/m³, 63% of which is relative to investment and 37% O&M. The Litoral region presents the highest crystalline groundwater costs, i.e., 0.105 U\$/m³, 42% above average. In sedimentary areas, groundwater is cheaper: 0.059 U\$/m³, 59% investment and 41% O&M. Sedimentary groundwater in the Parnaíba region costs 0.088 U\$/m³, the highest in Ceará, and in the Salgado region 0.042 U\$/m³, the lowest.

Costs of Alternative Water Sources

Water Reuse Costs

The possibility of wastewater reuse is strategic for water policy: at the same time it reduces fresh water pollution and fresh water demand. Reuse water costs were assessed by calculation of investment and O&M yearly costs divided by addition of input (sewage to be treated) and output (water available for reuse) flows. It was assumed that treatment and reuse volumetric costs are the same. Stabilization ponds were admitted as the wastewater treatment method, so 13 treatment plants (spatially distributed and serving a wide range of population, from 500 to 30,000) were used for computation (Barbosa 2000). Investment costs include land acquisition, pond construction and pumping facilities. O&M costs include maintenance, personnel and supervision expenses. Per-capita sewage flow, 100 L/day, was measured in three different areas that use stabilization ponds as treatment method. Regression equations were obtained (Eqs. 3 and 4) and a high linear correlation coefficient found (r^2=0.62 and 0.91, respectively). Pi is initial population served, Ci\$ the investment costs (US\$/m^3), and Co\$ the O&M costs (US\$/m^3). Wastewater reuse is economically feasible: for a served population over 25,000 inhabitants, reuse costs are compatible with surface water costs and environmentally cleaner.

$$Ci\$ = 2.375.Pi^{(-0.4107)} \tag{3}$$

$$Co\$ = 0.478.Pi^{(-0.4433)} \tag{4}$$

Rain-Water Tank („Cisternas“) Costs

The most susceptible population to droughts is the rural population, especially those who live far from sanitation services. This group of society lives unattended and, as soon as the rainy season ends, water quality in the small dams they mostly live on decreases. Besides, they have very little political power to force enhancement of their living standards. Due to this situation, several non-governmental organizations have started a program to build rain-water tanks, called *cisternas*. In the state of Ceará alone about 1000 *cisternas* have already been built in the driest rural areas, and the Brazilian Environmental Ministry plans to finance about one million of such tanks in the whole „drought polygon“ in the northeast of Brazil. Tanks can store 15 m^3 of good quality water collected from house tailings, which is used for drinking and cooking purposes only. Water quality in the tanks is much better than that in small dams, where the cattle also drink. According to data from NGO Esplar, one of the technical supporting groups for the *cisternas* program, the capital cost of one tank, including material and human labor, is about 195 US$. Maintenance is about 10 US$/year and operational costs are negligible. Considering that a tank is filled three times a year and that the costs pay a yearly interest rate of 6% (typical for agricultural infrastructure finance programs in Brazil), within its 20-year lifetime the total yearly cost of the water reaches 0.596 US$/$m^3$. This cost, though, is not high if one considers water quality, water supply guarantee (approximately 95% yearly) and water spatial distribution. Indeed, the official rural sanitation program of the state of Ceará (SISAR) supplies some rural communities with drinking water at an average cost of about 0.413 US$/$m^3$. However, water supply guarantee in the SISAR system is about 90%. According to Araújo (2000), surface water costs with 95% guarantee in the Brazilian Semiarid are 49.4% higher than those for 90% supply guarantee. In other words, water costs for rural sanitation systems and 95% guarantee is 0.617 US$/1000$m^3$, very close to the value obtained for the *cisternas*.

Costs of Distribution by Water Trucks

Whenever a drought occurs in Ceará, many villages and cities receive their water supply from water tanks. Campos et al. (1997) have estimated the cost of such water based on data from the Water Resources Secretariat for the 1993 drought: 787 truck routes for 143 municipalities (78%) were analyzed. Results indicate an average distance of 118,201 km/day, assisting 302,850 people with daily consumption of 6050 m^3, which means a per-capita demand of 20 L/day. The water sources were 82 dams, 47 wells and 14 regularized rivers. The lowest cost assessed was in Tejuçuoca, 0.910 US$/$m^3$, and the highest 11.140 US$/$m^3$ in Sobral. Average cost for Ceará is 6.650 US$/$m^3$, i.e., US$ 3.30 per person per month.

Water Demand Management Costs

Water policy cannot only be concerned with offering more water, but also of using it more efficiently. Therefore, water demand management should be encouraged, especially within the less efficient sectors such as irrigation (de Araújo 1998). This research proposes irrigation technological enhancement as an important management instrument. After visiting the most important irrigation areas in Ceará, four scenarios (23,400 ha) where technological improvement could reduce water demand considerably were identified (Joca 2001). Scenario C1 consists of changing sprinklers for micro-sprinklers for fruit irrigation; scenario C2 consists of changing furrows for sprinklers for grain irrigation (corn and beans); scenario C3 consists of changing rice irrigation for sugarcane irrigation with sprinklers; and scenario C4 consists of changing grain (corn and beans) irrigation with sprinklers for fruit irrigation with micro-sprinklers. Costs for technological enhancement ($\$C_A$, US$/ha) were analyzed for a reference area of 2 ha, which include: equipment cost ($eq), personal training cost ($pt=27.80 US$/ha), transaction cost ($tr=5% of $eq + $pt), and old equipment price ($vr=10% of similar new equipment, according to irrigators that recently sold their old equipment). In order to calculate realistic costs for a much larger intervention than a 2 ha area, scale factor fs had to be assessed (see Eq. 5 and Fig. 2).

$$\$C_A = fs.(\$eq + \$pt + \$tr - \$vr) \tag{5}$$

Irrigators cannot pay for all the costs alone, so a capacity-to-pay (CP) analysis based on de Araújo (1998) is performed. The state saves water and could complement the investment costs. The financing plan should work as follows: the state makes a deposit at a savings account in the name of the irrigator; the bank pays for the new technology and considers it as a loan for the irrigator; the irrigator deposits yearly the payment in the same account. After a 3-year term, the bank starts charging the irrigator for the loan (6% yearly interest rate) by taking the annual payment directly from the savings account for 12 years (these are the conditions of loans for agricultural investments in Ceará). Meanwhile, interest is deposited as a normal savings account. The state saves water and both the technological and social-economic profiles are enhanced. Results show that it is much cheaper to save than to increase water supply (Table 1). A field survey (Joca 2001) showed high acceptance of scenarios C1 and C2 by irrigators (100%), low acceptance of scenario C4 (25%), and no acceptance of scenario C3 due to market problems with sugarcane.

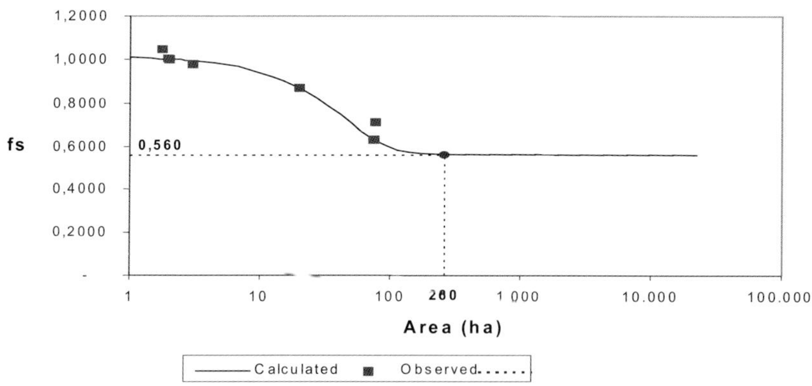

Fig. 2. Behavior of scale factor (fs) as a function of area

Table 1. Applicability of water demand management in irrigated areas in Ceará, 2000

Scenario	Area (ha)	$\$C_A$ (US\$/ha)	Capacity-to-pay (US\$/ha/year)	State deposit (US\$/ha)	ΔV (m^3/ha)	Global cost (US\$/m^3)	State cost (US\$/m^3)
C1	4709	851.26	38.09	531.88	58,200	0.023	0.014
C2	3983	350.25	21.06	173.67	22,200	0.025	0.012
C3	12,109	350.25	10.40	272.87	98,400	0.006	0.004
C4	2716	852.58	42.33	610.86	48,000	0.028	0.020
Global	23,427	508.59	21.44	346.96	71,810	0.015	0.010

„Valor" Algorithm for Water Cost Assessment in Ceará

In order to allow water management and planning institutions to benefit from the data and methodology developed in the context of this research, software was developed. VALOR is an algorithm (see Fig. 3) in DELPHI that accesses a database. Its objective is to assess investment and O&M costs for several water sources categories (surface bulk water Q_{90} and Qo, groundwater in crystalline and sediment, reuse water and water delivered by trucks) for the eleven planning regions of Ceará. Figure 4 shows the „start menu" on within Windows.

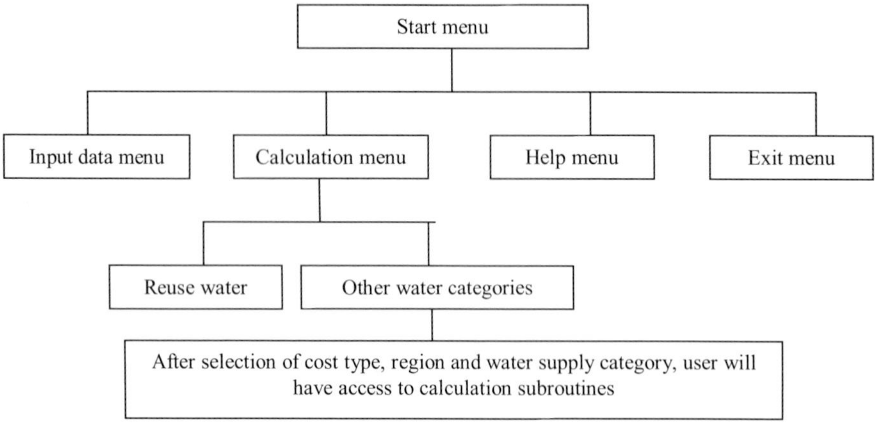

Fig. 3. Algorithm VALOR

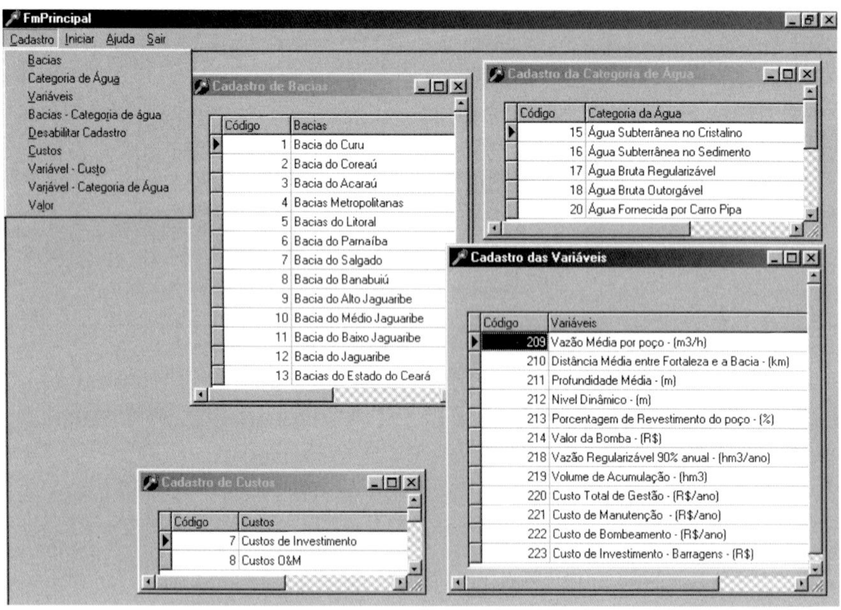

Fig. 4. Start menu with „Windows“

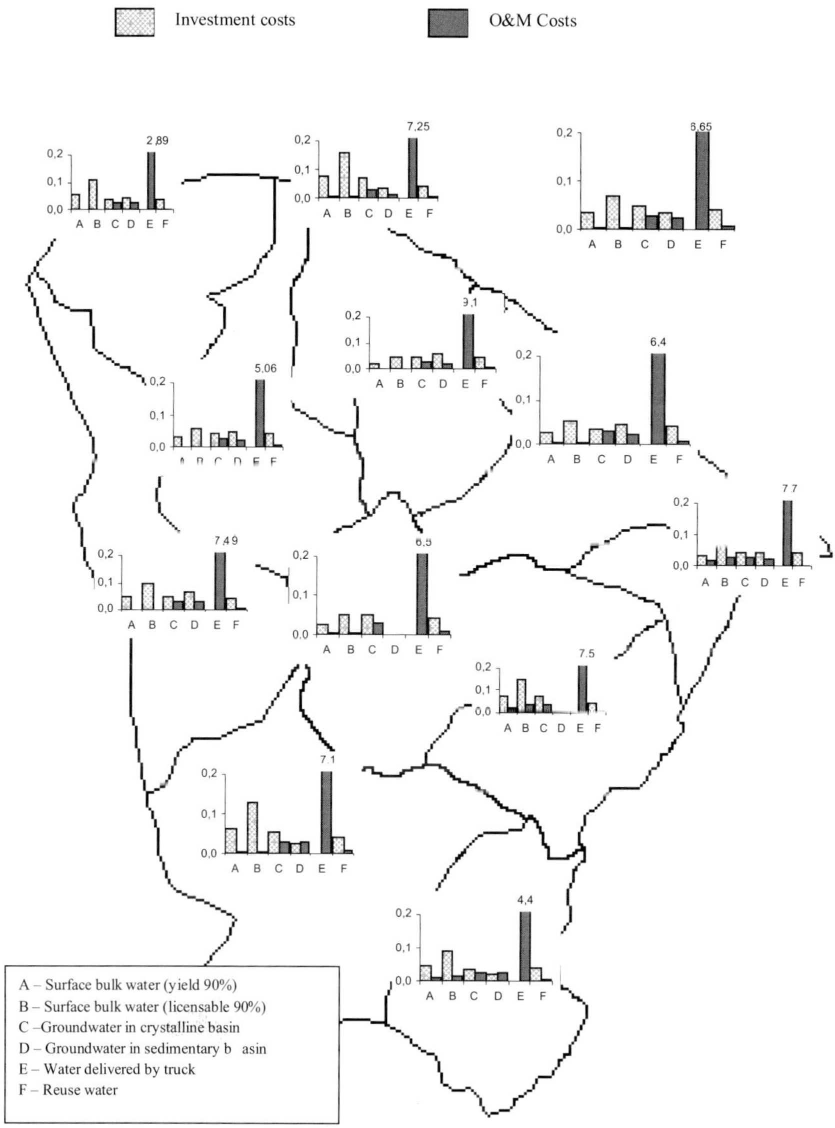

Fig. 5. Investment and O&M costs (US$/m^3) for several water supply categories in Ceará in relation to the location of the watershed

Conclusions

Different water supply costs (both investment and O&M) are computed for all planning regions in the state of Ceará (Tables 2 and 3). In terms of water supply possibilities, this research has assessed average investment costs for Ceará: bulk surface water (90% guarantee yield 0.035 US$/m^3 and licensable 0.067 US$/m^3), groundwater in crystalline (0.046 US$/m^3) and sedimentary (0.035 US$/m^3) basins.

Alternative water sources were also investigated, such as rain water tanks (0.567 US$/m^3) and reuse water (0.042 US$/m^3). O&M costs are usually lower: 90% guarantee yield surface water (0.002 US$/m^3), licensable 90% surface water (0.003 US$/m^3), groundwater in crystalline (0.027 US$/m^3) and sedimentary (0.024 US$/m^3) basins, water delivered by tanks (6.650 US$/m^3), rain water tanks (0.029 US$/m^3) and reuse water (0.006 US$/m^3). Figure 5 shows the distribution of average costs in Ceará. Demand management cannot be neglected as a major water policy tool. In this study it was shown (see Table 1) that irrigation method improvement would cost 509 US$/ha, or 0.015 US$ per m^3 of water saved. Considering irrigators capacity-to-pay, the state would have to pay 0.010 US$/m^3, about 30% of bulk water investment costs. In other words, it would cost less to save water by financing technological improvement in irrigation than to enhance water supply by means of dams. The data generated in this research provides the basis for water planning and water tariff policy in Ceará.

References

AraújoWC (2000) Avaliação dos Custos da Água Bruta Associados a Diversas Garantias na Bacia do Rio Paraíba. MSc Thesis, Federal University of Ceará, Fortaleza

Barbosa CP (2000) Avaliação dos Custos de Água Subterrânea e de Reuso de Efluentes no Estado do Ceará. MSc Thesis, Federal University of Ceará, Fortaleza

Campos JNB (1996) Dimensionamento de reservatórios. UFC, Fortaleza, 52 pp

Campos JN, Oliveira JB, Vieira V, Campos VR (1997) Custo do Fornecimento da Distribuição de Água através de Carro Pipa. Proc XII Brazilian Symp of Water Resources, vol 1. ABRH, Vitória

De Araújo JC (1996) Estudos de Tarifa d'água e Hidrológicos. Technical report. COGERH/CNPq, Fortaleza

De Araújo JC (1998) Avaliação da Capacidade de Pagamento de Irrigantes no Ceará. Technical Report. Dept Hydr Envir Eng. UFC, Fortaleza

Joca ELL (2001) Gestão da Demanda Hídrica na Agricultura Irrigada do Ceará: Financiamento do Incremento Tecnológico, MSc Thesis, UFC, Fortaleza

Mota FA (1995) Análise dos Custos do Volume Regularizado e da Eficiência Hídrica de Reservatórios do Ceará, MSc Thesis, Federal University of Ceará, Fortaleza

Tancredi AC (1996) Recursos Hídricos Subterrâneos de Santarém: Fundamentos para Uso e Protenção. PhD Thesis, Federal University of Pará, Belém

Table 2. Investment costs (US$/$m^3$) for several water supply categories and planning regions in Ceará

Water supply category	Alto Jaguaribe	Médio Jaguaribe	Baixo Jaguaribe	Salgado	Banabuiú	Metro-politan	Litoral	Curu	Acaraú	Coreaú	Parnaíba	Ceará
Surface water (yield 90%)	0.064	0.072	0.047	0.047	0.047	0.027	0.080	0.022	0.028	0.055	0.048	**0.035**
Surface water (licensable 90%)	0.126	0.141	0.064	0.093	0.050	0.052	0.157	0.043	0.055	0.108	0.093	**0.067**
Groundwater crystalline basin	0.054	0.066	0.044	0.039	0.049	0.036	0.075	0.044	0.042	0.039	0.047	**0.046**
Groundwater sedimentary basin	0.024	n.a.[⊕]	0.044	0.019	n.a.	0.043	0.038	0.058	0.049	0.043	0.059	**0.035**
Water delivered by truck	n.a.	n.a.	n.a.	n.a.	n.a.	n.a.	n.a.	n.a.	n.a.	n.a.	n.a.	**n.a.**
Rain water tanks (cisternas)	0.567	0.567	0.567	0.567	0.567	0.567	0.567	0.567	0.567	0.567	0.567	**0.567**
Reuse water[*]	0.043	0.043	0.043	0.042	0.043	0.041	0.042	0.042	0.042	0.041	0.042	**0.042**

[⊕] not applicable

[*] Reuse water costs depend heavily on population. Estimation for 20,000 inh. and climatic data for each planning region

Table 3. Operational, maintenance and management O&M costs (US$/$m^3$) for several water supply categories and planning regions in Ceará

Water supply category	Alto Jaguaribe	Médio Jaguaribe	Baixo Jaguaribe	Salgado	Banabuiú	Metro-politan	Litoral	Curu	Acaraú	Coreaú	Parnaíba	Ceará
Surface water (yield 90%)	0.002	0.019	0.017	0.008	0.002	0.003	0.003	0.002	0.001	0.002	0.002	**0.002**
Surface water (licensable 90%)	0.004	0.037	0.032	0.016	0.004	0.003	0.003	0.002	0.002	0.002	0.002	**0.003**
Groundwater crystalline basin	0.028	0.030	0.026	0.026	0.028	0.028	0.030	0.026	0.025	0.025	0.027	**0.027**
Groundwater sedimentary basin	0.028	n.a.(♰)	0.023	0.023	n.a.	0.022	0.016	0.020	0.023	0.023	0.029	**0.024**
Water delivered by truck	7.110	7.580	7.700	4.470	6.570	6.440	7.250	9.150	5.060	2.890	7.490	**6.650**
Rain water tanks (cisternas)	0.029	0.029	0.029	0.029	0.029	0.029	0.029	0.029	0.029	0.029	0.029	**0.029**
Reuse water(*)	0.006	0.006	0.006	0.006	0.006	0.005	0.006	0.006	0.006	0.006	0.006	**0.006**

(♰) not applicable

(*) Reuse water costs depend heavily on population. Estimation for 20,000 inh. and climatic data for each planning region

Part 5: Agricultural Productivity and Soils

An Information System for Land Resources in Piauí and Ceará

Thomas Gaiser, Frieder Graef, Thomas H. Hilger, Luiz G.R. Ferreira and Karl Stahr

Institute of Soil Science and Land Evaluation, University of Hohenheim, Emil-Wolff-Str.27, 70593 Stuttgart, Germany, tgaiser@uni-hohenheim.de

Introduction

The Brazilian federal states of Piauí and Ceará are characterized by a semiarid environment that is highly variable in terms of soil resources and climatic conditions. The assessment of areas with high potential for agricultural production as well as the identification and quantification of constraints to crop production are high priorities for the planning authorities of both states. Attempts towards land suitability classification have been made with different approaches at the regional scale for Ceará (SEARA 1988) and Piauí (Cordeiro 1985) and at the national scale for Brazil (SUPLAN 1979; EMBRAPA 2000). However, national classification approaches are not detailed enough for regional planning and, in some cases, lack climatic evaluations. The need for land resource information has been confirmed by a series of joint workshops with the participation of scientists of the research program WAVES and representatives of several ministries of the state of Ceará. For this purpose, the **S**oil and Land Resources Information System for the states of **PI**auí and **CE**ará (SPICE) has been established. The present paper focuses on applications of SPICE with special emphasis on the assessment of the spatial and temporal crop production risk. Further examples of applications are available on the SPICE homepage (http://www.uni-hohenheim.de/spice).

Material and Methods

The information system SPICE contains three main domains: the meta data domain, the data storage domain and the methods domain (Fig. 1). The *meta data domain* contains a catalogue of data and methods available together with explanations about the different data types, methods, sampling and analytical procedures. The *data storage domain* of SPICE stores raw data about climate, soil, terrain and crop requirements as well as specific input information for crop growth models.

Furthermore, it contains the geometry of spatial units (e.g., mapping units). In SPICE, the soil and terrain attributes are structured according to the SOTER (Soil and Terrain Digital Database) approach (FAO 1993; Gaiser et al. 2000a). Information on soil data of Piauí and Ceará was extracted from soil maps (SUDENE 1972; EMBRAPA 1983), their corresponding explanatory volumes (SUDENE 1973; EMBRAPA 1986), and additional data sources (SUDENE/BGR 1976; SUDEC 1980; Cordeiro 1985). Presently, the soil distribution is visualized at scales of 1:600,000 and 1:1,000,000 for Ceará and Piauí, respectively. More detailed soil information (1:200,000) is available in the two WAVES focus regions, Picos (PI) and Tauá (CE). Climate data were obtained from national sources (SUDENE 1990; DNM 1992). In the paper presented here, data sets reconstructed by a statistical model from historical climate data were used (Gerstengarbe and Werner 1997).

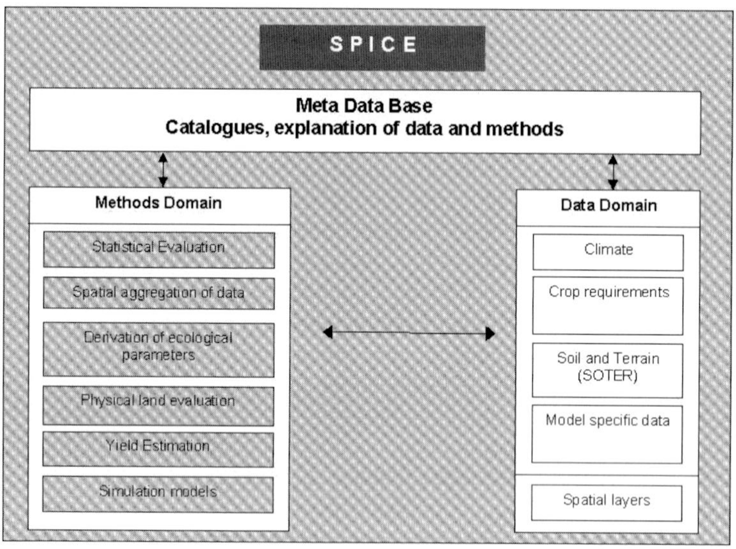

Fig. 1. Structure of the soil and land resource information system SPICE

Various applications are available in the *methods domain* (Fig. 1), such as derivation of soil ecological parameters (Gaiser et al. 2000c), yield estimation (Gaiser et al. 2000b) and coupling of soil and climate data with simulation models (Gaiser et al. 2000a). Land evaluation methods are based on the FAO framework (FAO 1976, 1983), refined with the help of crop-specific quantitative land suitability classification by Sys (1993). This method was tested and adapted for various crops, such as cowpea (*Vigna unguiculata*) in northeast Brazil as well as pearl millet (*Pennisetum glaucum*), maize (*Zea mays*) and rubber in other tropical regions (Van Ranst and Vanmechelen 1995; Van Ranst et al. 1996; Gaiser et al. 1999; Gaiser and Graef 2001; Igué 2000). Presently, land suitability evaluations can be carried out for 48 crops; however, only five of them have been tested intensively.

Under the high spatial and temporal climatic variability in northeast Brazil, the climatic suitability of alternative crops is of particular interest. Therefore, the spatial distribution of the climatic suitability of two crops (cowpea and millet) in four representative years has been evaluated. Following the statistical analysis of long-term precipitation series, four precipitation patterns were selected (Table 1):

- „normal“: years with annual rainfall corresponding to the long-term mean and regular distribution, represented by the year 1929
- „dry“: years with extremely low annual rainfall, represented by the year 1942
- „normal“ with dry spells: years with annual rainfall corresponding to the long-term mean, but with frequent dry spells („veranicos“), represented by the year 1957
- „wet“: years with extremely high annual rainfall, represented by the year 1974

Table 1. Mean annual amount of rainfall in Ceará from reconstruction of historical climate conditions (N=184)

	Rainfall Pattern			
	Normal (1929)	Dry (1942)	Normal with dry spells (1957)	Wet (1974)
Mean annual rainfall (mm)	1136	446	928	1838

The soil and terrain index and the climatic index are calculated separately by first specifying the minimum factor rating within different groups of ecological factors and then multiplying these ratings, representing the most limiting factor within each ecological group, according to Storie (1950):

$$I_{[c,s]} = A \times \frac{B}{100} \times \frac{C}{100} \times \frac{D}{100} \quad (1)$$

where $I_{[c]}$ is the climatic index, $I_{[s]}$ is the landscape and soil index and A, B, C, D are the ratings of the most limiting factors for climate and soil as defined below. The following groups of climatic factors were considered in the evaluation procedure: (a) rainfall (total amount of rainfall during the growing period, rainfall per month), (b) temperature (mean temperature during the crop cycle, monthly mean, minimum and maximum), (c) air humidity (relative humidity during the crop cycle), and (d) radiation (sunshine hours during the crop cycle). For the calculation of the soil index, six factor groups were defined: (a) topography (terrain slope), (b) wetness (terrain flooding risk, profile drainage conditions), (c) soil physical properties (soil depth, texture/structure, gravel, carbonate and gypsum content), (d) soil chemical properties (base saturation, sum of basic cations, pH, total organic carbon, CEC of clay fraction), and (e) salinity/alkalinity (electrical conductivity and exchangeable sodium percentage in the topsoil). Depending on the matching of land characteristics and crop requirements, ratings from 0 to 100 are attributed to the individual characteristics.

The calculation of the climatic index developed by Sys et al. (1991) involves modifying and adjusting to the local setting (Gaiser and Graef 2001). Compared with other calculation methods like the „maximum limitation method" (Sys et al. 1991), the „multiplication" approach (Eq. 1) proved to be the most suitable for quantification and correlation of crop yields with climate, soil and terrain factors, because it considers the synergistic interactions of the limitations (Gaiser et al. 1999). The soil and terrain suitability is calculated per representative profile set of each soil component. The results per profile are aggregated per soil association to the terrain component or terrain unit level. The final index is then correlated with four suitability classes (Table 2).

For the evaluation of the climatic suitability, the definition of the optimal planting time is crucial. February was the most suitable planting month for cowpea in the four selected years, whereas, for millet, March, and in 1974 April, was on average for the whole area of the state the best planting month.

Table 2. Key for the determination of climatic and soil suitability from the calculated indices

Suitability classes	Class symbol	Climatic index I_c or soil index I_s
Very suitable	S1	76-100
Moderately suitable	S2	51-75
Marginally suitable	S3	26-50
Not suitable	N	0-25

Results and Discussion

Evaluation of Climatic Suitability

The evaluation of the climatic suitability of the individual municipalities in Ceará for cowpea production indicates that in regular years the climatic conditions are moderately to very suitable (overall mean climate index=61), especially in the interior or „Sertão“ region of Ceará (Fig. 2). In the extremely dry year, the climatic suitability for cowpea decreased by at least one class. Exceptions are located in the northwestern littoral and the „Chapada do Ibiapaba“, where climatic suitability is lower in regular years, when cowpea is planted in February, due to the humid tropical conditions in both regions. Also, in some municipios in the Sertão, the climatic suitability does not decrease, probably due to a series of local convective events that are very common in this region. Such local events pose a serious problem when only one meteorological station represents the whole area of a municipality, because it is difficult then to regionalize measurements and suitability evaluations. In years with dry spells the climatic suitability for cowpea is somewhat similar to that in the extremely dry year, although the total annual precipitation was only reduced by 16% in this particular case. In the extremely wet year, the suitability for cowpea decreased in the coastal region and even in some municipalities of the Sertão region, due to the abundance of rain causing enhanced risk of disease infestations.

When the municipalities of Ceará were classified according to their suitability for pearl millet production, there were similar results in the normal, extremely dry and wet year (Fig. 3). However, in the year with frequent dry spells, pearl millet seemed to perform better than cowpea, which can be attributed to its adaptation to drought stress and slightly lower water requirements. From an ecological point of view, pearl millet should therefore be considered as an alternative to cowpea cropping or to cereals such as maize and rainfed rice. This information is important for the agricultural extension service (EMATERCE) in the municipalities, which were encouraged by the evaluation results to design field tests with different pearl millet varieties.

The importance of planting time is illustrated by comparing the climatic suitability of the municipalities in Ceará for cowpea cropping in a dry year, when planting is done either in January or February (Fig. 4). Planting in January led to total crop failure in almost all municipalities (mean climatic index=24), whereas planting in February increased the climatic suitability (mean climatic index=56). The evaluation procedure may thus be helpful in optimizing planting dates for various crops and regions, provided that reliable predictions of spatial and temporal distribution of precipitation for the growing period are available.

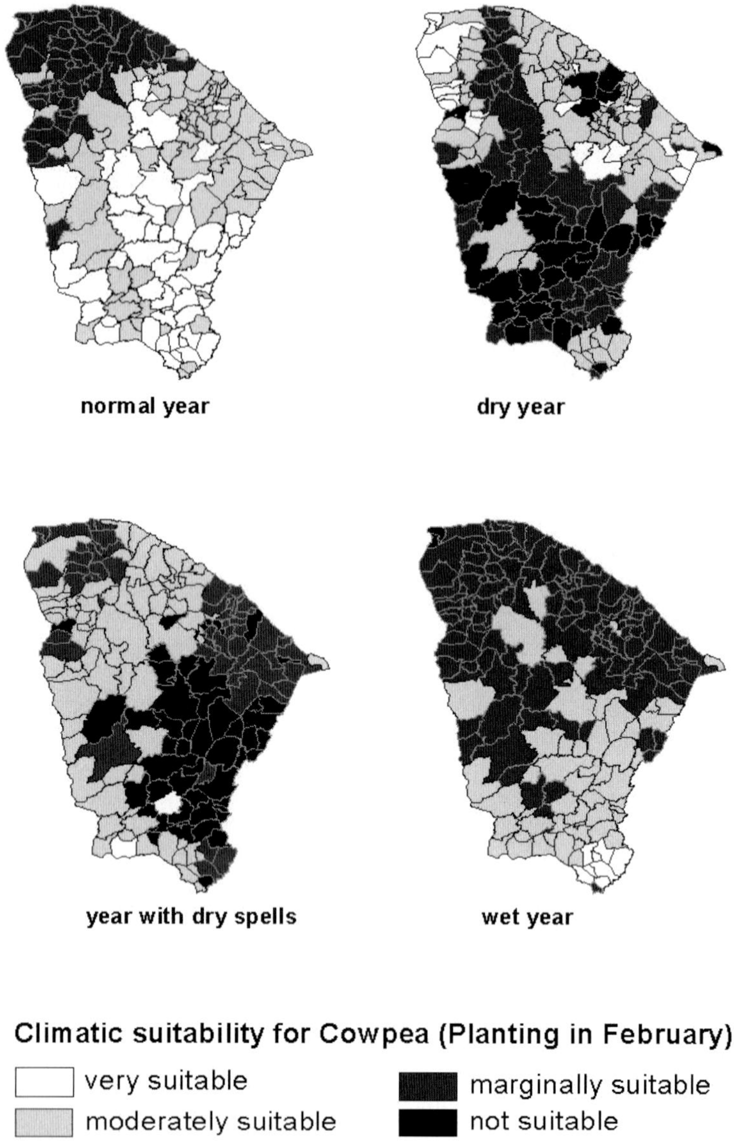

Fig. 2. Climatic suitability of the municipalities in Ceará for cowpea (*Vigna unguiculata*) production under four precipitation patterns. Cowpea was planted in February

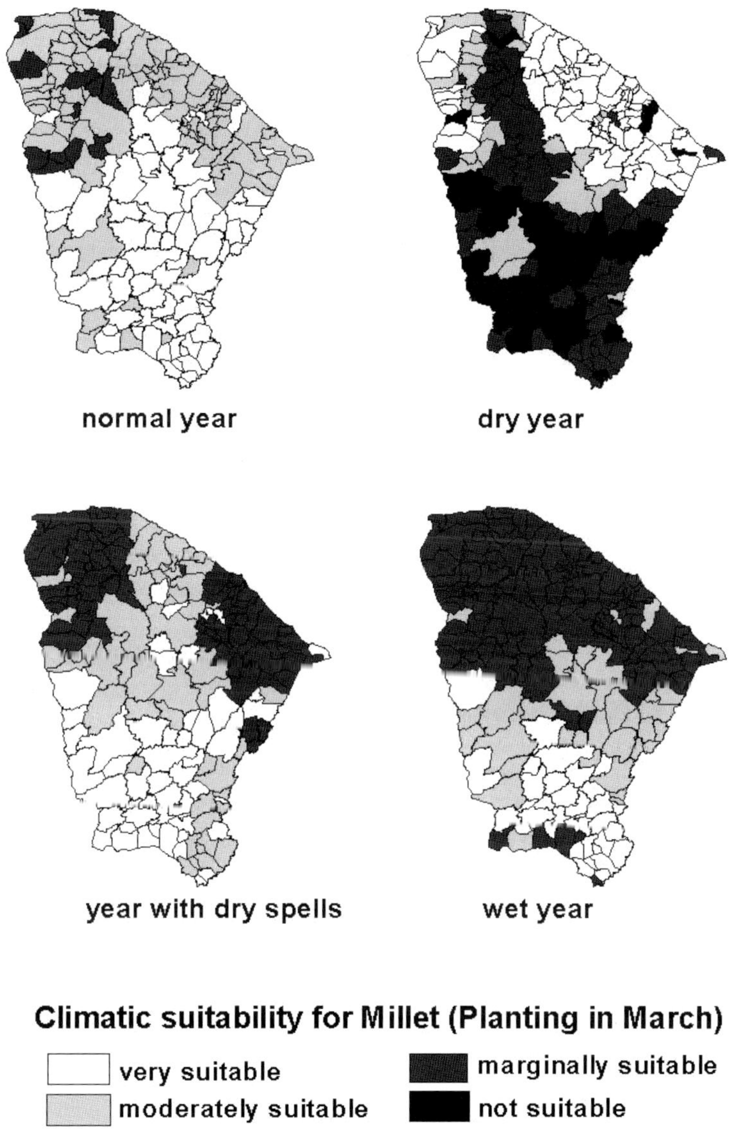

Fig. 3. Climatic suitability of the municípios in Ceará for pearl millet (*Pennisetum glaucum*) production under four precipitation patterns. Pearl millet was planted in March/April

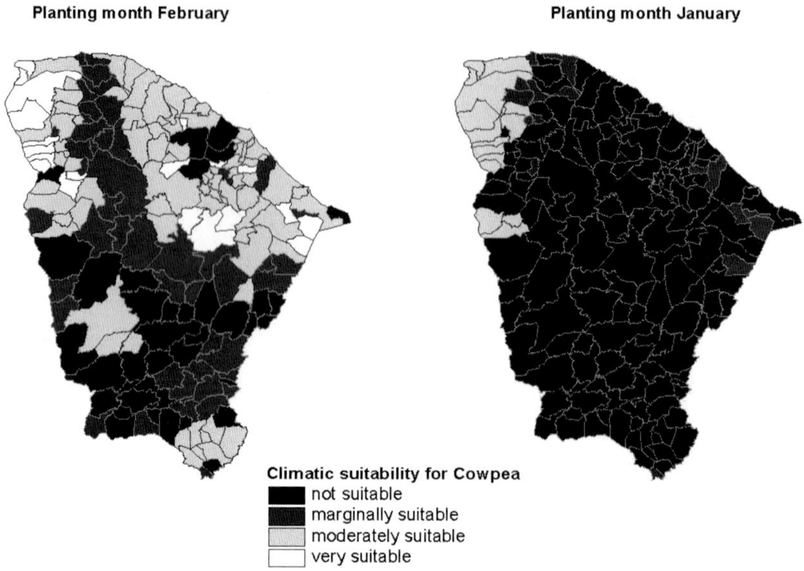

Fig. 4. Effect of two planting periods on the climatic suitability of cowpea in a dry year (1942)

Evaluation of Soil and Terrain Suitability

In many cases, crop production is not only limited by climatic constraints, but also by soil and terrain conditions. The soil and land resources information system uses the soil data that are available at the respective scales of the states to evaluate the soil and terrain suitability as illustrated by Figure 5 for cassava production in Piauí. The suitability is expressed as the mean over different soil components within each mapping unit. In Piauí, the most suitable sites for cassava production are located on plateau sites with sand cover where deep soils with predominantly sandy or loamy texture (LA) are found(Fig. 5). In spite of their acidity, which constitutes the most important constraint of these sites (Table 3), cassava production is not seriously hampered, as most cassava varieties are well adapted to low pH and high aluminum saturation. The standard deviation of the total soil index in these soil associations is generally low, due to the rather homogeneous soil distribution in the sand covered areas. The standard deviation is much higher in soil associations with predominantly shallow soils (R, Table 3) with coefficients of variance above 100%. In general, these soil associations are rated as not suitable. Nevertheless, there are some soil components that actually are at least marginally suitable. When comparing selected soil associations, the most limiting factors for cassava production in Piauí seem to be soil physical (texture, soil depth, gravel content) and soil chemical properties (organic matter, pH, base saturation). Salinity and wetness (drainage, flooding) are moderate problems in the soil association BV1 (predominantly Brunizems Avermelhados) that has soil components that tend to water logging. Soil depth and gravel content are limiting in the soil associations with predominantly shallow soils (R8 and R9, Table 3).

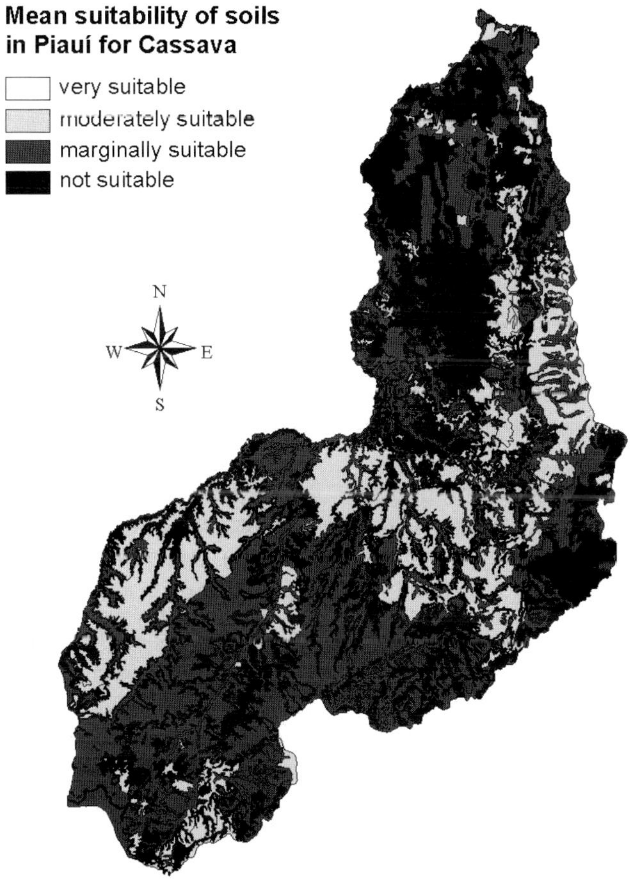

Fig. 5. Mean terrain suitability of soil associations in Piauí for cassava (*Manihot esculenta*) production

Table 3. Suitability of soil and terrain in different soil associations for cassava production (mean limitation indices and standard deviation of total soil index, 100 = no limitations, 0=complete limitation)

Soil association	Slope index	Wetness index	Index of soil physical properties	Index of soil chemical properties	Index of salinity	Total soil index	Std. Dev.	Suitability-class[1]
LA1	97	100	98	57	100	53	7,4	S2
BV1	66	62	63	64	78	33	12,8	S3
PV1	85	81	57	65	93	31	11,2	S3
PE15	90	81	74	73	98	42	4,3	S3
LA36	90	93	70	67	100	38	12,9	S3
R8	48	77	42	66	100	21	24,4	N2
R9	62	94	39	72	100	21	23,6	N2

[1] S2=moderately suitable, S3=marginally suitable, N2=not suitable

Conclusions and recommendations

The soil and land resource information system for the states of Piauí and Ceará (SPICE) provides useful information for planning authorities (state ministries for planning and for rural development) and extension services (EMATER) with respect to environmental constraints for crop production. In particular, the effect of temporal and spatial variability of precipitation on crop production can be assessed in relation to crop management and soil, extremely important parameters for the evaluation of climatic risk of crop production in individual municipios The evaluation results may encourage the introduction of new crops by the agricultural extension services and influence the allocation of investments into rural development by the state ministry for rural development. Crop status evaluations based on cumulative precipitation data of the ongoing rainy season are also helpful for deciding on emergency measures and the rapid verification of the necessity of financial support to individual municipios by the state government. However, considering the spatial variability of precipitation in Piauí and Ceará, the spatial resolution of the available soil and climate data is still too low for planning decisions within the individual municipios. For this purpose, monitoring of weather conditions and detailed mapping of soil distribution are required and recommended. In its present state, the soil information supplied by SPICE is suitable for large-scale evaluations. Applications at the municipality or watershed level necessitate maps of soil distribution at a more detailed scale. In Ceará, a soil inventory at the scale of 1:100,000 is planned and SPICE is an adequate framework to store, retrieve and use these data efficiently.

References

Cordeiro JC (1985) Classificação de terras para irrigação no semi-árido da bacia do Canindé. Fundação Universidade Federal do Piauí, Teresina, Brazil, 343 pp

DNM (1992) Normais climatológicas. Departamento Nacional de Meteorologia, Brasilia

EMBRAPA (1983) Mapa exploratório - reconhecimento de solos do Estado do Piauí. EMBRAPA, Rio de Janeiro

EMBRAPA (1986) Levantamento exploratório - reconhecimento de solos do Estado do Piauí, vol I and II. EMBRAPA, Rio de Janeiro, 782 pp

EMBRAPA (2000) Zoneamento agropédologico de Nordeste do Brazil. Empresa Brasileira de Pecuária e Agricultura, Escritório Regional de Pesquisa, Recife, Pernambuco (CD-ROM)

FAO (1976) A framework for land evaluation. Soils Bulletin 32. FAO, Rome, 72 pp

FAO (1983) Guidelines: land evaluation for rainfed agriculture. FAO soils bulletin 52. FAO, Rome, 237 pp

FAO (1993) Global and national soils and terrain digital databases (SOTER): procedures manual. World soil resources reports 74. FAO, Rome, 115 pp

Gaiser T, Graef F (2001) Calibration of a parametric land evaluation method for cowpea and pearl millet production in semiarid regions. Agronomie 21:705–712

Gaiser T, Marohn C, Bach B, Stahr K (1999) Standortsbewertung im semi-ariden Nord osten Brasiliens (land evaluation in the Northeast of Brazil). Mitt Dtsch Bodenkundl Gesell 91:981 984

Gaiser T, Ferreira LGR, Hilger TH, Römheld V, Stahr K (2000a) The land resources information system SPICE as a tool for integration of the results of the research area „Agroecosystems". In: Printz A (ed) Proceedings of the IIIrd international WAVES workshop in Freising-Weihenstephan. Technische Universität München, Institut für Landschaftsökologie, pp 46–49

Gaiser T, Hilger TH, Ferreira LGR, Herfort J, Barros I, Stahr K (2000b) The soil and terrain information system SPICE for estimating yield potentials at a regional scale in the states of Piauí and Cearà (Brazil). In: Knowledge partnership: challenges and perspectives for research and education at the turn of the millenium. Deutscher Tropentag on 14/15 Oct 1999 in Berlin. Humboldt Universität zu Berlin, Fachgebiet Tierzucht in den Tropen und Subtropen. Berlin, Germany (CD-Rom)

Gaiser T, Graef F, Cordeiro JC (2000c) Water retention characteristics of sandy soils with contrasting clay mineral composition in semi-arid tropical regions. Austr J Soil Res 38:523–536

Gerstengarbe F-W, Werner PC (1997) A method to estimate the statistical confidence of cluster separation. Theor Appl Climatol 57:103–110

Igué AM (2000) The Use of a Soil and Terrain Database for Land Evaluation Procedures - Case Study of Central Benin. Hohenheimer Bodenkundliche Hefte 58. Institut für Bodenkunde und Standortslehre, Universität Hohenheim. 235 p.

SEARA (1988) Zoneamento agrícola, Governo do Estado do Ceará, Secretaria de Agricultura e Reforma Agrária, Fortaleza

Storie RE (1950) Rating soils for agricultural, forest and grazing use. Transactions of the 4th international conference of soil science, vol I, pp 336–339

SUDEC (1980) Levantamento de reconhecimento semidetalhado dos solos da região dos Inhamuns -Salgado. Parte I (Iguatu). Superintendência do Desenvolvimento do Estado do Ceará, Fortaleza, 192 pp

SUDENE (1972) Mapa exploratório - reconhecimento de solos do Estado do Ceará (1:600.000). Superintendência do Desenvolvimento do Nordeste, Recife

SUDENE (1973) Levantamento exploratório - reconhecimento de solos do Estado do Ceará, vol I and II. Superintendência do Desenvolvimento do Nordeste, Recife, 803 pp

SUDENE (1990) Dados pluviométricos mensais do Nordeste. Série pluviometrica no 2. Recife, Pernambuco

SUDENE/BGR (1976) Pesquisas hidrogeológicas, pedológicas e agro-econômicas para aproveitamento da região centro-leste do Piauí. Parte IV, Anexos, vol 3. Métodos de trabalhos pedológicos e descrição dos perfis. Superintendência do Desenvolvimento do Nordeste, Recife

SUPLAN (1979) Estudos básicos para o planejamento agrícola, Aptidão agrícola das terras, no 8 e no 9. Ministério da Agricultura, Secretaria Nacional de Planejamento Agrícola, Brasília

Sys C (1993) Land evaluation in the tropics. Pédologie 43:117–142

Sys C, Van Ranst E, Debaveye J (1991) Land evaluation, part II. Methods in land evaluation. Agricultural Publ, General Administration for Development Cooperation, Brussels

Van Ranst E, Vanmechelen L (1995) Application of computer-based technology in assessing production potentials for annual crops in developing countries. Bull Séanc Acad R Sci Outre-Mer 41:613–636

Van Ranst E, Tang H, Groenemans R, Sinthurahat S (1996) Application of fuzzy logic to land suitability for rubber production in peninsular Thailand. Geoderma 70:1–19

Effectiveness of Fertilization and Other Agricultural Management Options in Picos and Tauá, NE Brazil

Joachim Herfort, Inacio de Barros, Luiz Gonzage Rebouças Ferreira, Thomas Gaiser, Leonhard Schneider, Bettina Bach and Thomas Hilger

Institute of Plant Production and Agroecology in the Tropics and Subtropics, University of Hohenheim, 70593 Stuttgart, Germany, jherfort@uni-hohenheim.de

Introduction

The socioeconomic development of northeast Brazil is highly related to the occurrence of droughts (Villa 2000). Therefore, a large part of northeast Brazil is also known as the polygon of drought. Frequency, duration and dimension of droughts seem to be closely linked to the ENSO (El Niño Southern Oscillation) phenomenon (Gommes et al. 1998), leading to a strong spatial and temporal inter- and intraannual variation in rainfall distribution (Conti 1995). Even in years with a high total rainfall amount, dry spells of up to 3 weeks and more may occur, reducing the agricultural production substantially. In more than one-third of the area of the northeastern Brazilian federal state of Piauí, the production of annual crops is strongly hampered by severe droughts (IBGE 1999). Additionally, the crop production of this area is often adversely affected by the widespread inherent low soil fertility. Nevertheless, agriculture is the most important economic activity in rural areas of northeastern Brazil and has to guarantee the subsistence of many small-scale farmers and their families. Climate changes, one of the key issues of the *Global Change* discussion, may exacerbate drought proneness of northeast Brazil (Nisbet 1994; Böhm et al. 2000).

Material and Methods

This study focuses on two regions in the Brazilian federal states of Ceará and Piauí, Tauá and Picos, respectively. Both regions were selected according to their climatic and environmental characteristics. Picos located at 7°1'S, 41°37'W has an average annual rainfall of 704 mm that varies from 80 to 1440 mm. The rainy season lasts from December to May (Table 1). The mean annual temperature is 30.5 °C. The altitude of the landscape ranges from 200 m in the valley up to

545 m on the chapadas (plateau). The natural vegetation is a transition between cerrado, a woodland savannah, and caatinga, a deciduous drought-resistant shrub land. The most intensively cultivated soil types are Alumi-haplic Acrisols or Alumi-Ferralic Arenosols on the plateau and Eutric Fluvisols or Chromic Luvisols in the lower areas and valleys (Halm 2000). The soils on the plateau are characterized by low pH levels, Al toxicity, and a low soil fertility, whereas the soils of the Rio Guaribas valley and the area between the plateau and the valley tend to be more fertile, with pH levels ranging from 6 to 7 and no toxic levels of Al (Table 2). Rather some locations in the valley with a high groundwater table tend to have higher sodium saturation (De Barros 2002).

Table 1. Annual rainfall for Picos, Piauí, and Tauá, Ceará, during 1998–2000 and its distribution during the rainy season

	Total amount	Dec.	Jan.	Feb.	Mar.	Apr.	May
				(mm)			
Picos[1] 1998	546	48.1[4]	159.2	117.6	166.5	28.0	0.0
1999	804	75.0	134.0	190.0	310.0	44.0	40.0
2000	664[3]	86.0	131.0	176.0	119.0	163.1	75.0
Tauá[2] 1998	236	18.0	0.0	141.4	8.2	0.0	0.0
1999	416	0.0	110.2	35.8	105.5	68.7	61.4
2000	579	28.7	56.5	67.5	196.0	84.7	50.9

[1] Measured in field trials on the chapada
[2] FUNCEME, Fortaleza-CE
[3] Measured till May 2000
[4] DHME, Teresina-PI

Table 2. Soil types in intensively used landscape units and soil types in Picos-PI and Tauá-CE and their top soil (0–20 cm) characteristics

Location		Soil types	pH(H_2O)	$CEC_{pot.}$ (cmol/kg)	Al-Saturation (%)	Avail. P (mg/kg soil)
Picos						
	Plateau	Alumi-Haplic Acrisol	4.0	4.9	54	5.0
	Basaltic plain[1]	Eutric Cambisol	6.4	10.9	0	10.0
	Valley	Eutric Fluvisol	6.6–6.8	21.5	0	22.2
	Valley	Chromic Luvisol	5.4–5.9	3.5-5.9	3	1.0–3.0
Tauá						
	Upper slope	Haplic Luvisol	6.1	6.2	0	4.0
	Valley	Eutric Fluvisol	7.4	16.4	0	10.5

[1] Between plateau and valley, limited extension

In Tauá, which is located at 6°0'S, 40°25'W, the average annual rainfall is 601 mm, ranging from 155 up to 1474 mm. In general, the rainy season starts in February and ends in May. The mean annual temperature is 26 °C. The altitude of the surrounding landscape ranges from 400 m in the lowlands (Depressas Sertanejas) up to 800 m in the mountains (Serra). The natural vegetation is a typical caatinga. Main soil types in the cultivated areas are Haplic Luvisols and Eutric Fluvisols. The pH ranges from 6 to 7 and in some places signs of alkalinization are evident.

In Picos, two field trials were established as main trials in a randomized complete block design with four repetitions on each of the two dominant landscape units – chapada (plateau) and vale (valley). One trial was installed to study the nutrient limitations of maize and cowpea in mixed cropping systems (termed fertilizer trial hereafter). The treatments focused in this paper are[1]: (i) control without fertilizer application and liming; (ii) complete fertilization with N, P, K without liming; and (iii) complete fertilization with liming. In the valley trial, no lime was applied due to the higher pH values of this site (Tables 3 and 4). In the year 2000, fertilizers were split-applied in two doses (Table 5). The maize variety BR 5039 (São Vincente) and the cowpea variety EPACE 10 were used in this trial. Both are improved varieties, adapted to semiarid conditions. Planting densities were 45×10^3 plants ha^{-1} for maize and 90×10^3 plants ha^{-1} for cowpea.

The second trial was installed to study the effect of various crop management practices on the yield performance of maize and cowpea in sole and intercropped stands (termed crop management trial hereafter). The treatments focused in this paper are[2]: (i) maize/cowpea intercropped with the crop management practice dibbling and no fertilizer application. The treatment represents the common small-scale farmers' practice of this region and is considered as the control treatment, and (ii) maize/cowpea intercropped with dibbling plus fertilizer application (Tables 3 and 4). For maize, the variety BR 5039 (São Vincente) was used, whereas for cowpea the traditional variety CEA 315 was included. Planting densities were 40×10^3 plants ha^{-1} for cowpea and 20×10^3 plants ha^{-1} for maize on the chapada, whereas densities of 60×10^3 plants ha^{-1} for cowpea and 30×10^3 plants ha^{-1} for maize were used in the valley due to the higher soil fertility there.

Complementary field trials were established on other sites in the Picos region to obtain information on crop growth and yields of different food crops and to study the effect of variety and planting densities. Thus, a trial was carried out on the chapada in 1998 and 1999 to study the influence of variety and planting density on cowpea yield. The treatments were two local varieties (Canapú comum and CEA 315) and a mixture of them (Table 6). The treatments were planted at two density levels: $60\times10^3$and 120×10^3 plants ha^{-1}. The trial was designed as a RCBD with three repetitions. Furthermore, growth and yield performance of cowpea, maize, cassava and rice were monitored on farmers' fields at various sites in the Picos region.

[1] For overview of all fertilizer treatments, see De Barros et al. (2001).

[2] For overview of all fertilizer treatments, see Hilger et al. (2000).

The studies were completed by a fertilizer application experiment similar to that established in Picos, which was installed in 1998 in Tauá. The trial was strongly affected by a drought in 1998 (Table 1) that led to a die-off of maize in its juvenile stage. The cowpea stand was maintained, but only by applying salvation irrigation of 126 mm. In 1999 and 2000, the trial was reinstalled at another location, and the treatments were modified: complete fertilization with manual weed control; complete fertilization without weed control; and no fertilization no weed control.

In all field trials, yield, dry matter production and distribution, crop growth rates and leaf area indices were determined. In this paper only yield data are presented.

Results

The trials of both focus areas heavily suffered from the adverse climate and site conditions. In the El Niño year 1998, droughts led to a total crop failure at the valley site in Picos (Tables 3 and 4). On the plateau site, however, yields were obtained in all three years. In Tauá, which is more arid, maize plants were also completely lost in 1998 (Table 4). At the valley site in Picos, trials suffered from several inundations from a nearby river in 1999 and 2000. Nevertheless, the trials revealed differences in the suitability of maize and cowpea under the different site conditions. Tables 3–5 and Fig. 1 indicate the enormous differences between the production levels of the plateau and valley sites in the focus region of Picos. This is mainly the effect of the strongly differing soil fertility (Table 2) of the two major landscape units in this area. Toxic concentrations of Al ions, a very acid soil pH of 4 or even less, and a low phosphorus availability strongly limit crop growth and do not permit a good yield performance of maize and cowpea under small-scale farmers' conditions on the plateau when no fertilizer is applied. Even in years with more favorable precipitation like 1999 and 2000 (Table 1), cowpea yield on the plateau remains at the level of 1998. The production potential of maize in the other areas of Picos and Tauá is higher compared with that of the plateau (Tables 4 and 5). In the former, the soils are more fertile and favorable for maize production.

Tabelle 3. Yield of intercropped cowpea in t ha^{-1}. Data were collected during 1998–2000 in Picos, Piauí and Tauá, Ceará, NE Brazil

		Picos				Tauá	
		Fertilizer trial		Crop management trial		Fertilizer trial	
		Control	+NPK -lime[1]	Control	+NPK[2]	Control	+NPK[3]
Plateau	1998	0.16	0.54	0.30	0.51	- -	- -
	1999	0.36	0.76	0.24	0.44	- -	- -
	2000	0.19	0.57	0.09	0.23	- -	- -
Valley	1998	0.0[4]	0.0[4]	0.26	0.37	0.28[5]	0.65[5]
	1999	0.07	0.71	0.37	0.34	- -	0.49
	2000	0.37	0.17	0.0[6]	0.0[6]	0.0[4]	0.0[4]

[1] 30 kg ha N, 40 kg ha P_2O_5 and 40 kg ha K_2O
[2] 26 kg ha N, 50 kg ha P_2O_5 and 35 kg ha K_2O
[3] 80 kg ha N, 70 kg ha P_2O_5 and 40 kg ha K_2O
[4] no yield due to water shortage
[5] salvation irrigation
[6] no yield due to inundation

Nutrient supply, however, is not the only limiting factor of crop production in the study area of Picos. Comparing the yield data from 1998 and 1999 clearly indicates a general yield increase at all test sites of Picos, although a yield decline was expected in the unfertilized plots in the second year of production due to the nutrient losses there (Tables 3–6). The low yield levels of 1998 were caused by a strong El Niño effect that led to a short, non-regular rainy season.

A similar observation was made for Tauá. In 1998, the trials had to face a total yield loss of maize (Table 4). Through a salvation irrigation it was possible to obtain a relatively good yield for cowpea. In Tauá, the subsequent cropping season was also dry. Though the total rainfall amount in both regions increased in 1999, in Tauá it was still low, when compared with the rainfall in Picos during 1998 (Table 1). In the last year of observation, total rainfall amount of Tauá increased again and the monthly rainfall distribution was also slightly better. However the rainfall events were concentrated to 2–3 rainy days per month and crop establishment was impossible in spite of several replanting attempts. This explains the low yield levels of the trials at Tauá during the entire observation period comparable to those obtained from the plateau sites at Picos (Tables 3 and 4), although the chemical soil properties are much better in Tauá (Table 2).

Tables 3 and 4 show that fertilizer applications increased the yields of cowpea and maize at the plateau site in Picos, whereas the effect observed in the valley was not that clear. In Tauá, fertilizer application improved cowpea yields in 1998, but this was only obtained in combination with a salvation irrigation (Table 3). Liming plays an important role at the plateau site in Picos (Table 5). When lime was applied, maize yields were increased reaching an acceptable level. Furthermore, Table 5 indicates that high fertilizer applications to cowpea on the plateau sites only had a beneficial effect when combined with liming. Under the site conditions of the valley, fertilizer application showed no improvement, even at high levels of fertilizer application due to the better natural nutrient availability (Table 2). In addition, the fertilizer trial in the valley elucidated that soil fertility may be highly variable within a small area caused by variations in Na concentration and rootable soil depth in plots within the same treatment (De Barros 2002).

Table 4. Yield of intercropped maize in t ha^{-1}. Data were collected during 1998–2000 in Picos, Piauí, and Tauá, Ceará, NE Brazil

		Picos				Tauá	
		Fertilizer trial		Crop management trial		Fertilizer trial	
		Control	+NPK –lime[1]	Control	+NPK[2]	Control	+NPK[3]
Plateau	1998	0.02	0.14	0.17	0.29	- -	- -
	1999	0.00	0.24	0.00	0.00	- -	- -
	2000	0.50	0.70	0.10	0.28	- -	- -
Valley	1998	0.00[4]	0.00[4]	0.00[4]	0.00[4]	0.0[4]	0.00[4]
	1999	0.84	0.82	1.44	1.37	- -	0.12
	2000	1.51	1.79	2.99	2.18	0.0[4]	0.00[4]

[1] 30 kg ha N, 40 kg ha P_2O_5 and 40 kg ha K_2O
[2] 26 kg ha N, 50 kg ha P_2O_5 and 35 kg ha K_2O
[3] 80 kg ha N, 70 kg ha P_2O_5 and 40 kg ha K_2O
[4] no yield due to water shortage

Table 5. Cowpea and maize yield in t ha^{-1} under two levels of fertilizer application as influenced by liming. Data were recorded in 2000 in Picos, Piauí, NE Brazil

		Level of fertilizer and lime application			
		+high NPK[1] +lime	+NPK[2] +lime	+high NPK[1] –lime	+ NPK[2] –lime
Cowpea	plateau	1.01	0.68	0.62	0.57
	valley	- -[3]	- -	0.21	0.18
Maize	plateau	2.81	1.26	0.82	0.70
	valley	- -	- -	1.79	2.45

[1] 30 kg ha N, 40 kg ha P_2O_5 and 40 kg ha K_2O; treatments with lime 3t ha only in the first year
[2] 100 kg ha N, 150 kg ha P_2O_5 and 120 kg ha K_2O
[3] no experiment or trials failed

The field trial installed to test the effects of variety and planting density on the yield performance of cowpea revealed again the positive response to more favorable rainfall distribution at Picos, as indicated by the strong yield increase from 1998 to 1999 (Table 6). This trial also showed that Canapú, an indeterminate cowpea variety with a long growing cycle, had almost always a significantly higher yield than CEA 315, a variety with determinate growth habit and medium growing cycle. The yield level of the mixture of both varieties was intermediate. In both years the shorter growing cycle of CEA 315 showed no advantage in yield performance. The two tested planting densities had no significant influence on the yield performance of cowpea. Planting density was at minimum nearly twice as high as under local farmers practice. This led in1999, when rainfall conditions were more favorable, to a higher cowpea yield compared with low plant densities, which is represented by the +NPK treatment in the crop management trial with 40,000 plants/ha (Hilger et al. 2000). This indicates the possibility of better yield performance of cowpea under small-scale farming conditions by increasing plant density, when access to small amount of fertilizers is provided.

Table 6. Cowpea yield in t ha^{-1} as influenced by variety and planting density. Data were recorded between 1998 and 1999 on the plateau in Picos, Piauí, NE Brazil

Variety	Planting density			
	60×10^3 plants ha^{-1}		120×10^3 plants ha^{-1}	
	1998	1999	1998	1999
Canapú[1]	0.37 a[2]	0.83 a	0.47 a	0.88 a
CEA315[1]	0.17 a	0.57 b	0.24 b	0.45 b
Canapú/CEA315[1]	0.23 a	0.62 ab	0.30 ab	0.72 ab

[1] 8 kg ha N, 47 kg ha P_2O5 and 24 kg ha K_2O
[2] Treatments followed by the same letter do not differ significantly ($P < 0.05$)

Considering the economic aspects, such as production costs for external inputs and labor, the benefit for small farmers by using fertilizer is small (Table 7). Most of the treatments showed only a small benefit or even a loss for the farmer, particularly when they depend on external farm workers. In 1998, when severe drought reduced the yields, the economic benefit was the highest due to the higher market price for agricultural products in that year. The best result was obtained when lime was applied combined with only a small amount of fertilizer. Higher levels of fertilizer application improve the yield performance, but do not produce enough to cover the higher production costs for the external inputs, as shown in Table 7 when benefits of low and high input in 2000 are compared.

Besides maize and cowpea, yield data of two other crops of local importance, cassava and rice, were collected from farmers' fields. Cassava produced more than 5.0 t ha^{-1} root dry matter without any fertilizer application on the plateau. In valley areas with regular flooding, rice produced up to 3 t ha^{-1} on fertile alluvial soils (Fig. 1).

Table 7. Benefit of fertilizer and lime application in R$ per ha in the intercropping cowpea/maize on the Chapada between 1998 and 2000

	Low input						High input	
	1998		1999		2000		2000	
Treatment	F.W.[1]	F.W. +S.E.W[2]	F.W.	F.W. +S.E.W	F.W.	F.W. +S.E.W	F.W.	F.W. +S.E.W
Control	103.2	–52.5	–23.3	–187.3	16.5	–164.7	–5.1	–189.3
Compl. Fert.[3]	766.9	579.9	211.1	9.2	50.9	–143.7	–47.8	–284.9
Compl. Fert. –lime[4]	625.1	445.3	–3.5	–176.3	5.8	–184.6	–436.1	–628.6

[1] F.W. family workers only
[2] F.W.+S.E.W. family workers plus seasonal external workers
[3] 30 kg ha N, 40 kg ha P_2O_5 and 40 kg ha K_2O; treatments with lime (3 t ha^{-1}) only in the first year
[4] 30 kg ha N, 40 kg ha P_2O_5 and 40 kg ha K_2O

Discussion

Results from the field studies in Picos and Tauá clearly showed that the actual land use practice of local small-scale farmers is strongly handicapped by adverse climate conditions. On the plateau sites of Picos, which are the predominant landscape unit in this region, the low soil fertility and the strong degree of acidification of soils are further limiting factors for crop production (Tables 3 and 4).

Moderate fertilizer applications can almost double the yields of maize and cowpea at this site. In addition, the production level of cowpea is raised above 0.25 t ha^{-1} (the average yield for intercropped cowpea in Piauí according to IBGE [1996]), which was achieved even under the poor rainfall distribution of 1998. High amounts of fertilizer are able to raise productivity level drastically, but it is necessary to apply lime (Table 5). It is uncertain, however, if local small-scale farmers have the financial conditions and technical facilities to apply the required amount of 2–3 t ha^{-1} lime. Considering our economic evaluation of all treatments, moderate fertilizer applications are feasible under small-scale farming conditions, especially when planting density is higher.

Maize cultivation on plateau sites is not recommendable. In years with good rainfall conditions and moderate fertilizer application, the average yield level for Piauí of 0.53 t ha^{-1} (IBGE 1996) may be reached, but this is still a very poor production level (Table 4). Considering that the maize produced in the field trials at the plateau site generally showed cob deformations due to water stress, it renders them unsuitable to be sold as freshly cooked or grilled cobs for human consumption. Therefore, they can be only used as fodder in animal production, but with a high production risk and poor economic return which does not allow the use of external inputs.

When the soil fertility is high, as in the valley sites, fertilizer application had no effect on the performance of maize and cowpea in the study period. Even under low or no input conditions, yield levels were already excellent for small-scale farming conditions, compared to the average yield level in Piauí. Other soil factors, like Na saturation or compacted subsoil layers, seem to be limiting factors to crop production (De Barros 2002).

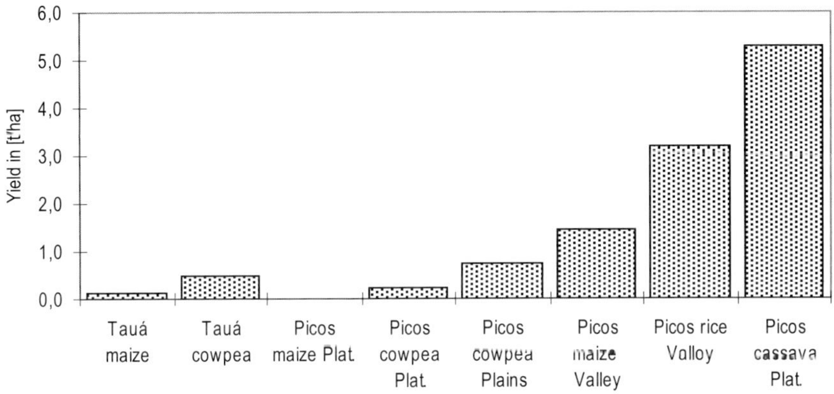

Note: Tauá maize = Tauá maize fertilizer trial; **Tauá cowpea** = Tauá cowpea fertilizer trial; **Picos maize Plat.** = Picos maize on the plateau crop management trial, intercropped without fertilizer; **Picos cowpea Plat.** = Picos cowpea on the plateau crop management trial, intercropped without fertilizer; **Picos cowpea Plains** = Picos cowpea farmer field, intercropped without fertilizer, on a plain fertile location below the plateau; **Picos maize Valley** = crop management trial at the valley, intercropped without fertilizer; **Picos rice Valley** = Traditional farmer field cropped to rice in the Picos valley, swamp land; **Picos cassava Plat.** = Traditional farmer field cropped to cassava at the plateau.

Fig. 1. Yield levels for various food crops at Tauá (Ceará) and Picos (Piauí), NE Brazil. Data were recorded in 1999

Apart from low soil fertility, the major production constraint for small-scale farmers is the lack of rainfall in some years or its erratic spatial and temporal distribution in all locations. In addition, areas close to rivers or brooks have a high risk of temporal inundation in the rainy season. Considering that with monthly amounts of less than 50 mm precipitation is insufficient for maize production, the situation in Tauá is severe. Table 1 shows that even in years with a high total amount of rainfall, e.g., in 2000, only one month had more than 100 mm of precipitation. In other years the rainy period is often split into two parts, separated by a temporary dry period or so-called „veranico“. This limits the number of crops that can be grown under these conditions and raises the risk for yield damage before the crops can be harvested.

In dry years, such as 1998, lack of adequate rainfall strongly reduced the crop production in both regions (Tables 3 and 4). In 1998, crop production mainly depended on the occasion of some rare rainfall events, which were limited to small land stripes. Farms within these strips were able to produce some crop yield, whereas farms outside of these favored areas faced total yield losses, as was the case in the field trials at the valley site.

Considering both, drought and low soil fertility, cassava is an alternative crop to cowpea on the chapada for rainfed agriculture. Under the adverse conditions during the growing periods of 1998 and 1999, cassava was able to produce about 5 t ha^{-1} root dry matter (Fig. 1). Even considering that cassava has a 2-year growing cycle, which reduces the yield to 2.5 t $ha^{-1}a^{-1}$, no other crop was able to reach this production level. The other advantage of cassava is that its starch and meal can be extracted by the farmers and sold on the local market.

In the valley and lowlands, which are more fertile, both maize and cowpea produce well under rainfed conditions when no severe droughts occur. However, for areas in the valley that tend to submerge, upland rice is an excellent alternative to maize (Fig. 1). There, cowpea development is strongly affected by oxygen deficiency in the soil. In Tauá, with its more arid climate, cowpea and especially maize cropping is restricted to floodplain areas in the valley.

Conclusions and Recommendations

Cowpea is well adapted to the growing conditions found in the regions of Picos and Tauá. However, improvements in drought resistance and low soil fertility should be promoted to guarantee security in production even in dry years, whereas maize can only be recommended on more fertile locations and when a minimum of water supply is guaranteed. Medium fertilizer applications can be recommended to increase productivity of cowpea on the plateau sites, even in dry years, especially phosphorus- and potassium-based fertilizers. However, to obtain best economic return, lime should be applied in a moderate rate. With fertilization higher planting densities are recommended (60–90,000 plants/ha).

On more fertile soils, fertilizer application should be considered when continuous cultivation is given. Then, nutrient replacement has to be calculated based on soil analysis and nutrient exportations. Under rainfed conditions, biannual or permanent low pH tolerant crops, like cassava, showed a higher drought resistance on the plateau sites with deep sandy soils.

Alternative crops for the regions might be groundnut, millet, shorgum or sesame, due to their drought resistance, but limitations through low soil pH have to be considered.

For improving yield stability in crop production, irrigation is an appropriate choice where applicable. Availability of irrigation water, however, is very limited in Tauá. In Picos, it is restricted to the valley of the Rio Guaribas River. However, it is most likely that the actual subterranean groundwater use in the municipality of Picos is higher than its recharge. If the present practice of furrow irrigation or flooding of rice continues, the water table will be further lowered, leading to severe problems in the future. It is therefore necessary to optimize irrigation techniques and to restrict them to cash crops, such as vegetables and fruits, which recompense high inputs of external resources. For vegetables under irrigation, the dry season has to be recommended as this period provokes less phytosanitary problems as the rainy season.

However, the agronomic improvements towards higher yields and more production security will only guarantee sustainability for small farmers if their performance is evaluated from an economic point of view. Therefore, the research results must be combined and integrated with economic knowledge as shown in the WAVES program. On one hand scientific progress must be available in a comprehensible form for small-scale farmers and, on the other, researchers should consider the demands and needs of the rural population to guarantee a successful implementation of new improved techniques.

References

Böhm U, Gerstengabe FW, Österle H (2000) Regional climate scenarios for Northeast Brazil. In: Printz A (ed) Proceedings of the IIIrd international WAVES workshop in Freising-Weihenstephan, Germany

Conti, J.B., 1995. Proposta de metodologia de estudo aplicada ao Nordeste Brasileiro. Tese de Livre-Docente apresentada ao Departamento de Geografia da Faculdade de Filosofia, Letras e Ciências Humanas da Universidade de São Paulo. São Paulo-SP, Brazil.

De Barros I (2002) Soil related limitations to crop production in the semiarid Northeast of Brazil. In: Kandeler E, Römheld V, Stahr K, Streck T (eds) Hohenheimer Bodenkundliche Hefte, vol 63. Universität Hohenheim, Stuttgart

Gommes R, Bakun A, Farmer G (1998) An El Niño primer. SD dimensions, Sustainable Development Department, Food and Agricultural Organization of the United Nations (FAO); http://www.fao.org/WAICENT/FAOINFO/SUSTDEV/EIdirect/EIan0008.htm

Hahn D (2000) Soil water balance in the semiarid northeast of Brazil – characterisation, simulation, evaluation and comparison of hydrological properties and processes in representative soils of the Picos region. In: Kandeler E, Kaupenjohann M, Stahr K, Balke KD (eds) Hohenheimer Bodenkundliche Hefte, vol 55. Universität Hohenheim, Stuttgart

Hilger T, Herfort J, Gaiser T, Leihner DE, Ferreira LGR (2000) Calibration of EPIC/ALMANAC in semi-arid environments of NE Brazil. In: Proceedings of the IIIrd international WAVES workshop, 13–16 March 2000. Freising-Weihenstephan, Germany

IBGE (1996) Censo Agroperuário 1995/96. Piauí, Rio de Janeiro, Brazil

IBGE (1999) Levantamento Sistemático da produção agicola 1999, Instiuto Brasileiro de Geographia e Estatística IBGE. Teresina PI, Brazil

Nisbet EG (1994) Globale Umweltveränderungen: Ursachen, Folgen, Handlungsmöglichkeiten, Klima, Energie, Politik. Spektrum Akademischer Verlag, Heidelberg

Villa MA (2000) Vida e morte no sertão – História das secas no Nordeste no séculos XIX e XX. Editora Ática, São Paulo, Brazil

Seepage and Groundwater Recharge in Sandy/Loamy Chapada Soils of the Picos Region, Piauí

Frank-Michael Lange, Dietrich Halm and Thomas Gaiser

University of Hohenheim, Institute of Soil Science and Land Evaluation, 70593 Suttgart, Germany, fmilange@gmx.de

Introduction

Water scarcity poses a serious problem in the semiarid northeast of Brazil since the region is affected by an irregular rainfall distribution within rainy seasons of 3–5 months duration (November–April) and also by regularly recurring El Nino events, causing strong decreases in rainfall (up to annual zero-rainfall at some locations) every 8–10 years (Halm et al. 1999). In addition, nearly 50% of the state of Piauí (mainly chapadas and serras) is covered by sandy soils, which generally show low water storage and plant-available water capacities but medium to high infiltration and percolation rates (EMBRAPA/SNLCS 1983, 1986).

Today, the mean annual rate of groundwater recharge (GR) for the sedimentary region of Picos is, according to Forster (1996), 10 mm. It is highly probable that the actual recharge is exceeded by extraction for human water supply and irrigation. However, safe yield concepts, e.g., for the region of Picos, have not been drawn up yet.

The investigation and quantification of seepage (SP) and GR are therefore high priorities for semiarid sandy regions such as the state of Piauí. This contribution presents model-based methods to assess SP and GR in sandy/loamy soils under different vegetation and land use.

Materials and Methods

Soils and Vegetation

Two sandy soils, developed from a 5-m thick sand layer above the Itapecurú sandstone formation and covered by different vegetation, were selected for the investigations and classified according to FAO (1988).

– A Ferralic Arenosol (ARo), used for cassava production.

– An Alumi-Haplic Acrisol (ACha), used for by maize/cowpea intercropping.

Both soils are located on the „Chapada do Frio“, 25 km west of the city of Picos, 600 m above sea level. The distance between both sites is 4 km in the E-W direction. They have four (ACha) and five (ARo) horizons, respectively. A detailed description is given in Halm (2000).

The texture of the ARo is, according to USDA (1992), loamy sand (LS) to sandy loam (SL) with a continuous increase in the clay content from the top to the subsoil by about 4.3% (7.8–12.2%, Table 1). The ACha shows a textural change at 27 cm depth from a BA-horizon with a clay content of 15.5% to a Bt-horizon with a clay content of 18.9%. Over the whole depth, the texture is SL.

Table 1. Depth distribution of horizons, texture, and clay content of investigated soils

Profile/horizon	Depth (cm)	Texture	Clay (%)
Ferralic Arenosol			
Ah	0–15	LS	7.8
AB	–33	LS	8.5
BA	–51	LS	10.4
Bw1	–84	LS	11.5
Bw2	–120	SL	12.1
Haplic Acrisol			
Ah	0–11	SL	15.6
BA	–27	SL	15.5
Bt1	–72	SL	18.9
Bt2	–120	SL	19.3

LS: Loamy sand; SL: sandy loam

Measurement and Simulation of Seepage Periods

The conditions for SP to occur are defined as periods where water suction values are below or equal to field capacity (FC). This implies the movement of soil water as quick draining water in coarse and medium pores, respectively. However, in the literature, the threshold for FC is reported to vary between 63 and 316 hPa (Lüken et al. 1970; Schachtschabel et al. 1989; Hartge and Horn 1999). Thus, the threshold of water suction in the ARo with sandy texture was set to 63 hPa (at a water content of 27 vol.%) since the hydraulic conductivity in sandy soils is strongly reduced between 63 and 316 hPa (Schachtschabel et al. 1989). This means that an effective water movement in sandy soils such as the ARo occurs at water suctions

≤63 hPa in pores with a diameter >50 μm and a hydraulic conductivity of $>10^{-4}$cm s^{-1}.

In the ACha, the threshold of water suction was set to 316 hPa (at a water content of 14 vol.%) since the hydraulic conductivity in loamy soils is significantly higher at 316 hPa than in sandy or clayey soils, leading to an effective water movement at lower water contents (higher water suctions) in pores with a diameter of >10 μm. This differentiation in defining the SP threshold for different textured soils seems to be important for soil hydraulic investigations under a semiarid climate, where the hydraulic conductivity is considered to be a more sensitive factor than under humid climate conditions. Furthermore, the amount and direction of the water flux also depend on the actual tension gradient.

Water contents were measured in both soils with time domain reflectometry (TDR) in the depth segments 30–60 and 90–120 cm. Three TDR probes were installed in each depth. The TDR measurements were carried out over two wet seasons but contrasting rainfall patterns (ARo: Nov. 1995–June 1996 and Nov. 1996–July 1997, ACha: Nov. 1999–Oct. 2000 and Nov. 2000–April 2001). Since the occurrence of SP was defined for a certain threshold of the water content at a corresponding water suction, the number of days with SP above this threshold could be calculated. Then, a simulation of the water contents was run to compare the measured SP periods with simulated SP periods.

For the simulations of the soil water dynamics in the ARo, the soil water balance model HILLFLOW (Bronstert 1994) was applied. For the simulation of the soil water changes in the ACha, the soil water balance model SIMPEL (Hörmann 2000) was applied. The approach of applying two model systems was chosen for a comparison of different model systems.

HILLFLOW solves the Richards equation by using van Genuchten parameters (VGNs). The VGNs were obtained from the literature (Bohne et al. 1993). The actual water content $\theta(\psi)$, the water suction $\psi(\theta)$, and the unsaturated hydraulic conductivity $k(\psi)$ for each depth layer, were determined according to Mualem (1976) and van Genuchten (1980). The determination of $k(\psi)$ requires the saturated hydraulic conductivity as input parameter. The values were taken from AG Boden (1996).

SIMPEL is a storage model with a vertical division of the soil into four compartments. Precipitation is added to and evapotranspiration is subtracted from the actual filling stage of the soil water store. When the actual filling stage of the upper compartment is higher than the storage capacity, the excess is directed to the lower compartment. In the stage of evapotranspiration, the compartments are depleted from top to bottom up to the threshold of the permanent wilting point. Reaching this threshold, the next lower compartment is depleted. Seepage is calculated according to the approach of Glugla (1969).

Evapotranspiration in both model systems is calculated according to the approach of Penman-Monteith (Monteith 1965).

Assessment of Groundwater Recharge in a Ferralic Arenosol

Using the HILLFLOW model, the GR was assessed with the subterranean water divide (SWD) approach for the ARo over the total investigation time of two hydrological years. Hoogmoed et al. (1991) confirm that modeling provides an alternative means of deep drainage calculation beneath lysimetry or groundwater piezometry. Since the simulated water contents showed high accuracy compared with the measured results, the empirical approach was extended to calculate the GR in the deep subsoil.

The SWD is a boundary in the subsoil without groundwater influence. Below that boundary water is directed to the deeper layers or to the groundwater (Renger et al. 1977). When the SWD is situated below the root zone, it can be used as integrational boundary to determine the GR. According to Schachtschabel et al. (1989) and AG Boden (1996), the SWD is defined as the soil layer where no withdrawal of water by plant roots occurs even in years with poor precipitation and where the water content does not fall below FC.

Since in this study the thickness of the sand layer above the sandstone plateau was 5 m, the total simulation depth in HILLFLOW was therefore increased to 5 m, divided into 18 depth layers. The SWD was met in that depth layer, where the water content did not fall below FC. The texture of the deeper sand layers was set to sandy loam (15 and 19% clay, respectively) in 120–150 cm and 150–300 cm depth, respectively, and sandy clay loam (25% of weight clay) in 300–500 cm depth. For each texture, different water retention curves were applied according to the respective VGNs (Fig. 1).

To compare SWD and GR under different vegetation, the HILLFLOW model was run with a vegetation file for actual cassava vegetation (mean root depth of 80 cm, maximum root depth 120 cm) and with another vegetation file for caatinga vegetation (natural drought deciduous shrub vegetation) in which the root depth was set to 170 cm (average value according to Cordeiro 1985).

Results and discussion

Seepage

The simulated courses of water contents in Figs. 2 and 3 show good correlation and high accuracy (ARo: $r > 0.7$, ACha: $r = 0.6$) after comparison with measured values. The ARo shows in the first period zero days (0) with SP when measured and 9 days when simulated (Table 2). Probably, the saturated conductivity was slightly overestimated as input parameter in HILLFLOW and, therefore, the simulated SP time is 9 days compared with a measured value of 0 days. In the second period, measured and simulated results are nearly identical (29 days measured and 23 days simulated).

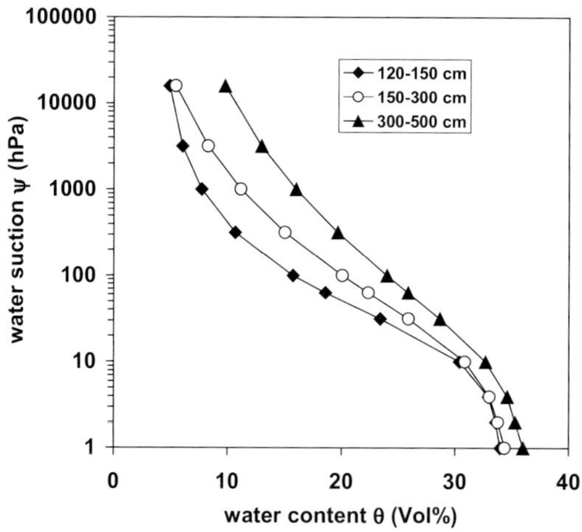

Fig. 1. Water retention curves of deep layers in a Ferralic Arenosol for the calculation of subterranean water divide and groundwater recharge

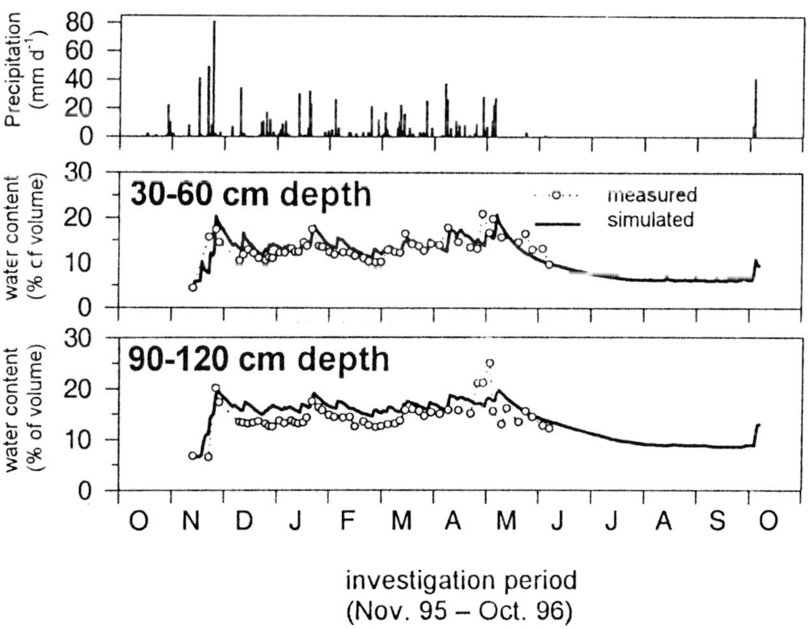

Fig. 2. Course of measured and simulated (HILLFLOW) water contents (r >0.7) in a Ferralic Arenosol for two depth segments in relation to rainfall distribution

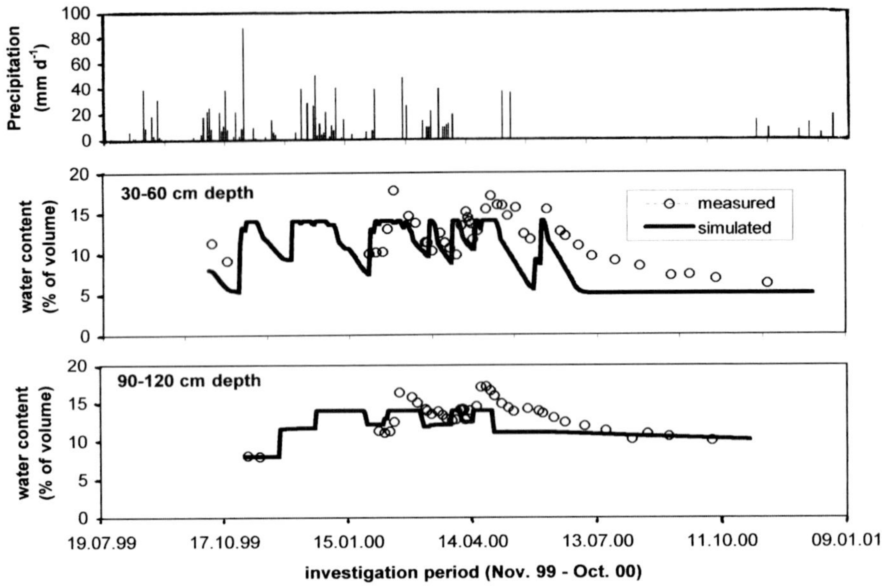

Fig. 3. Course of measured and simulated (SIMPEL) water contents (r=0.6) in an Alumi-Haplic Acrisol for two depth segments in relation to rainfall distribution

Table 2. Seepage times (in days) of two typical sandy chapada soils in 120 cm depth at the seepage threshold (ARo: water content of 27 vol.% at a water suction of 63 hPa, ACha: water content of 14 vol.% at a water suction of 316 hPa)

Soil	Period 1 (Nov. 1995–Oct. 1996)		Period 2 (Nov. 1996–Oct. 1997)	
	Measured	Simulated	Measured	Simulated
Ferralic Arenosol (Cassava)	0	9	29	23

Soil	Period 1 (Nov. 1999–Oct. 2000)		Period 2 (Nov. 1999–Oct. 2000)	
	Measured	Simulated	Measured	Simulated
Haplic Acrisol (Maize/Cowpea)	60	23	12	16

In the first year, the ACha shows 60 measured days with SP, which is 2.6 times higher than the simulated number of days (23). A sensitivity analysis of the seepage threshold showed, at a slightly higher water content at FC of 15 vol.% (instead of 14 vol.%), a reduction from 60 days with SP to 33 days. This shows that the SP threshold is very sensitive and cannot be considered as fixed within a narrow range of ±1 vol.% of water content. On the other hand, it has to be taken into account that SIMPEL could have underestimated the seepage days caused by setting a slightly overestimated value for FC as input parameter. Nevertheless, both model systems, HILLFLOW and SIMPEL, calculated very comparable SP periods related to measured values.

The calculated SP portion in the water balance in both soils is nearly identical (28 and 30% of precipitation, respectively, Table 3). However, in the ACha under maize/cowpea vegetation, a higher portion of precipitation is consumed by actual evapotranspiration (67%) than in the ARo under cassava vegetation (57%). Additionally, the ACha under maize/cowpea shows a lower interception of 5% than the ARo (13%) under cassava. However, the sum of interception and actual evapotranspiration is in both soils comparable (ARo: 70%; ACha: 72%) although cassava has a higher leaf area index (LAI, 1.0–3.5) than maize/cowpea (LAI 1.0–1.5), and a longer vegetation period during the year. This could be explained by the fact that cassava has a higher transpiration and a higher LAI causing a denser cover of the soil. Thus, the soil evaporation is considerably reduced compared to maize/cowpea intercropping. At the same time, less water is available in the ARo during the vegetation period for the transpiration of cassava due to the lower water storage capacity. From this it can be derived that the type of soil and vegetation has a strong influence on the proportioning of SP on the one hand and actual evapotranspiration and interception on the other.

Table 3. Seepage in 120 cm depth, interception, and actual evapotranspiration according to model calculations of two investigated chapada soils with different vegetation cover (mean values of two hydrological years in % of precipitation)

Balance component	Ferralic Arenosol (Cassava)	Haplic Acrisol (Maize/Cowpea)
Precipitation	100 (=1873 mm)	100 (=1969 mm)
Seepage in 120 cm depth	30	28
Interception	13	5
Actual evapotranspiration	57	67

Subterranean Water Divide in a Ferralic Arenosol

In the ARo, the obtained results of the calculated depth of the SWD under cassava and caatinga are different. According to the calculations by HILLFLOW, the SWD under cassava vegetation is reached in the depth segment 150–175 cm where the water contents do not fall below the FC threshold of 15.1 vol.% during the dry seasons in both years (Fig. 4). Under natural caatinga vegetation, the SWD is reached 225 cm deeper in the segment 350–400 cm (Fig. 5), where the water contents do not fall below the FC threshold of 19.6 vol.%. Since the roots of cassava penetrate to a mean depth of 80 cm and the roots of natural caatinga to a depth of 170 cm, it seems that the depth of SWD can be estimated as twice the root depth (Halm 2000).

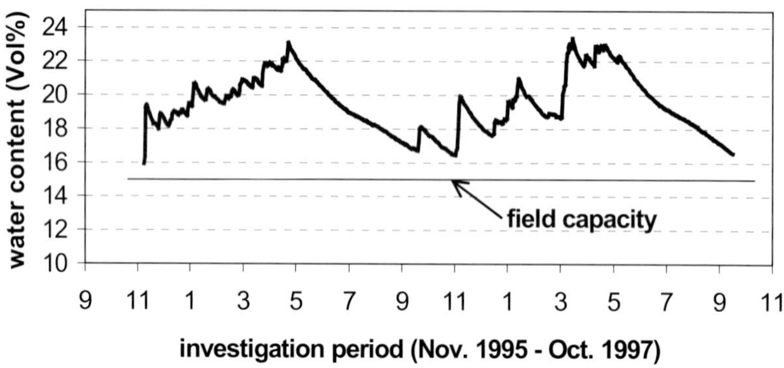

Abb. 4. Course of simulated soil water content over two hydrological years in the depth layer 150–175 cm for the determination of the SWD in a Ferralic Arenosol under cassava vegetation

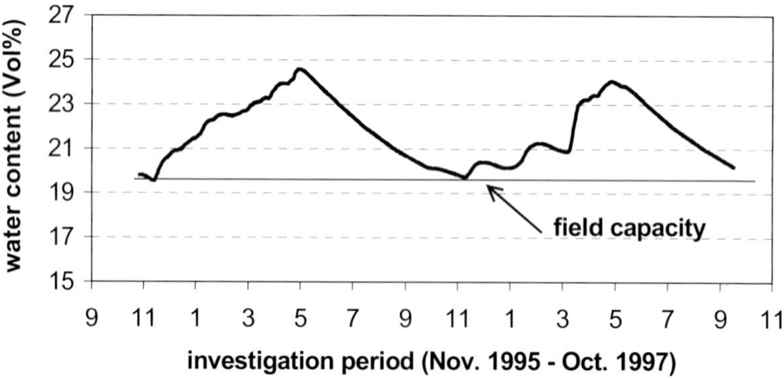

Abb. 5. Course of simulated soil water content over two hydrological years in the depth layer 350–400 cm for the determination of the SWD in a Ferralic Arenosol under caatinga vegetation

Groundwater Recharge

As consequence of higher water consumption by natural caatinga vegetation, the GR in 5 m is 6 and 7 mm in the first and second hydrological year, respectively (Table 4A). Under cassava vegetation it is 13 and 16 mm in the first and second year, respectively. GR is mainly generated when high rainfall intensities can infiltrate and percolate through the soil matrix. Stephens (1996) thinks it highly probable that GR occurs during phases of redistribution, especially in locations where the groundwater table is deep and precipitation events are brief. Redistribution is that hydraulic state in a soil where – after infiltration has stopped – the topsoil drys through evapotranspiration whereas at the same time the subsoil is wetted. This constellation was frequently met during both hydrological years (see Fig. 2a, precipitation diagram).

As was assumed for the ARo, the simulation of GR at a site with a large distance to the groundwater surface is only a small portion of precipitation (1.2%) in years with average precipitation (Table 4B). However, generally, simulated results have to be viewed critically. The crucial problem is the application of realistic van Genuchten water retention curves since GR shows high sensitivity to this parameter. Nevertheless, the obtained rates of GR indicate a level (or a limit), which facilitates comparisons of different types of land use. BGR (1977) and Forster (1996) reported a maximum rate of 10 mm a^{-1} for GR according to climatic balance calculations. These results are very comparable with the obtained mean of 10.5 mm a^{-1} under different vegetation over two investigated hydrological years (Table 4A).

Table 4. Calculated groundwater recharge in mm (A) and in % of precipitation (B) in a Ferralic Arenosol in dependence on different vegetation cover (single values for two hydrological years and mean value for both years)

	Hydrological year 1 (Nov. 1995–Oct. 1996)	Hydrological year 2 (Nov. 1996–Oct. 1997)	Mean of both years
A			
Cassava	13	16	14.5
Caatinga	6	7	6.5
Mean of both kinds of vegetation	9.5	11.5	10.5
B			
Total precipitation (in mm)	923 (=100%)	950 (=100%)	936.5 (=100%)
Cassava	1.4	1.7	1.6
Caatinga	0.7	0.7	0.7
Mean of both kinds of vegetation	1.1	1.2	1.2

Conclusions

For two chapada soils (sandy ARo and loamy ACha) of the semiarid Brazilian northeast, we have shown that the measured number of days where SP occurs is very comparable with the simulated number of days with SP obtained by different soil water balance models (HILLFLOW and SIMPEL) for two different investigation periods. Simulation results showed further that a maize/cowpea crop on an ACha shows a distinct higher evapotranspiration than a cassava crop on an ARo and, therefore, the portion of SP is nearly identical in both soils. This shows that SP is strongly dependent on soil and vegetation cover.

The subterranean water divide for an ARo was assessed in dependence on different vegetation (cassava and caatinga) by the application of the soil water balance model HILLFLOW. The subterranean water divide occurs in that depth which is approximately twice the root penetration depth. We could calculate mean rates of 14.5 mm (cassava) and 6.5 mm (natural vegetation) for the GR in 5-m depth at a sandlayer/sandstone border under an ARo, comparable to reports assessing the groundwater recharge according to climatic balance calculations. This shows that land use changes may lead to an increase in GR when climate conditions remain constant.

References

AG Boden (1996) Guidelines for soil mapping (in German). German Federal Agency of Geosciences and Natural Resources. Geological States Agencies (eds). 4th edn. Schweizerbart, Stuttgart, 392 pp

BGR (German Federal Agency of Geosciences and Natural Resources) (1977) Foundation and possibilities of a land development in dry areas of the state of Piauí/northeast Brazil (in German). German hydrogeologic mission in Recife. Final report. BGR, Hannover, 128 pp

Bohne K, Horn R, Baumgartl T (1993) Provision of van-Genuchten parameters for the characterization of hydraulic soil properties (in German). J Soil Sci Plant Nutr 156:229–233

Bronstert A (1994) Modelling of runoff generation and soil moisture dynamics for hillslopes (in German). IHW 46, 193 pp, University of Karlsruhe

Cordeiro JC (1985) Classificação de terras para irrigação semi-árido da Bacia do Canindé. Fundação Universidade Federal do Piauí - Convênio Banco do Brasil SA (FIPEC)/FUFPI, Teresina, 341 pp

EMBRAPA/SNLCS (1983) Mapa exploratório – reconhecimento de solos de Estado do Piauí. 1: 1,000,000. EMBRAPA/SNLCS, Rio de Janeiro

EMBRAPA/SNLCS (1986) Levantamento exploratório – reconhecimento de solos do Estado do Piauí. EMBRAPA, Rio de Janeiro, 782 pp

FAO (1988) Soil map of the world. Revised legend, with corrections. World resources report 60. FAO, Rome, 140 pp

Forster M (1996) An investigation of water quality, water availability, actual use, and a contribution to the development of a water management complex. In: Water availability in the semi-arid northeast of Brazil (in German). WAVES-status report, vol II. BMBF/PT-USF, Bonn, 35 pp

Glugla G (1969) Calculation procedure for the determination of the actual water content and gravity driven runoff in soils (in German). Albrecht Thaer Arch 13:371–376

Halm D (2000) Soil water balance in the semi-arid Northeast of Brazil – characterisation, evaluation, and comparison of hydrological properties and processes in representative soils of the Picos region, Piauí. Hoh Bodenkd Hefte 55, 205 pp

Halm D, Gaiser T, Stahr K (1999) Soil water balance modelling in the semi-arid northeast of Brazil (in German). Mitt Dtsch Bodenkd Ges 91:161–164

Hartge KH, Horn R (1999) Introduction into soil physics (in German), 3rd edn. Enke, Stuttgart, 304 pp

Hörmann G (2000) Homepage: http://www.pz-oekosys.uni-kiel.de/~schorsch

Hoogmoed WB, Klaij MC, Brouwer J (1991) Infiltration, runoff and drainage in the Sudano-Sahelian zone. In: Soil water balance in the Sudano-Sahelian zone. Proceedings of an international workshop held at Niamey, Niger, 18–23 February, 1991. IAHS Publ 199:185–198

Lüken H, Grüneberg F, Richtscheid P (1970) Land classification: a system to evaluate the qualification of soils in developing regions (in German). German Federal Agency of Geosciences and Natural Resources, Hannover, 78 pp

Monteith JL (1965) Evaporation and environment. Symp Soc Exp Biol 19:205–234

Mualem Y (1976) A new model for predicting the hydraulic conductivity of unsaturated porous media. Water Res 12:513–522

Renger M, Strebel O, Münnich KO, Sonntag C (1977) Methods to determine the rate of groundwater recharge (in German). Geol Jb Reihe C 19, 98 pp

Schachtschabel P, Blume H-P, Brümmer G, Hartge K-H, Schwertmann U (1989) Soil science (in German), 12th edn. Enke, Stuttgart, 491 pp

Stephens DB (1996) Vadose zone hydrology. Lewis Publ, Boca Raton, 428 pp

Van Genuchten MT (1980) A closed-form equation for predicting the hydraulic conductivity of unsaturated soils. Soil Sci Am J 44:892–898

USDA (1992) Keys to soil taxonomy. Soil survey staff, United States Department of Agriculture (USDA), 5th edn. Blacksburgh, Virginia, 541 pp

Part 6: Vegetation and Landscape Ecology

Sustainable Use of Natural Resources in the Municipality of Tauá-Ceará

Vládia Pinto Vidal de Oliveira, Andreas Printz, Stephan Schmidt and Carlos Lineu Frota Bezerra

Federal University of Ceará (UFC), Department of Geography, CEP 60451-970 Fortaleza-Ceará, Brazil, vlpinto@ufc.br

Introduction

The municipality of Tauá was selected as one of the WAVES focus regions because of its fragile, semiarid „Sertão“ environment, typical of northeastern Brazil. Over the course of a two and a half year study, the landscape ecology working groups were concerned with analyzing the complex interrelation of geo-ecological conditions and land use; the objective being to detect degradation and desertification processes and to define sustainable development strategies. In many ways, this study was breaking new ground. Basic data concerning site conditions and land use were surveyed for the first time. In addition, human interaction with the environment had to be assessed, which seemed ambitious, considering the lack of knowledge of human behavior within the reference areas. However, the careful synthesis of new findings with prior studies by expert knowledge, integration methods, geographic information systems and remote sensing technology led to a new quality of environmental analysis for the municipality of Tauá. Cooperation with the local authorities, the local affiliate of the state's university (UECE), and the municipal agricultural school were further important factors in making realistic assumptions and defining further applications.

Area Descriptions

The municipality of Tauá is situated in the administrative microregion of Inhamuns in the Southwest of Ceará , in the „heart of the Sertão“ (Fig. 1). It lies approximately 320 km from Fortaleza, the state capital (S 6°00'11'', W 40°17'34''). With an area of 3940.3 km^2 (IBGE 2000), Tauá is the second largest of Ceará's municipalities. In 1996, the reference year of the WAVES study, the total population of Tauá was 50,258, 52% of which lived in rural areas. However, there has been an increasing exodus from rural areas towards urban centers (mainly the cities of Tauá, São Paulo and Fortaleza). By the year 2000, the majority of the population of Tauá lived in urban areas for the first time: 51.44% of a total of 51,937 inhabitants (IBGE 2000). These emigration statistics are indicative of the quality of life of the rural population in the municipality of Tauá. Poor health conditions caused by substandard water quality and inadequate nutrition remain a problem for parts of the population. Underdevelopment is illustrated by several indicators: almost 85% of the population is living in total poverty (less than 50% of minimum income), and overall mean income in Tauá is approximately 25% of the national average. Furthermore, 48% of the population is illiterate (PNUD 1998).

Fig. 1. Position of Tauá – CE

Even though the city of Tauá has gained importance as a market place and service center for the surrounding rural areas, almost no alternatives to agriculture have been developed (in 1996, 18,426 persons were registered on 5544 farms; 70% of the total rural population). Economic activities are in most cases restricted to basic subsistence farming (extractivism and livestock breeding). Furthermore, 60% of the peasants cultivate farms of less than 10 ha (3% of the total agriculture area), whereas 16% of the peasants belong to large-scale land holders (farm size >100ha), owning 78% of the total agricultural area. In 1998, only 55 tractors were counted in Tauá (IBGE 1998b).

Because of this, the agricultural production system in Tauá suffers from an accentuated disparity of ownership, undercapitalization, and backward agricultural techniques (slash-and-burn is still a common cultivation method). Furthermore, the well-established, powerful traditional land-owner structure, coupled with a lack of development programs, are strong impediments to social and technological innovation.

Besides socioeconomic and cultural reasons, natural conditions place another strong restriction on agriculture; the climate in Tauá is typical of the semiarid Sertão: a rainy season of only 4 months and a mean annual temperature of 26 °C. Moreover, precipitation is often irregular: 155–1474 mm are historical extremes, with a mean annual precipitation of approximately 600 mm. At the center of the pediplain lies an arid area, with a highly negative annual water balance. Due to global climate change, extreme climatic events such as droughts are likely to occur with greater frequency and further accelerate the ecological degradation and depopulating processes.

The soils of Tauá are for the most part shallow, or have had their surface horizons eroded, and tend to be covered with emerging stones or coarse palaeopavements. They have predominantly low water-holding capacity, nutrient levels, and in some areas high levels of salinization. Therefore, suitable areas for cultivation are usually restricted to small stripes of alluvial soils in riparian areas, depending on stored capillary water and the accumulation of relatively fertile sediments.

Until the 1980s, drought-tolerant cotton served as a cash crop for the region. The oily straw that remained after processing was used as animal feed, providing a perfect complement. However, after losing its competitiveness on the global markets and suffering from pests, the role of cotton has been greatly reduced. Since then, the region has lacked a cash crop, and peasants have been forced to extend traditional animal husbandry practices. As in other semiarid regions, animal husbandry in Tauá is based predominantly on woodland grazing. Even if the caatinga shows an astonishing ability for regeneration, the tolerance for grazing can be quickly exceeded. Large tracts of land in Tauá are already affected by erosion and degradation problems, and this trend is further exacerbated near settlements and roads. Deteriorating environmental conditions are an important cause of the aggravation of socioeconomic problems.

Approximately 22% of the area is unused (IBGE 1998a), and remains more or less in a natural or conserved state. This is a fairly high percentage, considering that land allocated for woodland grazing is abundant. Primary forests are largely gone. Vegetation types are differentiated according to phenotype (density and height in particular). These characteristics reflect land use/land cover conditions, which range from dense, arboreal caatinga in the precipitation rich mountains, including remnants of „Mata Seca", to a more or less open shrub caatinga with herbaceous layer, abundant in the pediplain of Tauá.

The municipality of Tauá is dominated by the geo-ecological unit „Semiarid hinterland depression of the Inhamuns" (mean elevation of approximately 400 m above sea level). The geological strata is dominated by Precambrian crystalline, and includes other layers of smaller significance: acid dikes, diorites, gabbros and granite of the magmatic suite (RADAM 1981, see Fig. 2). According to Souza (2000), who describes morpho-structural units at the state level, the municipality of Tauá contains the following geological structures: Domains of sedimentary Cenozoic deposits with plains and fluvial terraces, domains of the Brazilian shield and of residual massifs, which comprise sectors of the „Sertão" depression. The residual massifs are found in border areas of the municipality at elevations of 550–800 m above sea level. In these areas, soil associations with Lixi-, Acri-, Phaeozems and Leptosols are abundant. The central valley has flat to undulating topography, dissected at the margins. The more mountainous areas contain soil types such as Luvisols, Leptosols, Lixi- and Acrisols, and Planosols, as transitory soils occurring in the alluvial stripes of valleys, all susceptible to degradation.

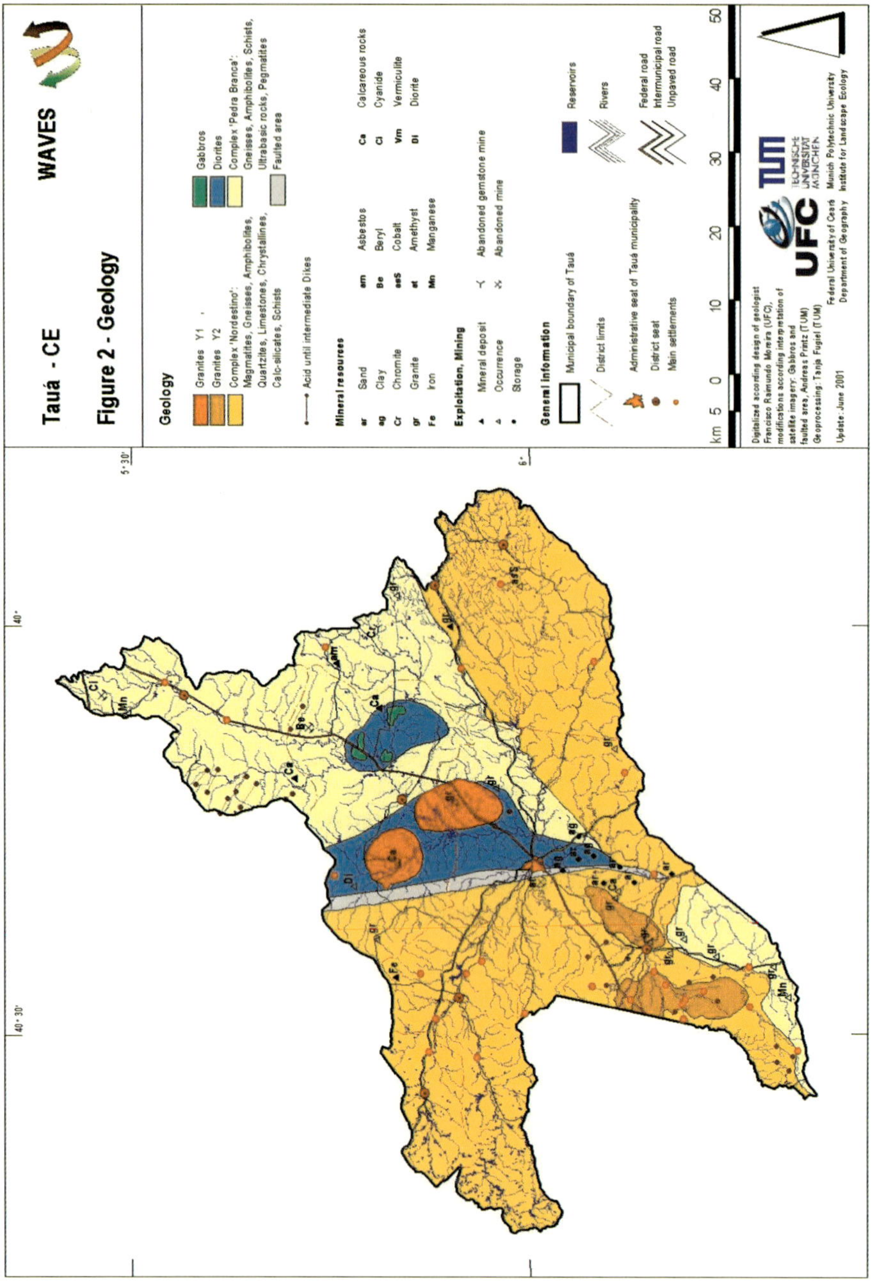

Fig. 2. Geology of the municipality of Tauá

The headwaters (e.g., Trici and Carrapateiras) of the Jaguaribe River, the largest and most important river in the Ceará region, are located in Tauá. The soils that developed on crystalline basement provide only low infiltration rates. Thus, the time and quantity of runoff is strongly correlated with intense precipitation events. All rivers in Tauá are intermittent and groundwater storage capacities are low (largely limited to fractures).

Methods

A four-step approach, analysis, aggregation, integration and application, combines both deductive („top down“) and inductive („bottom up“) methods (see Fig. 3):

1. Analysis phase: assessment of natural resources and anthropogenic impacts to provide a basic geo-ecological data set. Literature and cartographic studies lead to a series of thematic maps such as geology, relief, hydro-climatology and phytoecology.
2. Aggregation phase: selection and linking of the most important parameters. Erosion susceptibility is evaluated based on a detailed „Terrain Component“ map.
3. Integration phase: systematic integration of natural site and anthropogenic data to define geo-ecological units and degradation susceptibility.
4. Application phase: landscape assessment and development of strategies for sustainable land use.

In order to interpret the satellite image data, point and transect studies served as „ground checks“. Data was processed by a geographical information system (GIS) and linked to spatial units in order to provide the necessary data input for MOSDEL (see Printz and Lang, this Vol.). Results were then spatially represented.

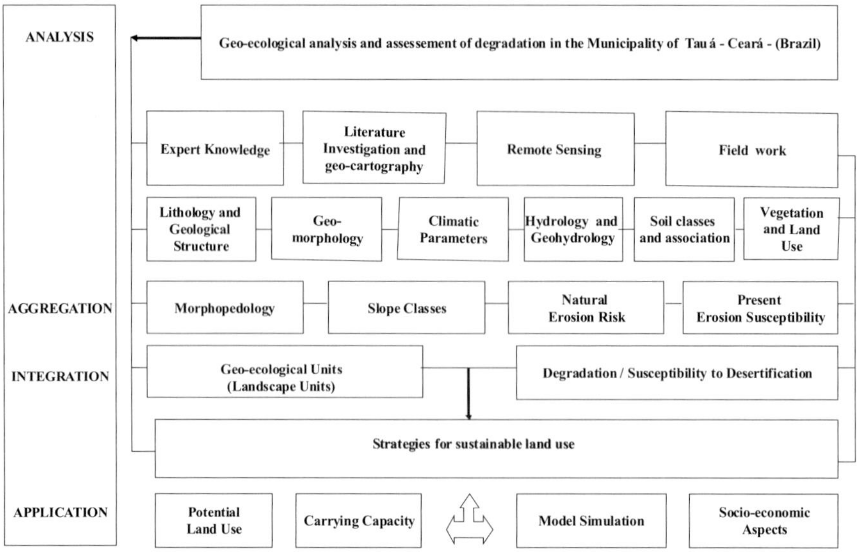

Fig. 3. Overview of the methodological procedure

Results and Discussion

The only comprehensive soil map for the state of Ceará has a scale of 1:600,000 (SUDENE 1972), not sufficiently detailed for a localized study. Another map of soil associations with higher resolution (SEAGRI 1987) was also available and seemed to be more appropriate for this purpose. In the legend, a detailed description of the soil types within the associations was given including their slope class. Using a digital elevation model, it was possible to disaggregate the overlaying SEAGRI map into slope classes. Thus, in cooperation with the soil science working group, a high resolution „Terrain Components“ map was generated (see Fig. 4). The number of mapping units in Tauá was increased from 9 (SUDENE 1972) to 42. During field surveys, simultaneously with the ground check of the satellite image interpretation, strong correlation of specific image signatures with „Vertisols“ was detected. The geometries derived by satellite image interpretation led to a further improvement of the soil map. This resulted in one of the best available soil maps at the municipal scale in the state of Ceará.

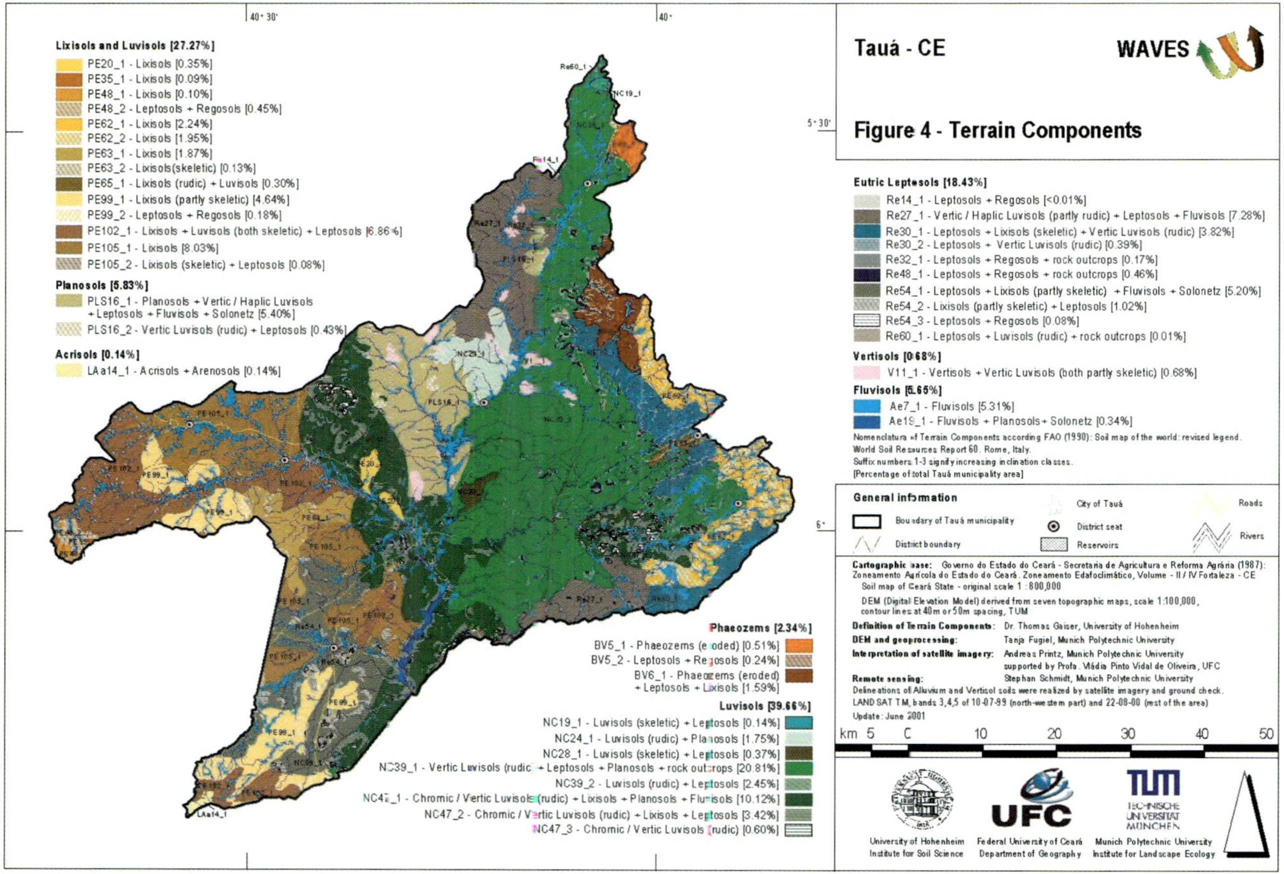

Fig. 4. Soil associations within the „Terrain Components" in the municipality of Tauá

Based on the spatially related „Terrain Components", inclination, and land use the erosion susceptibility was assessed. The general approach has already been described by de Oliveira and Printz (2000).

The „Terrain Component" database, SPICE (Gaiser et al., this Vol.), provides a k-factor value for each „Terrain Component". The k-factor characterizes the soil erosion susceptibility. Through cluster analysis, the „Terrain Components" were grouped into four classes. These classes were entered into a matrix together with four slope classes derived from SEAGRI (1987; see Table 1, Figs. 5 and 6).

Table 1. Natural erodibility [NE x], k-factors [K x] and slope classes [x %] matrix

k-factor	Class	Slope class	K 1	K 2	K 3	K 4
<0.2	K 1	0–3%	NE 1	NE 2	NE 3	NE 3
0.2–0.25	K 2	3–8%	NE 2	NE 2	NE 3	NE 3
0.251–0.32	K 3	8–20%	NE 2	NE 3	NE 4	NE 4
>0.32	K 4	>20%	NE 3	NE 3	NE 4	NE 4

The NE-values range from 1 to 4 (low, medium, high and very high). They were associated with the „Terrain Components" and were integrated into the natural erosion risk map (see Fig. 7).

The next step was to evaluate the actual erosion susceptibility by taking land use into account. This was done by overlaying natural erosion susceptibility with actual land cover units that were classified according to their ability to protect against erosion (see Table 2). Combining the matrix results (AE-values) with the „Terrain Components" map resulted in the identification of the spatial distribution of actual erosion susceptibility (see Fig. 8).

The spatial distribution of the classes is represented in Table 3. Obviously, the natural erodibility is already relatively high, as almost three-quarters of the area shows (very) high erosion susceptibility. Not surprisingly, current land use practices will further increase erosion susceptibility. However, the fact that more than one-third of the land lies in the highest class is a serious indicator of extreme ecological vulnerability. With the exception of steep mountain areas and intensely used alluviums, the most affected districts are those with the highest cattle densities.

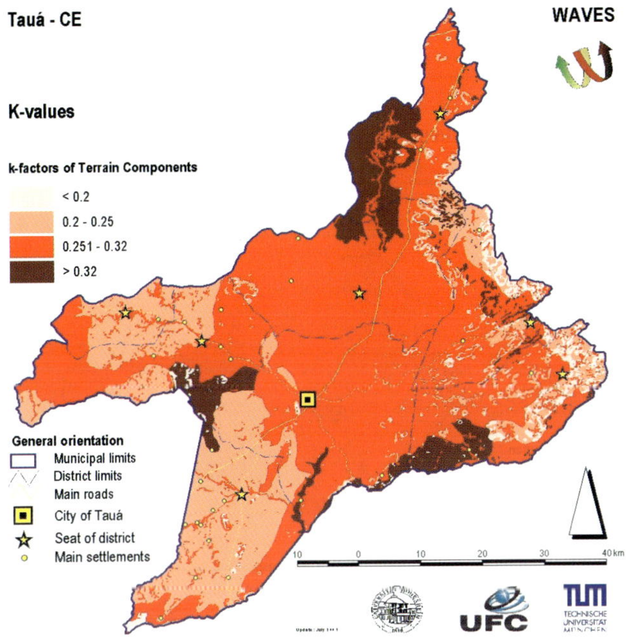

Fig. 5. Mean soil erodibility factor within the „Terrain Components"

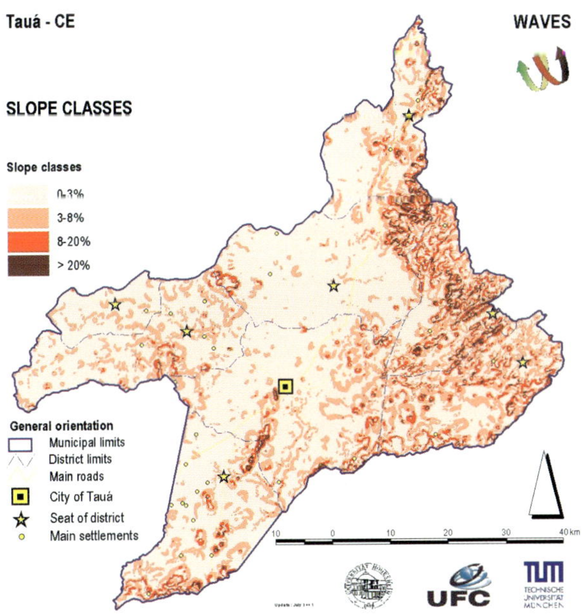

Fig. 6. Distribution of slope classes within the municipality of Tauá

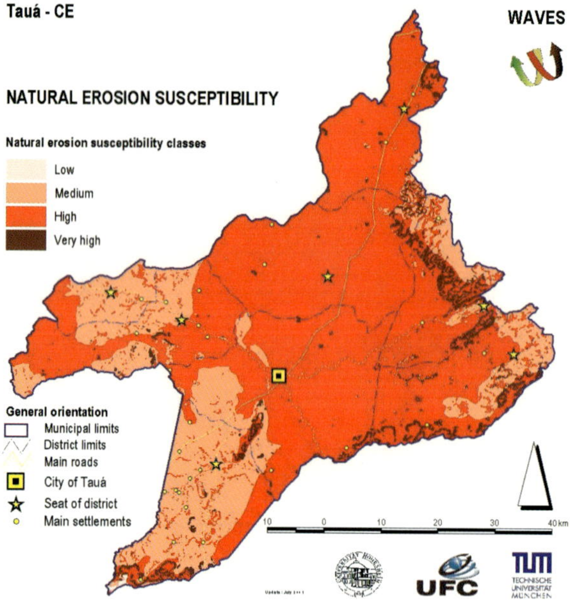

Fig. 7. Distribution of natural erosion susceptibility in the municipality of Tauá

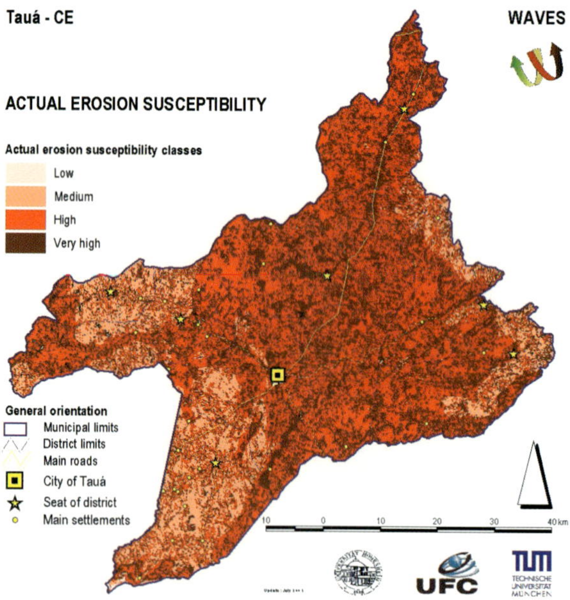

Fig. 8. Distribution of actual erosion susceptibility in the municipality of Tauá

Table 2. Evaluation of the vegetation units [classes V1–V4] and actual erosion [AE] susceptibility matrix

Vegetation class code	V 1	V 2	V 3	V 4		NE 1	NE 2	NE 3	NE 4
Dense tree caatinga with elements of Mata Seca	X				V 1	AE 1	AE 2	AE 3	AE 3
Dense tree caatinga	X				V 2	AE 2	AE 2	AE 3	AE 4
Tree-shrub caatinga		X			V 3	AE 3	AE 3	AE 4	AE 4
Dense shrub caatinga		X			V 4	AE 4	AE 4	AE 4	AE 4
Open shrub caatinga			X						
Alluvium				X					
Humid zone				X					
Succession or agriculture with dominant herbaceous vegetation				X					
Succession with dominant woody vegetation, plantation			X						
Bare soil with herbs, cactus and shrubs				X					
Bare soil with rocks				X					
Settlements and reservoirs		No data/ valorization							

Table 3. Distribution of area [naterclass: natural erosion susceptibility, acterclass: actual erosion susceptibility]

Description	Naterclass	km^2	%	Acterclass	km^2	%
Settlem./ reservoirs	nater_cl0	-	-	acter_cl0	16.9	0.4
Low	nater_cl1	3.0	0.1	acter_cl1	1.7	0.1
Medium	nater_cl2	1003.9	25.5	acter_cl2	705.2	18.0
High	nater_cl3	2775.0	70.6	acter_cl3	1783.4	45.4
Very high	nater_cl4	150.9	3.8	acter_cl4	1421.6	36.2

In order to gain a complete understanding of land use in Tauá, satellite imagery was used. Two scenes were necessary in order to cover all of Tauá (10.7.99 and 20.8.00; Landsat, bands 3,4,5, already rectified, 30×30m pixels). They were generously provided by IPLANCE (Fundação Instituto de Pesquisa e Informação do Ceará). After merging the two images, an unsupervised classification created 98 land use and vegetation classes. These were then aggregated through a supervised classification to create 13 final target classes (aggregated to 100×100m). Corrections were then made following ground truthing. Aggregating different „wet" classes along rivers allowed for the delineation of alluvial areas within Tauá. As a result, productive agricultural, as well as degraded areas (bare soil associated with shrub remnants, herbaceous layer, or rock), could be identified for the entire municipality. However, cultivated fields under rainfed agriculture showed the same spectral information as bare soil, whereas irrigated areas were clearly identifiable. This was due to the date of the imaging (July and August) being at the beginning of the dry season, and various interpretation problems emerged: the image is almost subdivided into two parts; the arid central plateau and the moist, mountainous regions, that had to be interpreted independently of the boarding mountains of the municipality. The vital chlorophyll reflection was quite different in this region and reflects the higher water availability for the vegetation at this time. This corresponds to the observed precipitation distribution in space and time, including a retarded harvest time in the mountain regions. Possible overestimations of the vegetation classes „Arboreal caatinga with relicts of Mata Seca" and the „Succession" classes should be checked in the future using multitemporal scenes.

The interpretation results served as an essential input for the „Model for sustainable development of land use" (Fig. 9; see also Printz and Lang, this Vol.). Highly degraded areas could be detected, the largest proportion of which were localized in the Carrapateiras district, where about 20% of the area belongs to the „bare soil" classes (Table 4), comprising more than one-third of the municipality's total classes. This corresponds to the greater number of cattle, erosion-susceptible soils, and low precipitation. The satellite interpretation served also to improve the soil and geology map. Irregular reflection and geometries were checked in the field and could partly be referred to the „Vertisol" soil class, which was in some cases correlated with the geological unit „Gabbro rocks".

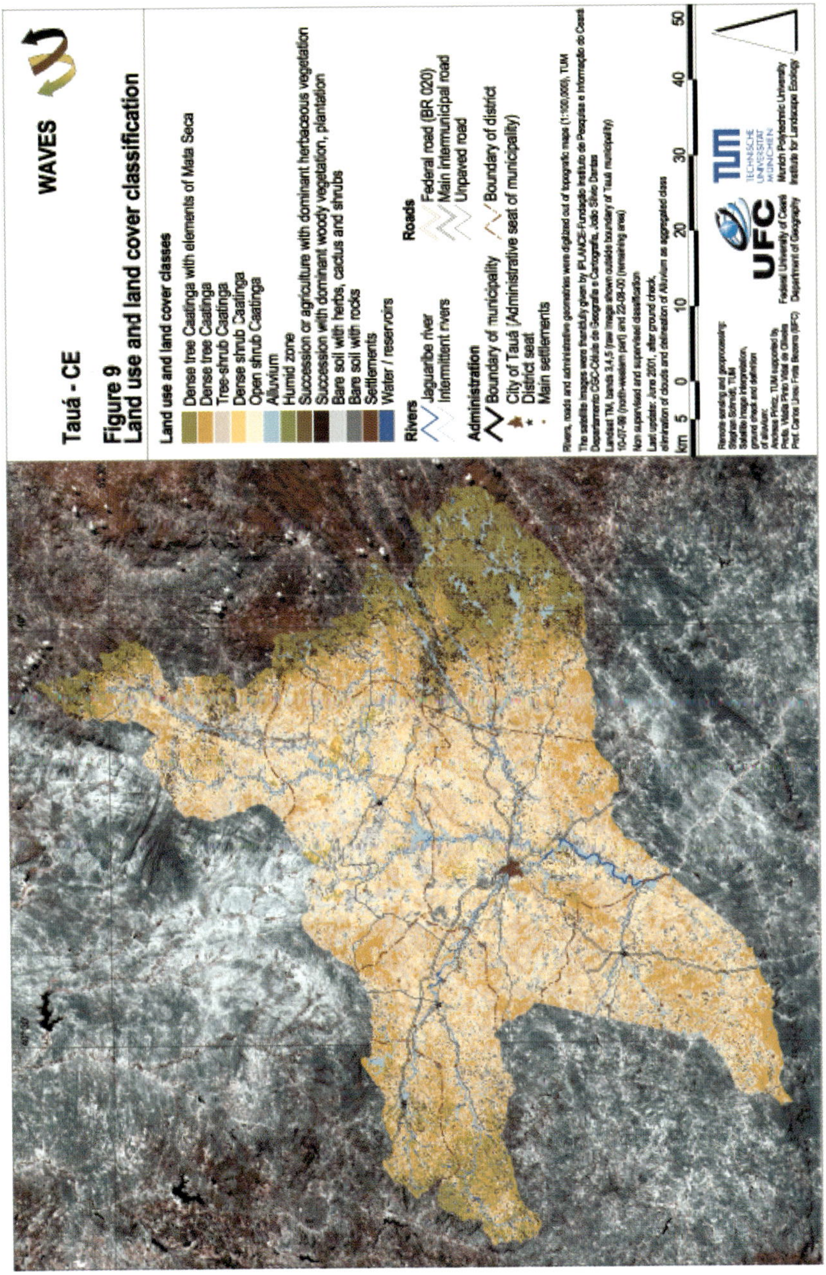

Fig. 9. Actual land use in the municipality of Tauá (Landsat TM image as background)

Table 4. Spatial distribution of vegetation and land use classes

Vegetation and land use classes	ha	%
Dense tree caatinga with elements of Mata Seca	44,470	11.31
Dense tree caatinga	75,670	19.24
Tree-shrub caatinga	81,575	20.74
Dense shrub caatinga	42,164	10.72
Open shrub caatinga	54,081	13.75
Alluvium	23,176	5.89
Humid zone	3,118	0.79
Succession or agriculture with dominant herbaceous vegetation	3,137	0.80
Succession with dominant woody vegetation, plantations	19,021	4.84
Bare soil with herbs, cactus and shrubs	30,399	7.73
Bare soil with rocks	14,766	3.75
Settlements	504	0.13
Reservoirs	1,216	0.31
Total	393,298	100.00

Geo-ecological units

Eight geo-ecological units were delineated and characterized by their dominant natural site characteristic (see Tricart 1977; de Oliveira 1990). Their potential and limitation for land use were evaluated and land use recommendations were given. Examples of two units are shown in Table 5 (see also Fig. 10).

Finally, a synoptic evaluation (see Table 6) of the geo-ecological units was generated based on the thematic numeric evaluation shown in Table 7.

Table 5. Example for characterization, valuation and recommendations for geo-ecological units

Geo-ecological units	Main natural characteristics	Environmental potentials and problems	Ecodynamic conditions/ evidence of degradation	Recomended use and sustainability	Degradation status
Occidental crystalline residual mountain of Pedra Branca	Dissected areas with elongated hills, moderately undulating until mountainous; interspersed with V-shaped valleys; shallow to moderately deep soils with high erosion risk; degradation of caatinga low to moderate.	Areas with Leptsols, Luvisols and Lixi- and Acrisols with low to medium natural fertility, steep slopes and high erosion susceptibility lead to high risks for agriculture.	Degradation of the vegetative cover may lead to accelerated erosion, causing soil loss, reduction of biodiversity and impair the land's production potential; transformation with instability effects; low recuperation ability of native vegeta-	Inadequate for agriculture, partially favorable for forestry; low sustainability.	Low degradation
Cipó-Carrapateiras intermountain depression („pediment corridor")	Plain surface, cutting various rock types of the Precambrian formation, dissected by low hills, gently undulating, interspersed with open valleys; shallow to moderately deep soils with middle erosion risk; moderately degraded tree and shrub caatinga.	On hill tops and ridges in gently sloped relief areas with Luvisols, middle to high natural fertility intermingle with Leptsols and rocky outcrops which means difficult agricultural conditions. Uncontrolled wood cuttings and extensive grazing may increase superficial runoff and degradation processes.	High instability; traditional livestock breeding and cotton cultivation lead to soil loss, reduction of biodiversity and agriculture production capacity.	Favorable for Sorghum cultivation and adapted pasture; melioration of water availability; increase of agricultural diversity; low sustainability due to accelerated processes of degradation with high risks for the future.	High degradation

Table 6. Evaluation rating of geo-ecological unities

Degradation Status	Value
Not degraded	8-11
Low degraded	12- 17
Moderate degraded	18-23
Highly degraded	24-30

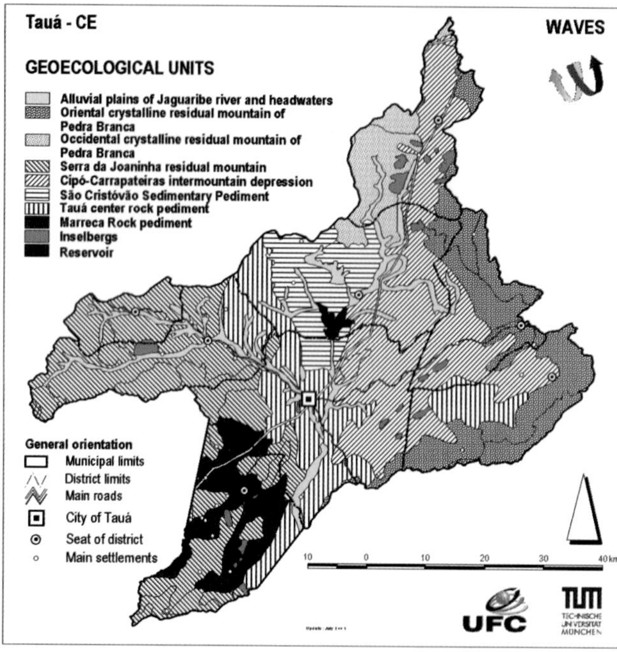

Fig. 10. Geo-ecological units

A public workshop to display the results to the public was organized by the Brazilian and German landscape ecology working groups, in cooperation with the local administration and agricultural school. In addition, all pertinent WAVES research was presented to approximately 300 citizens, including local representatives. Further cooperation and consultation of the administration was agreed upon for future municipal development. The landscape ecology groups were invited to show their methodology and results to the environmental committee of the state representatives, and discuss further implementation. The results of the workshop will be released in order to further distribute this model approach throughout Ceará. The WAVES program, and in particular the landscape ecology working groups, began initial steps to close the gap between scientists and policy makers in NE Brazil. The WAVES experience in Tauá showed that, with a common effort for application-oriented results and using integrated approaches, perspectives for sustainable development are becoming more feasible.

Table 7. Thematic numeric evaluation of geo-ecological unities (for interpretation of the total rating, refer to Table 6)

Geoecological unit	Oriental crystalline residual mountain of Pedra Branca	Occidental crystalline residual mountain of Pedra Branca	Serra da Joaninha residual mountain	Cipó-Carrapatei-ras intermountain depression	São Cristóvão sedimentary pediment	Tauá center rock pediment	Marreca rock pediment	Alluvial plains of Jaguaribe River
01 Vegetation								
Vegetation layer	1	2	2	4	3	3	3	1
Plant cover	2	1	2	4	3	3	3	3
Pasture intensity	3	2	1	4	4	5	3	3
02 Geomorphology								
Weathering	2	1	1	2	2	2	2	4
04 Hydric erosion								
Drainage density	3	2	2	3	2	2	2	2
Run off	3	2	2	3	2	3	3	2
05 Actual erosion risk								
Evidences	3	2	1	4	5	4	4	2
06 Soil characteristics								
Soil depth	1	2	1	3	3	3	4	1
Total	18	14	12	27	24	25	24	18
Degradation status	Moderate degradation	Low degradation	Low egradation	High degradation	High degradation	High degradation	High degradation	Moderate degradation

References

De Oliveira VPV (1990) Zoneamento Geo-Ambiental do Sertão de Quixeramobim-Ce. Coleção Mossoroense Série C – V (DXX), Mossoró

De Oliveira VPV, Printz A (2000) Analysis of geoambiental structures and degraded areas in municipality of Tauá. Proceedings of the IIIrd international WAVES workshop in Freising, Weihenstephan, 13–16 March 2000, Freising

IBGE (2000) Synopse preliminar do Censo demográfico 2000, vol 7. Data-CD: tabelas/Ce/Uf/tabela16.xls, Rio de Janeiro

IBGE (Fundação Instituto Brasileiro de Geografia e Estatística) (1998a) Censo agropecuário 1995–1996, no 9. Ceará, Rio de Janeiro

IBGE (Fundação Instituto Brasileiro de Geografia e Estatística) (1998b) Base de informações municipais (1996) [CD-ROM]. Rio de Janeiro

PNUD (1998) Desenvolvimento humano e condições de vida. Indicadores Brasileiros, Brasília

RADAM (1981) Projeto RADAMBRASIL, 1981. Levantamento de Recursos Naturais. Folha SA-24/25 Jaguaribe/Natal. Mme SG vol 23, Departamento Nacional de Produção Mineral, Rio de Janeiro

SEAGRI (Secretaria de Agricultura e Reforma Agrária) (1987) Zoneamento agrícola do Estado do Ceará. Zoneamento edafoclimático, vol II/IV. Mapa de Solo, Fortaleza

Souza MJN (2000) Bases naturais e esboço do zoneamento geoambiental do Estado do Ceará. In: Compartimentação Territorial e Gestão Regional do Ceará, part I. Funece, Fortaleza

SUDENE (1972) Mapa exploratório – reconhecimento de solos do Estado do Ceará (1:600.000). Superintendência do Desenvolvimento do Nordeste, Recife

Tricart J (1977) Ecodinâmica. Fibge-Supren, Rio de Janeiro

Cerrado and Caatinga in the Picos Area

Antonio Alberto Jorge Farias Castro, Andreas Printz, Maura Rejane de Araújo Mendes, Francisco de Assis Rodrigues Soares, José Orlando Soares Oliveira, Rigoberto Sousa Albino, Frank-Michael Lange and Ruth Raquel Soares de Farias

Universidade Federal do Piauí (UFPI), Núcleo de Referência em Ciências Ambientais do Trópico, Ecotonal do Nordeste (TROPEN), CEP 64049-550 Teresina-Piauí, Brazil, aajfcastro@uol.com.br

Introduction

Defining ecosystem vulnerability is a central focus of the WAVES program, and the ecology of plant systems is a key component in addressing this issue. Accordingly, the WAVES working group in landscape ecology has concentrated its research on the vegetation ecology of the Piauí region.

Research concerning the caatinga biome, especially within the regional context of the Piauí district, is still in its initial stages. For example, until now, only three botanical studies have been undertaken within the entire Picos region: a complementary floristic survey (Goergen 1983); a botanical economy study (Jenrich 1989); and a floristic and phytosociological study concerned only with the cerrado „sensu lato“ (Castro 1994), all of which were conducted in the municipality of Tanque de Piauí (until 1996, part of the municipality of Oeiras).

A pioneer study concerning the floristic and phytosociological aspects of arboreal-shrubby caatinga is needed for a better understanding of this ecosystem and its vulnerability. Therefore, four sampling areas within the WAVES focus region of Picos, concentrated around the Bocaina Reservoir, were selected. The results of this survey are presented here. In addition, this research contributes to the assembly of the herbarium (TEPB) of the Federal University of Piauí (UFPI), an important collection comprising 429 species (Tropen 2000).

Material Studied, Area Description and Methods

The WAVES focus region in Picos comprises an area of about 3000 km^2 (1.2% of the total area of the State of Piauí). It is composed of four municipalities (as of 1996), with a population of about 91,000 and with a demographic density of 30.5 inhabitants/km^2. This can be considered very high when compared with the mean of Piauí (11.25 inh./km^2). However, population and agriculture are concentrated mainly in or around the urban areas of Picos and along the fertile Guaribas valley (due to its good groundwater accessibility). The drier plateaus of the caatinga biome are mainly used for pasture, whose intensity diminishes as distance from roads increases.

The semiarid caatinga biome reaches its natural distribution limits in Picos, rapidly transforming as one heads westward through a transition zone into the subhumid cerrado biome. This is primarily due to a steep precipitation gradient, which is especially noticeable on the windward cliffs of the plateaus to the west of the Guaribas valleys, an area significantly represented by cerrado, „sensu lato".

The selection of the sample areas considered logistics (accessibility) as well as the ecological conditions, indicating low levels of anthropogenic influence (see characteristics of the sample areas in Table 1).

Table 1. Details of the four sample areas. A1 (heights of Sangradouro), A2 (Curral Velho Hill), A3 (Borges Hill), and A4 (Baixio Hill). (1) Semiarid without any water excess, fourth megathermic. (2) Semiarid with slight excess in the autumn, third megathermic

Parameters	Areas			
	A1	A2	A3	A4
Geographical coordinates	06°54'S, 41°18'W	06°54'S, 41°17'W	07°02'S, 41°32'W	06°51'S, 41°28'W
Altitude (m)	333	260	236	500
Municipality [1997]	Bocaina	Bocaina	Susssuapara	São José do Piauí
Vegetation	Caatinga Arboreal-Shrubby	Caatinga Arboreal-Shrubby	Caatinga Arboreal-Shrubby	Caatinga Arboreal-Shrubby
Minimum, average and maximum temperatures (°C)	21.6, 27 and 33.2	21.6, 27 and 33.2	21.6, 27.2 and 33.2	20.4, 25.6 and 31.8
Annual Precipitation (mm)	687.9	687.9	704.2	816.4
Climatic Typology	$Dd_1A'_4a'$ (1)	$Dd_1A'_4a'$ (1)	$Dd_1A'_4a'$ (1)	$Dd_2A'_3a'$ (2)
Soils	Litholic soils+yellow-red Podzolics	Litholic soils+red Latosols	Litholic soils	Litholic soils

The floristic survey was conducted with an unsystematic, preferential collecting method (Castro 1994) between January, 1999, and June, 2000. The phytosociological survey, in turn, was conducted with the quarter method (Cottam and Curtis 1956; Martins 1991), using a magnetic needle, precision altimeter, GPS, tape measure, string, galvanized wire handles, labels and wooden pickets. A total of 50 study points were allocated to each area. Each point was separated by 10 m and the points were sequentially distributed along a transect determined by the physiognomic state of the vegetation. For each quadrangle, only one ligneous individual (tree, bush, or liana) was selected. The orientation of the quadrangles was randomized with the help of a mobile orientation cross, located at the central point of each quadrangle picket. Individual plant species, with stalk diameter of greater or equal to 3 cm (at soil level), were included in the sample.

All individuals sampled were numbered and collected during the field examination. Samples with reproductive material were collected with the help of pruning scissors, and were then properly catalogued in the herbarium by specialists, according to standard techniques (Castro 1987, 1994). The botanical identifications were done at the Núcleo de Referência em Ciências Ambientais do Trópico Ecotonal do Nordeste (TROPEN), in the Graziela Barroso Herbarium (TEPB) and at other herbariums, through the examination of exsiccate. Voucher specimens were deposited in the TEPB of the Department of Biology of the UFPI. The orthography of the families and botanical genera followed Brummitt (1992), and the abbreviation of the authors of the plants' names followed Brummitt and Powell (1992).

The floristic representation of the samples was accompanied by the collector's curves, real and estimated (Mueller-Dombois and Ellenberg 1974; Castro 1987, 1994). The Whittaker's species diversity indices (WDIs) and Shannon's species diversity indices (SDIs) were also measured (Magurran 1988). The importance values (IVs) of the sampled species and the SDIs for the sample communities were calculated, based on the package FITOPAC, developed by Prof. George John Shepherd of the State University of Campinas (UNICAMP).

Information on climatic elements and soils was obtained from the SUDENE (1990), Jacomine (1986), Krishnan (1980), Nimer and Brandão (1985, 1989), Thornthwaite and Mather (1955), and Varejão-Silva and Reis (1988).

Results and Discussion

Altogether 58 species were sampled. In addition to this, 8 species were recorded (merely observed). A total of 62.1% of the sampled species were classified completely, 25.8% were classified to the level of genera, and 12.1% of the species were left without any taxonomic classification. This reflects the fact that the taxonomic knowledge of the plants typically found in the caatinga, and especially within transition areas, is not satisfactory, and suggests that more taxonomic studies need to be conducted.

As seen in Table 2, the species ***Combretum leprosum*** (Combretaceae), ***Croton sonderianus*** (Euphorbiaceae), ***Machaerium acutifolium*** (Fabaceae), ***Guapira*** sp. (Nyctaginaceae), ***Rollinia leptopetala*** (Annonaceae) and ***Tabebuia impetiginosa*** (Bignoniaceae) were the only species that consistently appeared in all areas.

Table 2. Sampled species of the four survey areas. NI (number of individuals sampled). A1 (heights of Sangradouro), A2 (Curral Velho Hill), A3 (Borges Hill) and A4 (Baixio Hill)

Species	Common names	NI			
		A1	A2	A3	A4
Agonandra brasiliensis Miers	pau-marfim	-	-	-	4
Alibertia rigida K.Schum. (3)	maria-preta	-	2	-	0
Amburana cearensis (Fr. Allem) A.C. Smith (1)	imburana-de-cheiro	0	-	-	0
Anadenanthera colubrina var. cebil (Griseb.) Altschul	angico-preto	1	-	2	4
Annona coriacea Mart. (1)	-	0	-	-	-
Apterokarpos gardneri (Engler) Rizzini	aroeira-brava	1	-	-	-
Aspidosperma pyrifolium Mart.	pereiro-preto	21	7	17	-
Aspidosperma cuspa (Kunth) Blake	pereiro-branco	8	13	5	-
Aspidosperma cf. refractum Mart.	pequiá	-	2	3	-
Aspidosperma sp.	pequiá	-	-	-	10
Bauhinia cheilantha (Bong.) Steud.	mororó	-	-	1	9
Bauhinia pulchella Benth.	mirorozinho	-	1	-	-
Bauhinia sp. nov.	mororó	-	-	-	9
Bignonia sp.	cipó-de-arco	-	3	-	-
Caeasalpinia bracteosa Tul.	catingueira	9	-	36	-
Caesalpinia ferrea Mart. ex Tul.	pau-ferro	1	-	-	-
Callisthene fasciculata Mart.	capitão-de-campo	4	-	-	-
Callisthene microphylla Warm.	carucim	-	15	-	-
Cenostigma gardnerianum Tul.	canela-de-velho	-	-	1	-
Cereus jamacaru DC. (2)	mandacaru	-	-	0	1
Cochlospermum vitifolium (Willd.) Spreng.	algodão-bravo	3	-	-	2
Combretum duarteanum Camb.	sipaúba	-	-	1	-
Combretum leprosum Mart.	mMofumbo	2	17	19	6
Combretum mellifluum Eichler	sipaúba	4	7	-	18
Commiphora leptophloeos (Mart.) Gillet	amburana-de-cambão	2	-	1	-
Copaifera sp.	pau-d'óia	-	1	-	-
Cordia alliodora Cham.	Frei-jorge	-	-	-	7
Coutaria hexandra K.Schum.	quina-quina	-	3	-	-
Croton sonderianus Muell. Arg.	marmeleiro-preto	116	49	84	8
Croton cf. zehntneri Pax ex Hoffm.	-	-	-	-	3
Dalbergia cearensis Ducke	birrote	2	21	-	-
Dyptychandra aurantiaca ssp. epunctata (Tul.) Lima, Carvalho & Costa	birro-de-caibo	1	3	-	-
Erythroxylum sp.1	carrasquinho	-	20	-	-
Erythroxylum sp.2	carocim	-	-	-	2
Godmania dardanoi (J.C. Gomes) A.H. Gentry	venta-de-tamanduá	-	-	-	2
Hancornia speciosa Gomez	mangaba	-	-	-	1
Helicteres baruenses Jacq.	guaxum	2	-	-	2

Table 2 (continued):

Species	Common names	NI			
		A1	A2	A3	A4
Hymenaea stigonocarpa Mart. ex Hayne (1)	jatobá	0	-	-	-
Hymenaea sp. (1)	jatobá	-	-	-	0
Machaerium acutifolium Vog.	violete-preto	11	11	4	4
Manihot sp.	maniçoba	-	-	-	2
Mimosa verrucosa Benth.	jurema-rasteira	-	1	-	-
Myracrodruon urundeuva Allemão (3)	aroeira	-	-	3	0
Piptadenia moniliformis Benth.	angico-de-bezerro	-	1	1	8
Piptadenia stipulacea (Benth.) Ducke	jurema-branca	-	4	2	3
Guapira sp.	pau-piranha	1	1	2	2
Poeppigia procera Presl.	-	-	-	2	-
Pseudobombax tomentosum Mart. & Zucc.) A. Robyns (1)	imbiratanha	0	-	-	0
Rollinia leptopetala (R.E.Fries) Safford	bananinha-brava	1	2	2	12
Senna cearensis A. Fern.	oca	2	-	-	43
Senna trachypus (Benth.) H. Irwin & Barn.	canafistinha	2	7	-	1
Senna trachypus (Benth.) H. Irwin & Barn.	canafistinha	2	7	-	1
Senna sp.	cabo-de-inchó	-	-	1	-
Spondias tuberosa Arruda (1)	-	-	-	0	0
Tabebuia impetiginosa (Mart. ex. DC.) Standl.	pau-d'arco-roxo	5	4	3	8
Tabebuia serratifolia (Vahl.) Nich. (1)	pau-d'arco-amarelo	-	-	-	0
Terminalia fagifolia Mart. & Zucc. (1)	chapada	-	-	-	0
Ximenia americana L. (3)	ameixa-brava	1	3	-	0
Unknown 1	-	-	-	7	-
Unknown 2	birro	-	-	-	10
Unknown 3	-	-	-	-	6
Unknown 4	-	-	-	-	3
Unknown 5	-	-	-	-	2
Unknown 6	-	-	-	-	1
Unknown 7	-	-	-	-	1
Unknown 8	-	-	-	-	1

(1) Observed species. (2) Observed species in A3. (3) Observed species in A4

In areas A1 and A2, 22 and 25 species were sampled, respectively. A total of 13 common species were found, 9 were unique to the area A1 and 12 to A2. In the two areas, 34 different species were sampled, including 26 genera and 17 families. Mimosaceae, Caesalpiniaceae and Apocynaceae were the most important families, with 7, 5 and 3 species represented, respectively. Bignoniaceae, Combretaceae, Fabaceae, Rubiaceae and Vochysiaceae appeared with two species each.

In area A3, 22 species were sampled, including 18 genera and 11 families. Floristically, Caesalpiniaceae, with 5 species, and Mimosaceae, with 4, were the most representative. Apocynaceae, Bignoniaceae and Euphorbiaceae were present with 3 species each.

In area A3, 33 species were sampled, including 21 genera and 17 families. At this site, Caesalpiniaceae and Euphorbiaceae (6 species each), Mimosaceae (5 species) and Bignoniaceae (4 species) were the families with the highest number of species. A species of „mororó“, sampled at Baixio Hill (A4), still remains under analysis for botanical determination (***Bauhinia*** sp., Caesalpiniaceae). This could indicate a new species (VAZ, 2000, pers. commun.).

Table 2 shows the distribution of all species, their common names, and the number of individuals sampled, listed according to species and area. Many of the species and families represented characterize the semiarid part of Piauí very well. However, the presence of species such as ***Agonandra brasiliensis*** (Opiliaceae), ***Annona coriacea*** (Annonaceae), ***Callisthene fasciculata*** (Vochysiaceae), ***C. microphylla***, ***Cenostigma gardnerianum***, ***Hymenaea stigonocarpa*** (Caesalpiniaceae), ***Combretum mellifluum***, ***Terminalia fagifolia*** (Combretaceae), ***Hancornia speciosa*** (Apocynaceae), ***Machaerium acutifolium*** (Fabaceae), ***Mimosa verrucosa***, ***Piptadenia moniliformis*** (Mimosaceae) and ***Pseudobombax tomentosum*** (Bombacaceae) are also indicative of the transition zone, as their presence implies the establishment of cerrado „sensu lato“ species and semideciduous forest.

In phytosociological terms, area 1 has 6 dominant species. Their IVs exceed 10%, taking into account their relative density, frequency and dominance (basal area). At area A2, 8 species are dominant. In the same way, at areas 3 and 4, 6 and 9 species, respectively, emerge as dominants, based on their IV. Table 3 shows these species, their IVs and their sociological positions (SPs – ranked importance value), based on these same values.

A variety of the quince tree, ***Croton sonderianus*** (Euphorbiaceae), was the only one that appeared in three of the four areas, occupying the same first sociological position. In area A4, ***Senna cearensis*** (Caesalpiniaceae) occupies the first position.

The minimum, average, and maximum heights of the sampled individuals were 1.5, 4.3 and 11.0 m in A1; 1.5, 4.2 and 11.0 m in A2; 1.6, 5.5 and 16.0 m in A3 and 1.6, 6.0 and 17.0 m in A4, respectively. The distribution of the height classes illustrates a progressive increase in the upper height classes from sample A1 to A4 (see Fig. 1). This will be further interpreted in the conclusions.

As an emergent species in the first layer, ***Anadenanthera colubrina*** var. ***cebil*** (Mimosaceae) is prominent in A1 and A3. The species ***Commiphora leptophloeos*** (Burseraceae) also appears as an emergent in A1. The species ***Dalbergia cearensis***, in this sense, is the only one that stands out in A2. In A3, ***Machaerium acutifolium*** (Fabaceae) appears to belong to the first layer. In A1 and A4, ***Tabebuia impetiginosa*** (Bignoniaceae) is also included in this category. The climbing bush, ***Combretum mellifluum*** (Combretaceae), appears as an emergent in A4.

Table 3. More important species, importance values (IVs) and sociological positions (SPs) of the species with base in the IV of the four survey areas. A1 (heights of Sangradouro), A2 (Curral Velho Hill), A3 (Borges Hill) and A4 (Baixio Hill)

SPECIES (1)	AREAS							
	A1		A2		A3		A4	
	IV (%)	SP	IV (%)	SP	IV (%)	SP	IV (%)	SP
Anadenanthera colubrina var. Cebil	-	-	-	-	15.33	5	-	-
Aspidosperma pyrifolium	28.73	2	-	-	29.89	4	-	-
Aspidosperma cuspa	-	-	13.65	7	-	-	-	-
Aspidosperma cf. refractum	-	-	-	-	-	-	11.39	6
Bauhinia cheilantha	-	-	-	-	-	-	10.14	9
Caesalpinia bracteosa	15.79	4	-	-	46.12	2	-	-
Callisthene microphylla	-	-	27.52	4	-	-	-	-
Combretum leprosum	-	-	23.05	5	35.14	3	10.72	8
Combretum mellifluum	-	-	10.41	8	-	-	30.50	2
Commiphora leptophloeos	13.56	5	-	-	-	-	-	-
Croton sonderianus	154.59	1	70.06	1	99.39	1	-	-
Dalbergia cearensis	-	-	36.29	2	-	-	-	-
Unknown 1	-	-	-	-	13.43	6	-	-
Unknown 2	-	-	-	-	-	-	10.87	7
Erythroxylum sp.1	-	-	29.31	3		-	-	-
Machaerium acutifolium	16.35	3	17.22	6	-	-	-	-
Piptadenia moniliformis	-	-	-	-	-	-	21.82	4
Rollinia leptopetala	-	-	-	-	-	-	16.80	5
Senna cearensis	-	-	-	-	-	-	59.74	1
Tabebuia impetiginosa	11.28	6	-	-	-	-	23.69	3

(1) Species with IV larger than 10%

The minimal diameters were determined by inclusion criteria (DNS greater or equal to 3 cm). The average and maximum diameters range from 10.2 to 43.9 cm in A1, from 9.7 to 33.4 cm in A2, from 11.7 to 62.1 cm in A3, and from 11.0 to 50.9 cm in A4.

The ranking of sample sites by increasing diameter classes (from A1 to A4) corresponds to the ranking of the height classes (see Fig. 2). With the exception of the general ranking interpretation (see Conclusions), the maximum diameter may also be associated with other factors, such as regrowth after cutting, or as a result of methodological limitations due to the estimation of absolute dominations (through basal areas) of the individuals sampled.

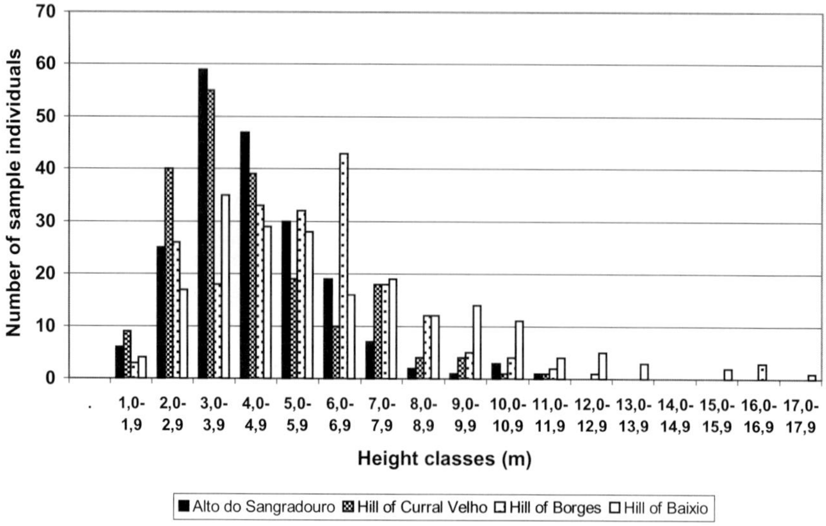

Fig. 1. Height classes of registered samples

The total densities for the areas sampled were 2245 individuals/ha for A1; 2532 individuals/ha for A2; 2900 individuals/ha for A3; and 1432 individuals/ha for A4. All areas consist of arboreal-shrubby caatinga but vary in terms of vegetation density; from A3 (most dense), A2, A1 to A4 (least dense).

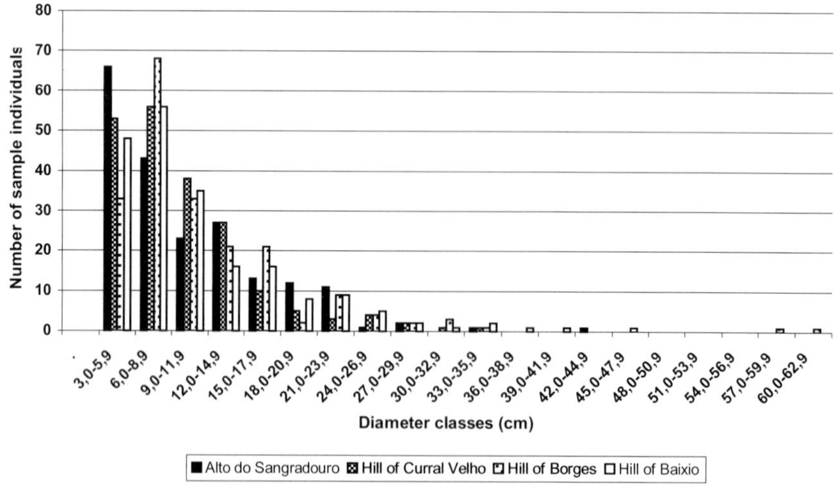

Fig. 2. Diameter classes of registered samples

The collector's curves indicate that at A1 the number of sample points taken were sufficient. In A2, more sample points should have been taken, but unfortunately this process was hindered by logistical problems. At A3 and A4, the 50 sampling points were sufficient. However, as these areas have been preserved, an increase in the number of sampling points is recommended.

Whittaker's diversity indices were 4.2, 4.7, 4.2 and 6.2 for areas 1, 2, 3 and 4, respectively. Shannon's indices of species diversity were 1.75, 2.63, 2.03 and 3.02 nats/ind, for the same areas and sequence. For the caatinga of Brazil's semiarid northeast, these numbers are relatively high, especially at A4 (Baixio Hill). This can most likely be attributed to better levels of conservation in the caatinga of Piauí.

Economically, 8 of the 20 dominant species (40.0%) are used intensively by the local population: ***Anadenanthera colubrina*** var. ***cebil*** (for construction, apiculture, and medicinal purposes), ***Aspidosperma pyrifolium*** (for furniture), ***Bauhinia cheilantha*** (forage), ***Combretum leprosum*** (firewood), ***Commiphora leptophloeos*** (furniture, workmanship), ***Machaerium acutifolium*** (construction, medicine), ***Piptadenia moniliformis*** (tools, apiculture) e ***Tabebuia impetiginosa*** (construction, apiculture, ornamental), according to Almeida et al. (1998) and Lorenzi (1998).

Conclusions

Although the dominant vegetation type in Picos is caatinga, the region is characterized by an ecotone situation due to the existence of a state of ecological tension in Piauí (Rivas 1996). This study reveals a continuous east-to-west ecosystem transition from caatinga to semideciduous woods to cerrado „sensu lato". The ecological analyses of the four sampled areas reflect this situation.

The ranking of the areas showed an almost linear increase in height, diameter and diversity for the areas 1 to 4 (from east to west). These findings further correlate with higher rates of precipitation, as A3 and A4 are situated within the near side of the windward cliffs along the western plateaus. Moreover, higher elevation and lower mean annual temperatures also contribute to a greater presence of semideciduous woods and cerrado „sensu lato" species (which generally surpass caatinga species in respect to height and diameter). The increasing transition influence also leads to greater natural diversity, which may be eclipsed by (anthropogenically induced) succession effects (e.g., A2).

Anthropogenic disturbances decline as precipitation increases between A1 and A4. A4, in particular, revealed a mature and undisturbed situation. In contrast, nearby A1 and A2 (which border the Bocaina Reservoir) the highest impact rates (clearcuts and grazing) were recorded. In A2, the disturbances are older and the values recorded reflect the recuperating effects of successional growth.

Ecological analysis methods prove to serve as a valuable tool for mapping vegetation types as well as ecological degradation. However, long-term research is needed to survey a greater number of sites and to collect typical mean values.

The studies of the arboreal shrubby caatinga (including the transition zones) show a considerable range of species diversity. The proximity of a variety of different biomes of large area and low elevation is common for Piauí, but unique for northeastern Brazil. In particular, the marginal caatingas of Piauí are considered unique because they grow on sedimentary soils (in contrast to the crystalline home domain of the Caatinga) as well as the ecotone (transition) zone situation.

The amount of anthropogenic influence is significant. The historical development and the influence of uncontrolled and unplanned settlement dynamics in the Picos region indicate that ecological planning at the landscape scale is urgently needed. Strategies and zones for conservation, cultivation and sustainable development have to be defined. The benefits and importance of biodiversity need to be addressed and incorporated into public policies that directly involve and affect the local population, especially in the light of how unique and fragile the Brazilian caatinga is.

Because of the direct influence of the WAVES program and the „Workshop about the evaluation and identification of priority actions for the conservation, sustainable utilization and distribution of the benefits of the biodiversity of the caatinga biome", which took place in Petrolina, Pernambuco, in May 2000, this region has been selected as one of several priority areas for the conservation within the caatinga biome.

References

Almeida SP, Proença CEB, Sano SM, Ribeiro JF (1998) Cerrado: espécies vegetais úteis. Embrapa/CPAC, Planaltina

Brummitt RK (1992) Vascular plant; families and genera. Royal Botanic Gardens, Kew

Brummitt RK, Powell CE (1992) Authors of plant names. Royal Botanic Gardens, Kew

Castro AAJF (1987) Florística e fitossociologia de um Cerrado marginal brasileiro, parque estadual de Vaçununga, Santa Rita do Passa Quatro – SP. MSc thesis. Unicamp, Campinas

Castro AAJF (1994) Comparação florístico-geográfica (Brasil) e fitossociológica (Piauí – São Paulo) de amostras de Cerrado. PhD thesis. Unicamp, Campinas

Cottam G, Curtis JT (1956) The use of distance measures in phytosociological sampling. Ecology 37:451–460

Goergen J (1983) Critérios ecológicos para o desenvolvimento de modelos de aproveitamento agrícola adaptados a regiões das chapadas no Piauí central, Brasil. DNOCS/GTZ, Teresina (mimeographed)

Jacomine PKT (coord) (1986) Levantamento exploratório de reconhecimento de solos do Estado do Piauí, vol 2. Embrapa/SNLCS, Rio de Janeiro. SUDENE/DRN, Recife

Jenrich H (1989) Vegetação arbórea e arbústea nos altiplanos das chapadas do Piauí central; características, ocorrências e empregos. DNOCS/GTZ, Teresina

Krishnan A (1980) Agroclimatic classification methods and their application to Índia. In: Virmani SM, Silvakumar MVK, Rosenberg G, Kumble V (eds) Climatic classification: a consultant's meeting. ICRISAT, Patanchern

Lorenzi H (1998) Árvores brasileiras: manual de identificação e cultivo de plantas arbóreas nativas do Brasil, 2 vols. Plantarum, Nova Odessa
Magurran AE (1988) Ecological diversity and its measurement. Princeton University Press, Princeton
Martins FR (1991) Estrutura de uma floresta mesófila. Edunicamp, Campinas
Mueller-Dombois D, Ellenberg H (1974) Aims and methods of vegetation ecology. John Wiley, New York
Nimer E, Brandão AMPM (1985) Balanço hídrico anual a partir de valores normais e tipologia climática. Rev Brasil Geograf 47:373–416
Nimer E, Brandão AMPM (1989) Balanço hídrico e clima da região dos Cerrados. IBGE, Rio de Janeiro
Rivas MP (coord) (1996) Macrozoneamento geoambiental da bacia hidrográfica do rio Parnaíba. Série estudos e pesquisas em geociências 4. IBGE, Rio de Janeiro
SUDENE (1990) Dados pluviométricos mensais do Nordeste; Estado do Piauí. Pluviometria 2. SUDENE/DPG/HME, Recife
Thornthwaite CW, Mather JR (1955) The water balance. Publ Climatol 8:1–104
Tropen (2000) Núcleo de Referência em Ciências Ambientais do Trópico Ecotonal do Nordeste, 2001. Relatório de Atividades 2000. Tropen, Teresina (mimeographed)
Varejão-Silva MA, Reis ACS (1988) Agrometeorologia e climatologia tropicais, Módulo 1.1. Abeas, Brasília, pp 86–89

Climate and Desertification in the Picos Region

M. C. L. Sales, V.M. Ramos and J.G.B. de Oliveira

Universidade Federal do Piauí (UFPI), Núcleo de Referência em Ciências Ambientais do Trópico, Ecotonal do Nordeste (TROPEN), CEP 64049-550 Teresina-Piauí, Brazil, mclsales@uol.com.br

Introduction

Currently, the desertification process is gaining increasing attention in the context of the global environment, due to the extension of the affected areas, as well as the promotion of the desertification of the productive system, that is causing the usual strong social and economic impacts that extend far beyond the boundaries of the regions directly affected.

The consolidation of the concept of desertification as the degradation of lands in arid, semiarid and subhumid dry areas resulting from various factors, including climatic variation and human activities, allows us to treat the environmental problems in these zones from the perspective of desertification. This way, these studies find the theme of desertification a privileged *locus* to shelter all sets of information about the natural, social, cultural and economic bases related to them, allowing the framing of the different arrangements resulting from the complex society–nature interaction in the dry regions of the planet.

The northeast region of Brazil (that comprises an area of 1,556,000 km^2) can be characterized from a geoenvironmental point of view by the diversity of its landscape. The climate constitutes a remarkable element in the northeastern landscape, especially if we consider the spatial and temporal variations of the precipitation. It is a semiarid climate of azonal character, marked by the frequent occurrence of droughts that impose deep limitations on the region's productive system.

According to Nobre and Moura (1985), this variability is directly linked to anomalous circulations and to the distribution of the sea surface temperatures (SSTs) of the tropical oceans, especially the Atlantic and the Pacific. Brabo et al. (1993) also refer to many investigations that relate the influence of the SSTs of the Pacific and the Atlantic tropical oceans to the regime of rainfall in the northern sector of the Brazilian northeast, especially the studies that relate the coupling ocean–atmosphere in years of the occurrence of the El Niño-Southern Oscillation (ENSO) phenomenon and the anomalous standards observed in the atmosphere in various areas of the globe.

As a result of these complex ocean–atmosphere interactions, associated with the actuation of other mesoscale meteorological systems, a high pluviometric variability is registered in the semiarid Brazilian region, with a concentration of rainfall of 60–85% in the months of February, March, April and May, resulting in an average of 8 dry months, as well as the occurrence of frequent droughts (Souza et al. 1999).

In the state of Piauí, the climatic conditions depend mainly on the actuation of two atmospheric systems: the ITCZ – the intertropical convergence zone, representing the northern system; and the western system with „lines of tropical instability“ and the predominance of W and NW winds that reach the state, which, in turn, cause rainfall, especially during summer and autumn (Nimer and Brandão 1989).

Specifically for the Picos region, the actuation area of the WAVES program in the state of Piauí, Souza (1998), studying the climatic dynamics, registers the principal atmospheric systems acting in the area: the ITCZ, the Atlantic tropical and equatorial continental air mass and the cold front from the south. The importance of the knowledge of these systems stems from the fact that it can be used to predict droughts, as well as for planning of the management and use of the land.

Although recognizing that desertification studies require multi- and interdisciplinary approaches, we discuss here the climatic conditioners of desertification for this region, considering, above all, the aridity and the thermic regime of the climate, the high rates of evapotranspiration, the irregularity of the temporal and spatial distribution of rain, besides the occurrence and persistence of droughts and torrential rainfall.

This work intends to discuss the implications that arise from these climatic regional conditions, above all, the high pluviometric variability, as conditioners of the management and use of the land and its implications for the process of desertification.

Material and Methods

For the climatic characterization of the southeastern portion of the territory of Piaui, more specifically, the area of influence of the Bocaina Reservoir, we utilized data from a climatological station in Picos and a pluviometric station in Bocaina (Table 1). The scarcity of meteorological stations in the region, as well as their poor spatial localization, achieves the study of the climate. Although recognizing the limitations of the climatic database for the area of interest, we applied some indices frequently suggested in climatic studies of the semiarid regions: the aridity index (IA) according to the United Nations Environment Program (UNEP 1991), the interannual variability coefficient (IVC) and the indices associated with the water balance of the soil–atmosphere, that can inform us about the severity of the local aridity, as well as the climatic instability to which the area is submitted. On the other hand, water balance indices allow us to determine the behavior of the annual hydric regime, indicating the existing potentialities and limitations for the biological activity, which are also useful for ecological and agroclimatological studies.

Table 1.

Municipality	Data collecton site	Lat.	Long.	Alt.	Series
Picos	Climatological station	7:05 S	41:28 W	195	1911–1985
Bocaina	Pluviometric post	6:45 S	41:19 W	250	1963–1985

The calculation of the water balance for Picos and Bocaina was carried out according to Thornthwaite and Matter (1955, 1957) using the monthly averages of rainfall along a historical series of data obtained from BRASIL/SUDENE (1990) and monthly mean temperature data estimated by multiple linear regression according to Cavalcante and Silva(1994), using the latitudes, longitudes and altitudes of the pluviometric station locations. To complete the existing gaps in the pluviometric data of Picos, we utilized the weighted regional method described by Tucci (1993). The water balance was computed both for the mean monthly rainfall and sequentially for the whole historical series.

Results and Discussion

The aridity index utilized here was proposed by UNEP for climatic susceptibility analysis of desertification and is calculated as the annual average rainfall divided by the annual potential evapotranspiration (P/ETP), calculated by the Thornthwaite and Matter method (Thornthwaite and Matter 1955). For UNEP, the potential areas susceptible to desertification present an aridity index that varies between 0.05 and 0.65, corresponding to conditions where the available heat for evaporation is greater than the quantity of water precipitated in a year. In the area studied, the values found were 0.43 and 0.45 for Picos and Bocaina, respectively (Table 2).

According to UNEP, regions that present an aridity index between 0.20 and 0.50, as is the case for the area studied, correspond to 17.5% of the emerged land on the globe. On these lands grazing is the main activity that may be carried out in a sustainable way, since the agricultural use of them is highly susceptible to their annual and interannual climatic variability (Hare and Ogallo 1993). The interannual climatic variability is one of the most remarkable characteristics of all arid and semiarid regions of the world. The coefficient of variation of rainfall shows interannual variability of the precipitation, and is above 25% in most parts of dry regions and exceeds 40% along the majority of desert margins (Hare 1992).

Table 2. Aridity indices (UNEP) for Picos-PI and Bocaina-PI

Municipality	P (mm)	ETP (mm)	IA (P/ETP)
Picos	704	1645	0.43
Bocaina	723	1603	0.45

The values found for the coefficient of variation (CV) for the studied area were 37.0 and 40.0 for Picos and Bocaina, respectively (Table 3). The historical series of pluviometric data for Picos, a period of 74 years, shows annual values that vary from 200 mm/year (1953) to 1433.9 mm/year (1985), with a coefficient of variation of 37.0%, indicating conditions of high climatic instability and, consequently, of ecological instability. Associated with this variability there are still years of droughts that correspond to a sequence of 2 years in which the rainfall is 60% below the annual average (Sant'anna; Neto and Zavantini 2000). According to Hare (1992), the prolonged droughts have greater important environmental impact than the interannual variability inherent to these regions.

Table 3. Average, standard deviation and interannual variability coefficients for Picos-PI and Bocaina-PI

Municipality	Average (mm)	Standard deviation	CV (%)
Picos	704	266	37.0
Bocaina	723	292	40.1

The water balance calculation for Picos and Bocaina (Tables 4 and 5), through the analysis of average monthly precipitation and temperature, shows the seasonal variation of the hydric regime in the area. For this calculation we used a field capacity of 100 mm, following Nimer and Brandão (1989), as the value that is more representative of the soils of the region.

The average monthly temperatures are elevated and do not present much variation, the lowest values occurring in February/March and the highest in October/November. Factors such as latitude, continentality and altitude are responsible for these conditions of air temperature, which, in turn, induce high values of potential evapotranspiration. The highest value of ETP also occurs in October/November, while the lowest value is registered during the months of February/March.

Table 4. Annual water balance (in mm) according to Thornthwaite and Mather (1955), for Picos-PI (CAD 100mm)

Month	T (°C)	P	ETP	P-ETP	Soil storage	Change	ET	Sur-plus	Deficit
Jan.	26.3	117.3	138	–21	0	0	117	0	21
Feb.	25.6	142.7	111	32	32	32	111	0	0
Mar.	25.8	189.5	124	66	97	66	124	0	0
Apr.	26.1	82.9	124	–41	66	–31	114	0	10
May	26.2	19.9	128	–108	22	–44	64	0	64
June	26.4	5.9	127	–121	7	15	21	0	106
July	26.6	1.4	137	–136	2	–5	6	0	131
Aug.	27.1	1.2	144	–143	0	–2	3	0	141
Sept.	28.4	3.7	151	–147	0	0	4	0	147
Oct.	28.8	16.9	161	–144	0	0	17	0	144
Nov.	29.0	45.9	161	–115	0	0	46	0	115
Dec.	27.1	80.7	152	–71	0	0	81	0	71
Year	26.9	708.0	1658	–950	226	0	708	0	950

The precipitations register an annual average of 708 mm for Picos and 723 mm for Bocaina, with highest values in March and lowest in August. The result of the combination of high temperatures with precipitations that concentrate mostly in January, February and March is a strong water deficit in the soil for the greater part of the year. In Picos, the water deficit occurs during 10 months and totalizes 950 mm through the year; in Bocaina, there is a accumulation record of 9 months of water deficit totaling 892 mm/year. Both Picos and Bacon do not present any water surplus.

With regard to the sequential water balance, this allows perception of the behavior of the climate throughout a historical Seri and to establish a climatic typology. The results that are presented in Table 6 show the characteristics of precipitation and the climate type according to the effective humidity index of Thornthwaite and Matter (1955). Picos, for example, throughout the 74 years that make up the studied series of precipitation, presented in 16.0% of the years a dry subhumid climate, in 56.0% of the years a climate of the semiarid type, and in 26.7% an arid climate. Therefore, during 98.7% of the period observed, the climate can be considered as belonging to the dry type.

Table 5. Annual water balance (in mm) according to Thornthwaite and Mather (1955), for Bocaina PI (CAD 100 mm)

Month	T (ºC)	P	ETP	P-ETP	Soil storage	Change	ET	Sur-plus	Deficit
Jan.	26.4	136.0	140	–4	0	0	136	0	4
Feb.	25.6	126.7	112	15	15	15	112	0	0
Mar.	25.4	158.3	118	40	55	40	118	0	0
Apr.	25.8	125.5	119	7	62	7	119	0	0
May	25.9	27.9	124	–96	24	–38	65	0	59
June	25.9	10.4	119	–109	8	–16	26	0	93
July	25.9	1.1	123	–122	2	–6	7	0	116
Aug.	26.7	0.6	141	–140	1	–1	2	0	139
Sept.	27.9	2.2	148	–146	0	–1	3	0	145
Oct.	28.8	18.1	161	–143	0	0	18	0	143
Nov.	28.4	34.6	155	–120	0	0	35	0	120
Dec.	27.6	81.3	155	–74	0	0	81	0	74
Year	26.7	722.7	1615	–892	166	0	723	0	892

Bocaina, throughout the 23 years that make up the studied series of precipitation, presented in 21.7% of the years a dry subhumid climate, in 52.2% of the years a climate of the semiarid type, and in 21.7% an arid climate. Therefore, during 95.6% of the observation period, the climate belonged to the dry type.

Conclusions

Although we should not compare Picos and Bocaina due to the difference in the length of their historical series of precipitation data, the indices regarding both localities allow us to identify some climatic parameters of the region and their importance as conditioners for land use and environmental degradation/desertification.

Table 6. Climate analysis in Picos and Bocaina

	Picos (7°05' S, 41°28' W) 195 m (Altitude)		Bocaina (6°54' S, 41°19' W) 250 m (Altitude)	
Characteristics	Precipitation (mm)	Dry months	Precipitation (mm)	Dry months
Average of historic series	704	9	723	9
Annual minimum of historic series	200	5	320	5
Annual maximum of historic series	1435	12	1283	12
Coefficient of variation (%)	37.0	18.4	40.0	22.4
Type of climate according to Thornthwaite and Mather (1955) based on historic series (% of years)				
Super-humid A	-		-	
Humid B4	-		-	
Humid B3	-		-	
Humid B2	-		-	
Humid B1	-		-	
Humid subhumid	1.3		4.3	
Dry subhumid	16.0		21.7	
Semiarid	56.0		52.2	
Arid	26.7		21.7	

Normally, the desertification begins with the removal of the plant cover that accelerates the erosion processes of soils due to the action of water and wind, bringing, as consequences, among others, a decrease in water infiltration and retention by the soil, an increase in runoff, loss of soil fertility and decrease in bioproductivity. From the biodiversity point of view, there is a decrease in the diversity of species and of plant biomass and, as a feedback process, an even greater decrease in the protection that the natural plant cover offers to the soil against erosion. This situation leads to the desertification process that then promotes drastic changes in the microclimates of the affected areas, with their consequent alterations of the whole ecological system.

According to William and Bailling (1995), the ecosystems of dry lands have a high degree of resilience and use a series of strategies to face the low and sporadic precipitations, the high temperatures, the poor and sometimes salty soils, besides the extreme climatic conditions such as droughts or floods.

The climatic conditions in the Picos and Bocaina regions show themselves to be extremely adverse. The high temperatures allied with a high pluviometric variability submit the ecosystems in this region to strong climatic and environmental stress. Despite the degree of resiliency that these ecosystems may present, other factors such as the characteristics of soils and the system of use of these lands constitute further conditioners to the process of desertification.

Table 7 presents a general view of the limitations for the use of the land in the regions of Picos and Bocaina. The soil associations presented here were compiled from the map elaborated by EMBRAPA (1983). The slope classes result from the slope gradient chart elaborated by Ramos and Sales (2000) in which they were subdivided into class A (0 to <3%); class B (3 to <15%), class C (15 to <30%); and class D (> 30%). Based on Table 7, it is possible to say that the soils found in areas of favorable relief positions (classes A and B) present reasonable potentialities of use. Their limitations are mostly low natural fertility, drainage and texture. The soils found in areas with pronounced inclination (classes C and D) present restricted potentialities of use and need adequate management practices.

Regarding the use of land, there is a predominance of agricultural activities and livestock grazing activities developed in small establishments with low level of technology, having as their main form of organization family labor.

In the case of the regions studied here, there is a conjunction of factors such as: adverse climatic conditions, natural fragility of soils, inappropriate use of natural resources, low level of technology without the use of conservative practices, allied to precarious socio-economic conditions of the local population. This makes the region extremely vulnerable to the process of desertification.

The climate has an important influence over the process of desertification, because of its effects on the soils and the vegetation on the hydrological cycle and on the use that human beings make of these lands (William and Bailling 1995). It is difficult to establish the effects of environmental degradation of dry lands in the meso- and macroscale climate, although its role in altering microclimates is undoubted. Therefore, any model of development proposed for this region must consider the adversity of climate and recognize the climatic irregularity as being the standard to be considered in land management strategies.

Table 7. Associations of soils, slope classes, use of land and limitations of the use of areas under influence of the Bocaina Reservoir, Piauí

Associations of soil	Slope classes	Use of land	Limitations of use
Latossolo Amarelo+Podzólico Verm.-Amarelo Concretionário+Solos Lithólicos	A	Extensive use by grazing animals, subsistence and cashew cultivation	Low natural fertility and high acidity
Latossolo Amarelo+ Areias Quartzosas	A	Cashew cultivation, corn, bean and manioc cultivation and extensive use by grazing animals	Low natural fertility and high acidity
Latossolo Verm.-Amarelo+Latossolo Amarelo+Latossolo Vermelho Escuro Eutrófico+Podzólico Verm.-Amarelo Concretionário	B	Extensive use by grazing animals, corn, bean, manioc and fruit cultivation	Low natural fertility and high acidity
Solos Aluviais+Cambissolos+Podzólico Verm.-Amarelo	A	Extensive use by grazing animals, extrativism, rice, corn, bean, manioc and cotton cultivation	Periodic droughts or floods or salinization
Areias Quartzosas+Latossolo Amarelo+Podzólico Verm.-Amarelo Concretionário+Solos Lithólicos	B	Extensive use by grazing animals, corn, bean and manioc and fruit cultivation	Sandy texture, excessive drainage and low fertility
Areias Quartzosas+Solos Lithólicos+Podzólico Verm.-Amarelo	A	Extensive use by grazing animals and cashew cultivation	Sand texture, excessive drainage and low fertility
Podzólico Verm.-Amarelo, Bruno Não-Calcico, Solos Lithólicos	B	Cotton, corn, bean and fruit cultivation, extensive use by grazing animals	Stony
Podzólico Verm.-Amarelo+Plinthossolos+Planossolos +Latossolo Amarelo	B/C	Manioc, corn, bean, fruit cultivation and extensive use by grazing animals	Stony
Podzólico Verm.-Amarelo+Cambissolos+Solos Aluviais	A/B	Corn, bean and rice cultivation, cultivated pastures	Stony
Solos Lithólicos+Areias Quartzosa+Rock outcrops	B/C	Extensive pasture and small subsistence cultivation	Low fertility, high inclination, stony
Solos Lithólicos+Podzólico Verm.-Amarelo+Latossolo Amarelo	C/D	Pasture and subsistence cultivations	High inclination, shallow, stony
Solos Lithólicos+Podzólico Verm.-Amarelo+Rock outcrops	C/D	Extensive cattle raising, pasture, subsistence cultivation	High inclination, stony
Solos Lithólicos+Podzólico Verm.-Amarelo Concretionário+Bruno Não-Calcico	C/D	Extensive cattle raising and corn and cotton cultivation	High inclination, stony

References

Brabo JM, Repelli CA, Melo NS (1993) A pré-estação chuvosa do setor norte do nordeste brasileiro e sua relação com a temperatura dos oceanos adjacentes. Rev Brasil Meteorol 8:22–30

BRASIL/SUDENE (1990) Dados pluviométricos mensais do nordeste, Estado do Piauí. Recife-PE

Cavalcanti EP, Silva EDV (1994) Estimativa da temperatura em função das coordenadas locais. Anais do VIII Congresso Brasileiro de Meteorologia, Belo Horizonte-MG

EMBRAPA (1983) Mapa exploratório – Reconhecimento de solos do Estado do Piauí. Empresa Brasileira de Pesquisa Agropecuaria, Rio de Janeiro.

EMBRAPA (1986) Levantamento exploratório de reconhecimento de solos do Estado do Piauí. Rio de Janeiro, EMBRAPA/SNLCS. Recife, SUDENE/DRN

Hare FK (1992) Desertificação, causas e conseqüências . Fundação Calouste Gulbenkian, Lisboa

Hare FK, Ogallo LJ (1993) Climate variations, drought and desertification. World Meteorological Organization, vol 653

Nimer E, Brandão AMPM (1985) Balanço hídrico a partir de valores normais e tipologia climática. Rev Brasil Geogr 47:373–416

Nimer E, Brandão AMPM (1989) Balanço hídrico e clima da região dos cerrados. IBGE, Rio de Janeiro

Nobre P, Moura CA (1985) Planetary sclae circulation anomaloies associated with droughts over Northeast Brazil. Trop. Ocean Atmos. Newsletter, March 1985, p.13-15.

Ramos VM, Sales MCL (2000) Análise da capacidade de uso da terra, com base na declividade e nas características dos solos, nas áreas de influência do Reservatório de Bocaina-PI. CNPq/Programa WAVES, Ano II, Relatório Final

Sant'anna Neto JL, Zavatini JA (2000) Variabilidade e Mudanças Climáticas, Maringá, EDUEM

Souza CF (1998) Dinâmica climática e as chuvas no nordeste brasileiro no eixo Barra do Corda/MA-Natal/RN (Relações com o fenômeno El Niño). USP-FFLCH - Departamento de Geografia. Tese de Doutorado, São Paulo

Souza EB, Alves JMB, Xavier TMBS (1999) Distribuição mensal e sazonal da precipitação no semi-árido nordestino durante os anos de predominância de aquecimento ou resfriamento observados em toda a Bacia do Atlântico Tropical. Rev Brasil Meteorol 14:1–9

Thornthwaite CW, Mather JR (1955) The water balance. Publications in climatology, vol VIII(1). Centerton, New Jersey

Thornthwaite CW, Mather JR (1957) Instructions and tables for computing potential evapotranspiration and the water balance. Publications in climatology, vol X(3). Centerton, New Jersey

Tucci CEM (org) (1993) Hidrologia: ciência e aplicação. Ed. da Universidade, ABRH, EDUSP, Porto Alegre

UNEP (United Nations Environment Programme) (1991) Status of desertification and implementation of the United Nations plan of action to combat desertification. Draft Report, Nairobi

UNEP (United Nations Environment Programme) (1992) Atlas of desertification. Arnold, London

Williams MAJ, Balling RC Jr (1995) Interacciones entre la desertificación y el clima: una visión general. Boletín de control de la desertificación, no 25. Programa de las Naciones Unidas para el Medio Ambiente

Part 7: Economic and Socio-Cultural Analyses

Quality of Life and Migration – Concepts and Results of the Socio-economic Survey in Tauá and Picos

Dagmar Fuhr with contributions from Martin Grebe, Andreas Döring, Francisco Matias da Rocha and Ernst-Dieter Lantermann

Center for Environmental Systems Research, University of Kassel, Kurt-Wolters-Strasse 3, 34109 Kassel, Germany, fuhr@usf.uni-kassel.de

Introduction

The semiarid northeast of Brazil is one of the most problematic regions in the world. Social structures were established in the last century under difficult natural and climatic conditions. These social structures are not very apt to deal with social and ecological crises in this region. Coping mechanisms that were developed for such natural and social problematic living conditions are highly questionable as individual and social solution concepts. This is especially the case for migration activities that can no longer be classed traditional compensatory migration or seasonal labor migration.

Therefore, the central tasks for the Working Group Socio-cultural Analyses (SOLAM) in the context of WAVES was to describe and explain the interaction between quality of life and adaptive behavior in the rural areas of northeast Brazil, in particular, for the federal states of Piauí and Ceará. As we are dealing with a problem-oriented approach, the set of action strategies is restricted to migration and direct alternatives, such as self-organized enhancement of local living conditions. The main objective is to evaluate possible future development plans of rural regions in the study area based on reference scenarios and simulation model results on the basis of differences in quality of life.

This chapter gives an overview of the main activities of the project „Socio-cultural Analyses", especially about (1) the methods to identify the socio-economic and socio-cultural options and restrictions for the improvement of the quality of life, and (2) their integration into a space-related migration model. This included (i) the formulation of an integrative concept for quality of life, (ii) the outline of functional relationships, and (iii) interviewing householders in the research area. The model results are presented for the present situation and for different reference scenarios up to the year 2025.

Method

To reach the goals described above, it is necessary to formulate a theory-driven actor model reflecting the dynamic interdependence of internal and external states. Different adaptive action strategies have to be implemented in this model to realize feedback functions of behavior on external (e.g., characteristics of the social surrounding) and internal variables (e.g., degree of satisfaction with social relations). The set of variables used and their interrelations should be derived from theoretical considerations. Survey data are needed to check the validity of the proposed model.

In order to develop a theory-based dynamic migration model the structure of the actor model has to be integrated into a space-related model of regional dynamics. This procedure allows calculation of aggregated effects of individual decisions.

Quality of Life as an Integrative Concept

Migration is interpreted as a multidimensional phenomenon. Therefore, an integrative concept is needed to connect the different thematic spheres. Possible causes for migration are located in the environmental sphere (lack of natural resources), in the social sphere (missing social support), and in the personal sphere (education). Quality of life integrates these different phenomenological spheres.

The concept of quality of life indicates the matching between external and internal state variables and serves as a global indicator for the stability of interrelated systems. Furthermore, the structure of WAVES demands the coupling of different types of data sets. On the basis of spatial distributions of natural resources and the availability of infrastructural supply, probabilities of various adaptation strategies should be calculated by the integrated model.

Development of an Adaptive Action Model

The general concept of the postulated action model can be described in the following steps:

1. calculation of the fitting of a set of theory-guided deduced needs and related possibilities/opportunities to meet them;
2. selection of different adaptive action strategies with reference to the specified needs and empirically derived competencies;
3. maximization of positive changes in the satisfaction of needs.

A set of needs (existence, security, freedom, flexibility, affiliation) was derived from Max-Neef's works (1991) intended to build up a set of human requirements with intercultural validity. Furthermore, two resource types were differentiated: infrastructural (medical support, education) and agricultural resources (soil quality, time since latest drought, water availability).

The set of adaptive action strategies is reduced to migration, innovation and persistence. The theoretical background for the selection of different adaptive action strategies is the protection-motivation theory formulated by Riptoe and Rogers (1987). It postulates the probability of specific strategies depending simultaneously on the perceived personal ability (the actor's competencies to act, e.g., education) and the perceived situational threat (described by the actual need state and the availability of action resources).

Satisfaction with actual living conditions is the central result of this coping process. The postulation of a general goal to maximize this satisfaction leads to the actor's preference to select those strategies that are supposed to result in an optimal need satisfaction regarding competencies and action resources.

As a first empirical check, a survey of about 200 participants was carried out in the two municipalities of Tauá (Ceará) and Picos (Piauí). For this purpose, structured interviews were held with family heads to yield data to test the hypotheses outlined above.

One central assumption was that general satisfaction with conditions of life leads to a reduced tendency to migrate. This tendency is measured by the individuals' desire to leave the region. General satisfaction is measured by two different sets of indicators (judgements) representing different aspects of contentment:

1. contentment related to social aspects (e.g., living in the countryside, mayor); and
2. contentment related to materialistic aspects (living space, water supply, usable goods).

In the case of migration, social competence influences the probability to leave the region in a way that it lowers the barriers (e.g., age, education, communication). In contrast to social competence, materialistic competence (e.g., farm size, land ownership, personal goods) works against migration tendencies.

MigFlow: Central Ideas

The conceptualization of migration processes in MigFlow (i.e., **Mig**ration **Flow** Model) follows the assumption that migration tends to equalize disparities of quality of life because migrants utilize opportunities to improve their living conditions. Therefore, migration is affected by opportunities in potential target regions and the competence to overcome obstacles of moving as well as to benefit from these opportunities for the potential actors.

These findings converge with the theoretical assumptions of the „classical migration theory" (Ravenstein 1885; Lee 1966):

- The degree of migration inside a given territory varies directly with the diversity of regions included in that territory.
- The migration rate between two locations is inversely related to the obstacles for overriding the distance.

MigFlow: Modules

MigFlow consists of three main modules: *demography*, *quality of life* and *migration*.

The submodel *demography* calculates the total population for each municipality in cohorts of 5 years based on age-specific and sex-specific rates of mortality and fertility (IBGE 1997). The trends for mortality and fertility follow scenario assumptions: We assume a decrease down to 60% (fertility) and 80% (mortality) of 1996 rates in 2025.

Quality of life is modeled as an integrated indicator composed of:

- the aggregated *family income* (HDI: PNUD, IPEA, FJP, IBGE 1998),
- an indicator for *educational infrastructure* (secondary education and upwards) (SOLAM), and
- an indicator for *health infrastructure* (costs) (SOLAM).

On the one hand these indicators were selected as a result of the survey (see below) and, on the other, because official data for these themes on a municipal level were available.

These indicators are further made dynamic by scenario assumptions (see Table 1; for the general scenario assumptions, compare also Döll et al. 2000).

MigFlow models Ceará and Piauí as a system of interacting spatial cells – the municipalities – with specific properties. The communication with external territories beyond the model boundaries is represented in one virtual cell (being represented by São Paulo which is a primary migration target outside of the survey region, see the results of the survey). Migration is calculated in the following steps:

- computation of disparities between the quality of life of the different spatial cells;
- computation of migration costs based on spatial distances between the cells (linear distance based on UTM coordinates);
- computation of potential migration to cells with a higher quality of life: potential profit in the quality of life (cost benefit calculation) multiplied by competence factor and the affected population (households with income of less than half of the minimum wage – Salário Mínimo);
- recalculation of migration based on emigration and immigration restrictions;
- integration of migration in the demographic model.

For a more detailed description of the model, the reader is referred to Grebe et al. (2000).

Table 1. Assumptions of the growth rates per year for the reference scenario A (RS A – Coastal boom and cash crop = Globalization) and the reference scenario B (RS B – Decentralization)

Scenario regions Growth rates of income resources	RS A %	RS B %
Brazil	2.5	2.5
Teresina	3.9	3.2
Metropolis Fortaleza and Pecém	4.4	3.4
Coastal region	4.6	3.2
South of Piauí	2.8	3.0
Area of large potential water resources - Piauí	2.8	3.1
Area of large potential water resources - Ceará	2.8	3.1
Area of small potential water resources - Piauí	1.4	2.4
Area of small potential water resources - Ceará	1.4	2.4
Growth rates of per-capita health and education		
Brazil	2.5	2.5
Teresina	2.0	3.2
Metropolis Fortaleza & Pecém	2.0	3.4
Coastal region	2.0	3.2
South of Piauí	2.0	3.0
Area of large potential water resources - Piauí	2.0	3.1
Area of large potential water resources - Ceará	2.0	3.1
Area of small potential water resources - Piauí	2.0	2.4
Area of small potential water resources - Ceará	2.0	2.4

Results

The following section gives an overview of the results of the survey in both examination areas, Picos and Tauá, as well as the simulation results with the space-related migration model MigFlow.

Results of the Surveys in Picos and Tauá

About 100 households (102 families in Tauá and 104 families in the focus area of Picos, this includes Picos and eight municipalities around Picos) were questioned in March 1999 in each of the two examination areas, Picos (Piauí) and Tauá (Ceará), with respect to their socio-economic situation, to their experience with migration, and to their satisfaction with the farming situation.

One person was questioned as a representative of each household, in general the head of the household or his/her partner; 87 men and 17 women were questioned in Picos and 54 men and 48 women in Tauá. The average age of the interviewees in Picos was about 48 years and in Tauá about 45 years. This is similar to the average age of the head of households in this region.

At first we give an overview about the situation regarding farms, since nearly all of the questioned households have their main source of income in farming. Whereas two-thirds of the questioned farms in Picos are situated in the fertile river plain and one-third on the less-fertile plateau (the so-called chapada), the share between both forms is in balance in Tauá.

In both examination areas, more farms are property owned; however, only 50% of the questioned families have a definite property title.

The average farm size of the questioned farms in Picos is rather small (19 ha) compared with the farm size in Tauá (95.5 ha). In Picos, annual crops are grown on 5.16 ha on average, whereas in Tauá 7.57 ha are used for this purpose. On one-third of all farms, perennial crops are grown in Tauá as well as in Picos. In Tauá, 3.3 ha are used for this purpose and 9.4 ha in Picos.

As can be seen from Fig. 1, in Tauá live-stock farming is predominant. More than half of all questioned farmers own pasture land with an average area of 43.9 ha. The average area of pasture land in Picos is only 11.6 ha.

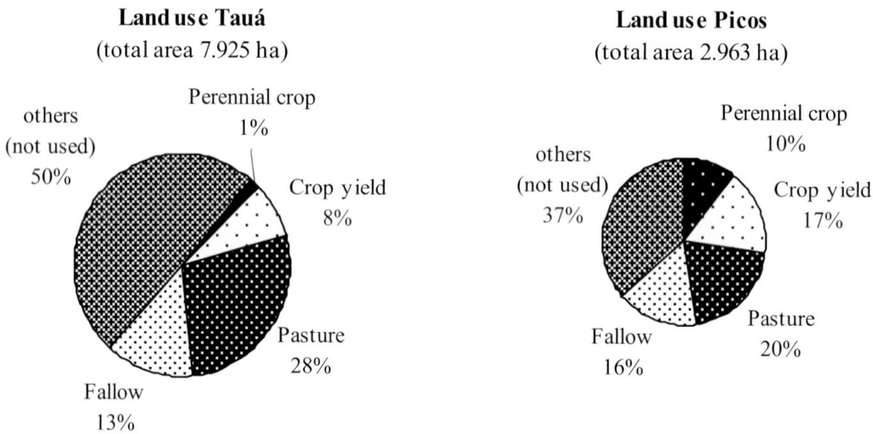

Fig. 1. Land use in Picos and Tauá (results from the survey 1999)

Satisfaction with Agricultural Life

The answer to the question as to how people feel with life in agriculture in general was „good“ from 13% of the questioned persons in Tauá, whereas 57% gave the answer „bad“. In contrast to this result, in Picos almost 50% gave a positive answer and only 10% had a bad impression of their life in agriculture. The remaining persons gave an indifferent answer. These judgements are the same for all farms independent of their size.

The main limitations to farming mentioned most frequently were: (1) the absent or irregular precipitation, (2) the desperate fight against pests, (3) the lack of high-quality seeds, and (4) the absence of a price guarantee for agricultural products.

In Picos, 74% of the farmers use almost the whole harvest for their own needs. In Tauá, this was the case for 60% of the farmers. This means that these families earn no or almost no money from their agricultural work. Accordingly, a third of the questioned persons in Picos and 44% in Tauá have a job, mostly as traders or craftsmen, in addition to their job in the agricultural sector. However, at the end of the year for the farmers in both municipalities, almost no money is left over for saving: only 21% of the questioned people in Picos and 32% of those in Tauá said that they were able to put some money aside.

Further, we were interested in whether the families were financially supported by family members who had already moved away. This is the case for about one-third of all questioned households in both municipalities. Furthermore, about 15% of the questioned households in Tauá and Picos receive financial support from the government retirement scheme. During the drought in 1998, there was additional governmental emergency aid available. These job programs or food aid were claimed by about two-thirds of the questioned families in each of the municipalities.

Although the general judgement about rural life was „not good“, 94% of the questioned persons in Tauá are satisfied with their job in agriculture; 37% even state that they would not like to do anything else but agriculture. In Picos, the satisfaction with living in agriculture is at a similar level (92% of the questioned persons). However, the willingness to change jobs is lower: more than 60% stated that they could not imagine working in a profession other than agriculture. The situation is somewhat different when people were asked what profession they wish for their children. In both regions, only five parents wished their sons to become farmers. For their daughters, agriculture was seen as a desirable profession only in three cases in Picos and not at all in Tauá. In addition to the socio-economic situation, migration was the second main interest in the evaluation. Questions related to personal experiences with migration as well as to intentions of family members to migrate.

In 27 of the families questioned in Tauá, one or more persons plan to migrate in the near future; in Picos, 18 families had members with the intention to migrate. Out of the group of questioned people, 280 persons (out of 96 families) had already migrated in Picos. In Tauá this was the case for 222 persons out of 61 families. The average age of people who plan to migrate is about 26 in Tauá and 28 in Picos. Looking at the difference in behavior of the two sexes, it was found that most migrants were male (two-thirds male compared with one-third female). However, there is a considerable difference between the two regions evaluated: in Picos, 75% of the persons who plan to migrate are male, whereas, in Tauá, the intention to migrate seems to be similar for both sexes (see Fig. 2).

We were further interested in the motivation to migrate and which town people preferred as a migration target. The main target is still São Paulo since the main motivation for the majority of migrants is to find a new job. This is valid for 87% of the migrants who move to São Paulo. The second very important reason why people migrate is reunification with family or plans to start a new family (16% of the motivation to migrate). The latter reason often leads to a migration within a smaller area when compared with the more economic motivation. As a third reason, people hope for an improvement in their general living conditions (including, secondary education or college education with 5% of the motivation to migrate); these improvements are mostly expected to happen in the south of Brazil. The results of our interviews differ in this considerably from official migration statistics. According to the Federal State Agency for Statistics (IBGE 1997), the migration within Brazil between 1991 and 1996 was 56% for Ceará, and 33% of the overall migration for Piauí. Migration within the northeastern region was 65% for Ceará and 50% of the overall migration for Piauí. In contrast, migration to São Paulo from Ceará was only 18% and from Piauí 23% of all migrations. This discrepancy between official statistics and our interviews can be explained by the fact that moving to the neighboring community is not always perceived as migration by the interviewees and thus not indicated in our questionnaires, whereas the official statistics register every single move into a new municipality.

It is further remarkable to note from the questionnaires that neither the size of the farm nor the type of ownership seems to play an important role in the decision to migrate. Migration is practically equal from all size and types of farms, which can be explained by the fact that it is generally not the head of the family who decides to migrate, but rather the sons and daughters.

Finally, we wanted to know where people prefer to live in the countryside or the city. The answers show a clear opinion: people prefer living in the countryside since this is seen as more quiet and peaceful, and solidarity with others plays an important role. People are afraid of life in the city because of crowding and criminality.

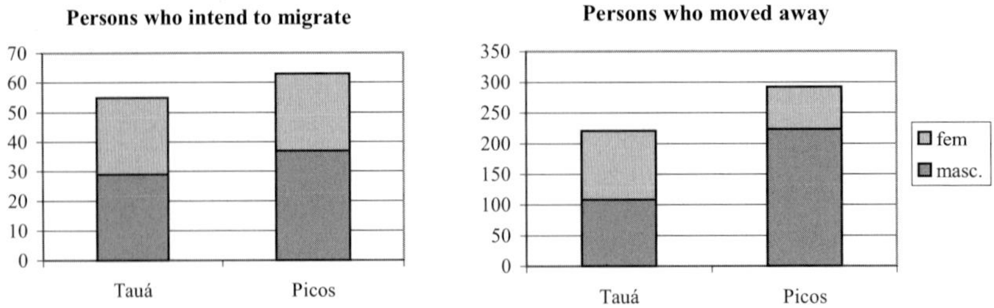

Fig. 2. Migrants by sex and study area (results from the survey 1999)

Results of the Simulation with MigFlow

Figures 3 and 4 show the differences in the results of the simulation with MigFlow in quality of life between the scenarios RS A Globalization (Fig. 3) and RS B Decentralization (Fig. 4) for 2025. One can see the higher and more disparate development in the globalization scenario: the index for quality of life ranges between 0.17 and 2.17 points. The more equal development of the rural area can be seen in scenario B (index from 0.19 to 1.68).

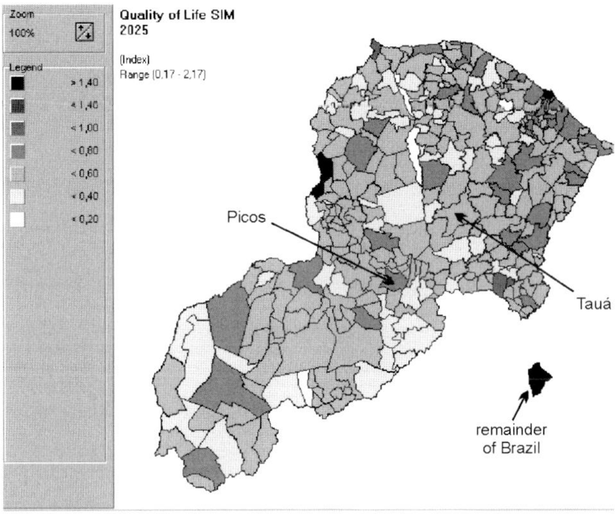

Fig. 3. Simulation results: quality of life RS A

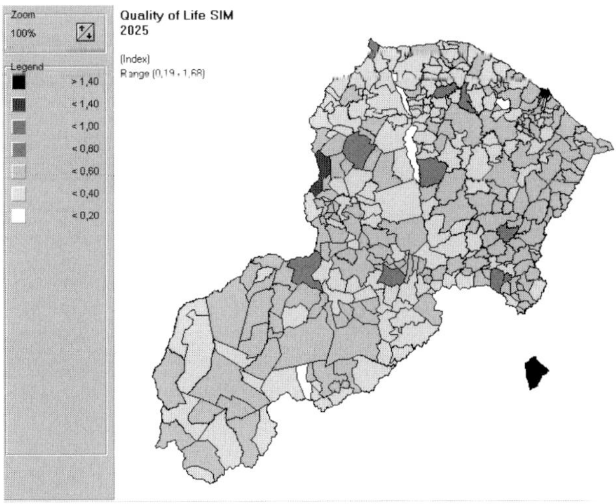

Fig. 4. Simulation results: quality of life RS B

Figures 5 and 6 show the results of migration for the two scenarios. Positive net migration rates in relation to the population of 1996 are found only for a couple of municipalities, mainly, the current urban centers but also in some up-coming areas, e.g., Picos, one of the examination areas of our studies.

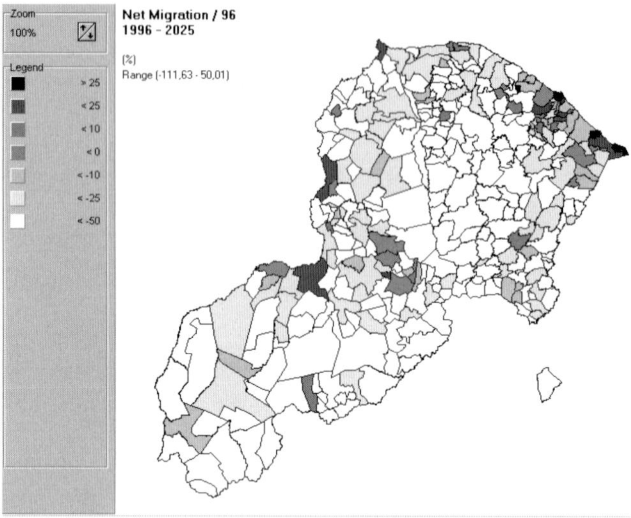

Fig. 5. Net migration 1996–2025, scenario A

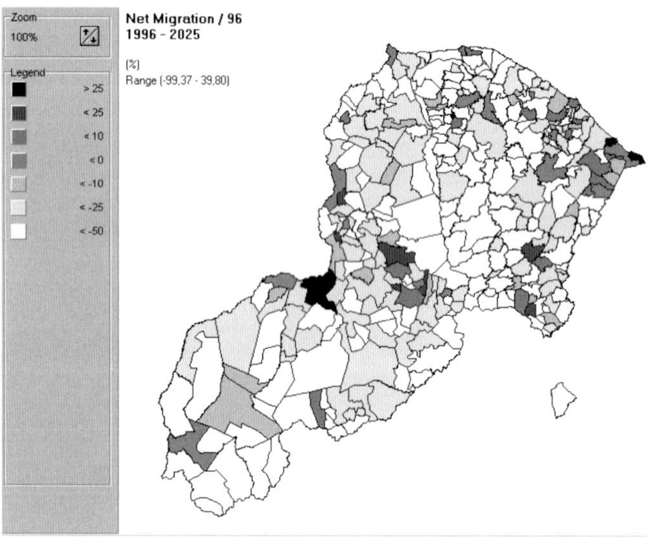

Fig. 6. Net migration 1996–2025, scenario B

In order to validate our results we have compared the results of our simulation of overall population with IBGE census data. For the year 2000, we have preliminary results of the census data on the municipality scale (IBGE 2000). Altogether, the simulation results of population for the whole examination area (the states of Piauí and Ceará) are lower in both scenarios than the results from the IBGE census (RS A and RS B 3.8%). This may be the result of our assumptions about the decrease in fertility rate for both scenarios.

More interesting is the comparison of the spatial distribution of population in each scenario region: Table 2 shows the comparison between simulation results for scenarios A and B and the results from the national census.

The comparison shows for both scenarios reasonable agreement between simulation results and census data for the regions with large potential water resources in both states. For the rural regions with small potential water resources, for the coastal area and for the south of Piauí, the comparison shows for both scenarios medium differences from the census results: here, all the scenario results are lower than the census results. For the Metropolis Fortaleza, the comparison of census data with the results of RS B is very good. However, for the other capital, Teresina, the results are worse. This could probably be attributed to our scenario assumptions in which we overrate the central role based on the infrastructure of Teresina.

Table 2. Fraction of total population of Piauí and Ceará in each scenario region [%]. Comparison of population data of IBGE (2000) with simulation results for reference scenarios A and B with MigFlow

Scenario region	Inhabitants census (IBGE)	Inhabitants MigFlow scenario A	Inhabitants MigFlow scenario B	Differences to IBGE in %	
	2000	2000	2000	RS A	RS B
Teresina	6.97	7.34	7.33	+5.18	+5.05
Metropolis Fortaleza & Pecém	28.06	28.63	28.23	+2.02	+0.61
Coastal area	13.04	12.74	12.72	–2.27	–2.40
South of Piauí	2.87	2.77	2.77	–3.34	–3.31
Area of large pot. water resources – PI	13.52	13.52	13.58	–0.04	+0.43
Area of large pot. water resources – CE	18.94	18.91	19.08	–0.13	+0.75
Area of small pot. water resources – PI	2.67	2.59	2.64	–2.95	–1.30
Area of small pot. water resources – CE	13.93	13.51	13.65	–3.06	–2.02

Conclusions

The results of the survey about satisfaction with agricultural business suggest an upcoming of structural change, especially for future generations of the very small farm holders, who do not seem content with farming as their sole profession. The lack of other jobs in rural areas is the main reason for migration; climatic changes influence migration only indirectly (through economic impacts).

Furthermore, quality of life in rural areas can only be improved through integrated measures, including, agricultural assistance, improvement to infrastructure, and creation of jobs in non-agricultural sectors in the rural area.

The concept used for migration modeling has shown to be useful. The space-related migration model connects demographic and socio-economic data in order to calculate an index for quality of life. Furthermore, the model provides quantitative estimates of migration processes on the community level in the northeast of Brazil. The comparison of simulation results with actual census data allows new insights into the sensitivity of adaptive mechanisms.

References

Döll P, Hauschild M, Fuhr D (2000) Scenario development as a tool for integrated analysis and regional planning. In: German-Brazilian workshop on neotropical ecosystems. Achievements and prospects of cooperative research. 3–8 Sept 2000, Hamburg. http://www.usf.uni-kassel.de/waves/english/vorl_ergebnisse/scenarios.pdf

Grebe M, Matias F, Döring A, Fuhr D (2000) Migration und Lebensqualität im semi-ariden Nordosten Brasiliens. Statusbericht der ersten Hauptphase. Kassel. http://www. usf. uni-kassel.de/ waves/vorl _endbericht/13_solam.pdf

IBGE (1997) Contagem da População. IBGE, Rio de Janeiro

IBGE (2000) Censo demográfico 2000 – sinopse preliminar. http://www1.ibge.gov.br/ibge/estatistica/populacao/censo2000/default.shtm

Lee ES (1966) A theory on migration. Demography 3:47–57

Max-Neef MA (1991) Human scale development. Conception, application and further reflections. Apex Press, New York

PNUD, IPEA, FJP, IBGE (1998) Atlas do desenvolvimento humano no Brasil. Belo Horizonte, Brazil

Ravenstein EG (1885) Die Gesetze der Wanderung 1 und 2 (1885/1889). Reprint: Széll G (ed) (1972) Regionale Mobilität, Nymphenburger Verlagshandlung, München, pp 41–94

Riptoe PA, Rogers RW (1987) Effects of components of protection motivation theory on adaptive and maladaptive coping with a health threat. J Pers Soc Psychol 52:596—604

Economic Analysis of Different Farming Systems on the Individual Farm Level in the Northeast of Brazil

Bettina Bach, Maria Irles de Oliveira Mayorga and Jürgen Zeddies

Institute of Agricultural Economy, University of Hohenheim, 70593 Stuttgart,Germany, bachb@uni-hohenheim.de

Introduction

One main scientific goal of the research program WAVES is the analysis of possible reactions of farm enterprises to changing frame conditions in the context of global change processes in semiarid rural areas in the northeast of Brazil concerning long-term ecological, social and economic sustainability of agricultural production systems.

To a great extent the area of Brazil's northeast is part of the semiarid hinterland called „Sertão", an area characterized by extreme uncertainty in the precipitation regime with long-lasting and recurrent drought periods with severe consequences for the rural population. Agricultural production and animal husbandry farm activities have to cope with an irregular rainfall regime resulting in high production hazards that lead to enormous rural income risks as well as to difficulties in guaranteeing the subsistence needs of the farmers' families. Therefore, the uncertainty of the precipitation regime is one of the major problems of agricultural enterprises in the investigated semiarid regions putting at risk the existence of rural families. Other important problems to be analyzed in the context of sustainable farming systems in the investigated regions are the continuous decrease in soil fertility, the degradation of natural pasture areas by overstocking and the low liquidity of capital resources for financing agricultural activities during the rainy season (Bazin 1993).

A necessary requirement for achieving plausible results in this research context is the analysis of agricultural production and farming systems on the individual farm level. In this specific case, the WAVES focus regions, the microregion of Picos in Piauí and the municipality of Tauá in Ceará, were selected for a case study based on primary data collection. These research activities on the individual farm level were really necessary, as a database for a detailed analysis of farming systems for the mentioned regions does not exist.

As main research objectives the case study focused on the following topics:

1. Analysis and classification of farming systems in the WAVES focus regions (Tauá – CE and Picos – PI) with the aim of analyzing the economic and social situation of different farming systems by a holistic approach.

   In Brazil it is more usual to analyze the economic benefit or the partial profitability of selected cropping systems or special animal husbandry systems without considering whether these systems match with the organization of the existing farming systems, or where they should be introduced to improve the economic situation on the individual farm level. Therefore, in this case, the research activities focused primarily on the analysis of real existing farming systems in the research areas to ensure a good basis for mathematical simulation models for analyzing possible future reactions and perspectives of these farming systems under changing climatic or economic frame conditions.
2. Analysis and simulation of possible adaptation strategies of farming systems to global change processes by means of linear programming models on the individual farm level.

   The mathematical algorithm used by linear programming is suited for a simultaneous analysis of all farm activities realized by the farmer considering all existing constraints such as, for example, the suitability of farm area for different cropping or animal husbandry systems, the availability of water for irrigation purposes, or the organization of family labor and hired labor used on the farm. Based on the data acquired for the analysis of farming systems, all processes related to agricultural and animal husbandry activities can be simulated as a reference for the actual existing farm situation. The mathematical model for the reference situation can be used for further analysis and simulation of possible adaptation strategies of farming systems to global change processes by changing the constraints or introducing new constraints.
3. Analysis and simulation of possible future sustainable development paths for the farming systems in the WAVES focus regions on the individual farm level by means of linear programming models.

   The same mathematical models simulating the reference situation can be used as a basis for further analysis and simulation of possible adaptation strategies of farming systems introducing innovative farm activities that can help the farmer to overcome difficulties on the farm caused by global change processes.

Material and Methods

Data Survey

The primary data survey was carried out in the microregion of Picos in the state of Piauí and in the municipality of Tauá in the state of Ceará. Both areas are part of the semiarid hinterland called „Sertão" in the northeast of Brazil and characterized by extreme uncertainty in the precipitation regime with long-lasting and recurrent drought periods.

By means of standardized questionnaires, detailed information was made available on the individual farm level relative to the farmer's familiar and economic situation. The data survey was carried out on 40 farms in both regions during the year 1998; during the year 1995 on 20 farms in the microregion of Picos – PI and on 10 farms in the municipality of Tauá – CE. The survey comprised three parts to obtain the following information:

- Data survey relative to the situation of the farmer's family (members of the family, age, gender, education, employment in the non-agricultural sector, housing situation, water resources for domestic and agricultural purposes, electricity, type and amount of energy for cooking).
- Data survey relative to the available farm resources (land and its suitability for the realization of different farm activities, labor and capital) and the actual situation related to the machinery and irrigation equipment on the farm, the market situation, the utilization of extension services and agricultural credit services.
- The most important data to be collected was relative to the actually realized farm activities concerning the cultivated crops and animal husbandry activities necessary for gross margin calculation for each farm activity.

On the one hand this database permits a partial economic analysis for each cropping and animal husbandry system realized on each farm by gross margin calculation per ha or unit of manpower and, on the other, allows, as mentioned previously, a holistic approach analyzing each farm or farming system by mathematical models like linear programming considering all activities realized on the farm and all economic, ecological or social constraints limiting agricultural production or influencing the farm management.

Calculation of Farm Income as a Basic Characteristic for Farming Systems Classification

Farming systems classification is an ingenious tool as a prerequisite for setting up linear programming models on the individual farm level to obtain generally accepted statements about the actual situation and the possible future development of farm types in an investigated region.

In a first approach, the collected primary data was used for the calculation of farm income. Besides other important characteristics like site conditions, water resources and socioeconomic conditions of each farm, income is an important feature to be considered for farming systems classification in the investigated regions. The importance of considering farm income for farming systems classification is related to the fact that, for implementing agricultural credit programs, the farms will be distinguished on the basis of farm income to decide which credit conditions will be applied on each case. In Brazil, and especially in the northeast, credit policy for the agricultural sector is an important tool for the development of rural areas applying different financial service programs for very small, small, middle and large-scale farmers. To distinguish these different categories, the financial institutions use the farm income found by estimation. Depending on the category, different items exist that can be financed for short- and long-term credit, different interest rates and limits of total credit amount to be received by each farm category. Analyzing special adaptation strategies of farming systems to changing economic and ecological frame conditions by mathematical models it is necessary to consider the different financial service programs offered by the financial institutions in the regions. For example, implementing new, innovative farm activities to facilitate the adaptation of agricultural activities to changing frame conditions in the investigated regions generally needs long-term investments which the farmer is not able to realize from his own capital resources but only with additional support of borrowed funds.

In the case study, the gross income was calculated for each farm based on different economic indices. Figure 1 gives an overview of the economic indices used and how the gross income was calculated.

Starting from the marketable production of agricultural and animal husbandry activities, a part of this production is sold at local markets or to regional merchandisers, a second part is used to fulfill the subsistence requirements of the farm family and a part of the agricultural production is transferred as forage to animal husbandry activities. To obtain the real gross cash income only the sold products at local markets are multiplied by local market prices. The subsistence requirements of the farm are considered as non-sold products, because they are exclusively destined for alimentation and therefore cause only variable production costs. From the sold agricultural yield are subtracted all variable costs that occur in relation to crop cultivation and animal husbandry farm activities to obtain the gross income contribution of all farm activities. To obtain the gross farm income destined to the remuneration of the family labor, fixed farm costs are subtracted from the sum of income contribution of all farm activities.

Table 1 gives an overview of the items that were considered to calculate the variable costs occurring in relation to agricultural and animal husbandry activities. The most important items of plant production activities are the costs for hired labor and hired machinery. Especially in periods with high demand for labor (e.g., preparation of the area for planting, planting, weeding or harvesting), most of the farmers have to pay for these services because, in most cases, there is not sufficient family labor available on the farm to carry out all services in a determined period of time. Variable costs for own machinery are not so important, as well as the costs for seeds, chemical and organic fertilizer and chemicals for plant protection, because they are not used to any great extent in the investigated regions.

Looking at the animal husbandry activities the most important item causing high variable costs, especially in drought periods, are bought-in feedstuffs. In many cases the forage base to fulfill all livestock needs is not sufficient in drought periods and makes it necessary to buy supplementary feedstuffs or hire additional pasture areas at high costs.

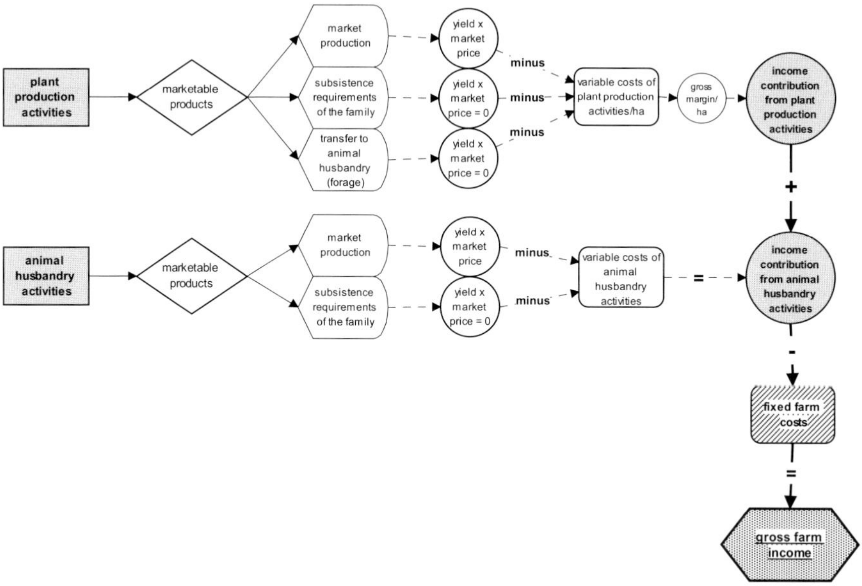

Fig. 1. Rules used for gross income calculation for each investigated farm in the research areas

Table 1. Items considered for the calculation of variable costs

Variable costs of plant production activities	Variable costs of animal husbandry activities
- Seeds	- Bought-in feedstuffs
- Chemical fertilizers	- Animal hygiene
- Organic fertilizers	- Hired pasture
- Chemicals for plant protection	- Supplemented mineral nutrition
- Own machinery	
- Hired machinery	
- Hired labor	

Fixed costs occurring at the farms comprises the following items:

- average costs per year for own machinery
- average costs per year for own irrigation equipment
- average costs per year for animal traction equipment
- average costs per year for hand tools
- average costs per year for water transportation systems
- costs for feeding draft animals
- yearly salaries for permanent hired labor

The average costs per year were calculated with the following formula:

$$AC/\text{year} = \frac{A - RV}{n} + \left[(A - RV) * \left(\frac{q^n}{q^n - 1} - \frac{1}{n(q-1)} \right) + RV \right] * \frac{i}{100} + \left(\frac{\sum_{t=1}^{n} R_t}{n} \right) \quad (1)$$

AC	=	average costs
A	=	acquisition value
RV	=	residual value
n	=	service life in years
i	=	rate of interest
q	=	$1+i$
R	=	repairs
t	=	years

The average cost per year consists of the value of depreciation, a factor for the interest and the yearly occurring average repairs of the technical equipment.

Description of the Research Areas

The microregion of Picos in the state of Piauí and the municipality of Tauá in the state of Ceará are localized in the semiarid hinterland called „Sertão“ in the northeast of Brazil (Fig. 2). The main economic activity in this region is characterized by agriculture and animal husbandry.

The mean annual rainfall regime in the microregion of Picos is about 700 mm. In the municipality of Tauá, the mean annual rainfall is lower with a mean rainfall regime of about 400 mm. In both regions the rainy season lasts about 4 months from approximately January/February until April/May (SUDENE 1990a,b; FUNCEME 1995). During this period it is possible to realize the cultivation of rainfed crops such as corn (*Zea mays* L.), cowpea (*Vigna unguiculata* L.), cassava (*Manihot esculenta* L.) cashew (*Anarcadium occidentale* L.) and forage crops used for animal husbandry activities. The mean annual temperature is about 25 °C. The rainfall regime is characterized by a high variability during each production period and/or between different agricultural years.

A principal difference between the two regions is the availability of water resources used for domestic water use, irrigation purposes and animal husbandry activities. The microregion of Picos is part of the sedimentary basin of the Parnaíba River where fossil groundwater resources can be made available for domestic water use and agriculture. The region is characterized by the exploration of these fossil groundwater resources since the year 1976, but without any governmental or institutional regulation. The water is transported in deep tube wells (port.: poço tubular) by means of electric pumps to the surface. The region is characterized by different geologic and ecological site conditions, the high „chapadas“ and the „valleys“ of small non-perennial rivers. At the „chapadas“, the tube wells are drilled up to 600 m deep, but they only supply small amounts of water

(1–5 m^3/ha). In the „valleys", the tube wells are drilled less deep up to 180 m providing satisfactory amounts of water (up to 45 m^3/ha). The water quality allows use for drinking water and irrigation purposes. The microregion of Picos is dominated by agricultural exploration. The main crops cultivated are cowpea, cassava, cashew and maize at the „chapadas" and irrigated vegetable in the „valleys" (tomato, bell pepper, garlic, onion, green maize and green cowpea).

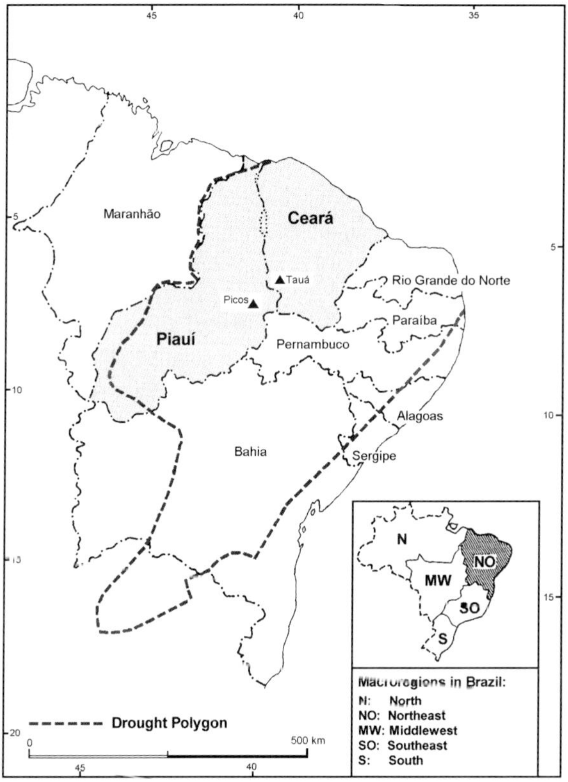

Fig. 2. Geographic location of the research areas Picos – PI and Tauá – CE in the northeast of Brazil

In contrast, the municipality of Tauá is part of a region with crystalline basement rock and, because of this specific geologic condition, the water resources in this area are more scarce. In the municipality of Tauá, fossil groundwater resources do not exist. The predominant water resources used for domestic water consumption and agricultural purposes are small storage lakes (port.: açudes), draw-wells (port.: poço cacimbão) and large water reservoirs (port.: barragem). Tube wells drilled in this region are only up to 60 m deep and the water provided is characterized by a high salinity that limits the use of this water for drinking water purposes. The municipality of Tauá is a region dominated by animal husbandry. Agricultural activities are focused on the production of forage crops and corn for animal production, as well as on the cultivation of cowpea for the fulfillment of the subsistence requirements of farmers.

The main problems of the farmers in the two investigated regions are:

- Coping with the high variability and uncertainty of precipitation regime which has an enormous influence on the production potential in crop cultivation and animal husbandry,
- coping with a low liquidity of own capital resources for financing agricultural activities during the rainy season, and
- coping with insufficient own capital resources for necessary investments on the farm, for example, investments to improve the basis of water resources on the farm or to improve the mechanization level.

Results

In this section we present the results of the calculation of farm and household income for three different farms located at different site conditions in the municipality of Tauá in Ceará.

As typical site conditions in this municipality, three different situations can be described. Agricultural enterprises exist situated at the *„Beira Rio"* site condition that are characterized by relatively good water resources due to the localization of some fields near small rivers producing on alluvial soils with relatively high fertility and good contact to the ground water level. Dried up river-beds can be used to dig water-holes in seasons of water shortage providing drinking water and low amounts of irrigation water for horticultural production activities in small gardens.

A second group of agricultural enterprises is located mainly in a hilly region of the municipality of Tauá called *„Serrra*". The specific site conditions in this area are characterized by a slightly more humid rainfall regime in comparison with other parts of the municipality.

As a third regional component of different site conditions, we can distinguish the *„Depressão Sertaneja"*. In this region most of the farms are not located near small rivers and therefore the situation concerning water resources is more precarious in comparison with other farms in the municipality (ESPLAR 1993).

Besides the differentiation of farming systems at a first level according to the site conditions, at a second level different mechanization levels can be distinguished in dependency of the farm size. Generally, large-scale farms use tractors for soil tillage. The utilization of implements for planting and weeding in combination with tractors is not widespread in the municipality of Tauá. Small- and medium-scale farms mainly use animal traction systems or hand tools. Paying for the service of mechanized soil tillage with tractors and for seasonal manpower for planting, weeding and harvesting is common for small- and medium-scale farmers.

In the following sections some preliminary results concerning income analysis are presented for three selected farms in the municipality of Tauá.

Income Analysis for a Farm Located at the Site Condition „Beira Rio"

The farm located at the ecological site condition „Beira Rio" has a total area of 600 ha. The focus of agricultural production at this farm is on animal husbandry. The farm has a large area with planted pasture and primary vegetation, which is normally used in the semiarid „Sertão" as natural pasture for cattle and small ruminants. Therefore, the herd and flock sizes with cattle and sheep on this farm are quite high. The area of utilized arable land is, at 70 ha, quite large. The predominant crop cultivated on this area is corn used for cattle and sheep feeding. The mechanization level of this farm is very high, but the farmer is working with equipment that is more than 20 years old. Soil tillage for the preparation of planting is done by tractor, weeding by animal traction and all other operations are realized by hand tools. The farm is characterized by the employment of day laborers in labor-intensive periods like planting, weeding and harvesting, as well as the employment of permanent manpower to care for the herd of cattle and the flock of sheep.

Table 2. Annual gross farm income of a farm at the „Beira Rio" in the years 1998 and 1995

	Drought year 1998	Normal year 1995
Plant production activities	–5063.00 R$	+5888.00 R$
Animal husbandry activities	+34,030.00 R$	+17,193.00 R$
Fixed costs	–7000.00 R$	–2247.00 R$
Total gross farm income	**+21,967.00 R$**	**+20,834.00 R$**
Income, non-agricultural activities	0.00 R$	0.00 R$
Total household income	**+21,967.00 R$**	**+20,834.00 R$**

Source: own calculations 1995 and 1998

Analyzing the income of the farm in a drought year, as it was in 1998, there is no income from plant production activities, because of the total loss of marketable yield of corn and cowpea. The income from animal husbandry, in this case the production of beef, milk, cheese and butter, provides a relatively good income for the farmer's family. The forage base produced for animal husbandry under this site condition permits the production of relatively drought-tolerant species such as *Cenchrus ciliaris* L. (Buffel grass) and, at the shores of small storage lakes and small rivers, the cultivation of *Brachiaria mutica*, even in a dry year due to the better surface near groundwater supply of the soil. The non-marketable biomass of corn production can be transferred to animal husbandry activities for feeding cattle and sheep and thus supporting, to a large extent, the survival of the herd of cattle and the flock of sheep during the time of the drought.

Subtracting the fixed farm costs of about 7000.00 R$, the total farm income in the drought year 1998 is about 21,967.00 R$. As the farmer has no further job activities in the non-agricultural sector, the household income is the same.

In a normal year, as it was in 1995, the situation of the farm is different. From plant production activities the farmer can earn about 5888.00 R$. A part of the harvested plant products, in this case principally corn, is not sold but transferred as forage to animal husbandry for the production of meat, cheese and milk.

The income contribution from animal husbandry in the normal year is only half the income contribution in the dry year. One reason is that in this case the farmer intensified the production of beef and milk from cattle as well as meat production from sheep. A second reason is the different price level in the two years: in a drought, prices normally are higher than in a normal year and, when the production level is nearly the same, or perhaps higher, as it was in this case, it is logical that the income contribution in a drought year can be higher than in a normal year.

Fixed farm costs in the normal year are lower, because the farmer does not employ as much permanent labor to care for the forage base for animal husbandry.

Comparing a normal and a drought year, farm and household income of a farm located at favorable site conditions for agricultural production and animal husbandry is quite stable. Also, due to the different price levels, farm income in drought years in some cases can be higher than in normal years (see Table 2).

Income Analysis for a Farm Located at the Site Condition „Serra“

The farm located at the ecological site condition „Serra“ has a total area of 74 ha. The focus of agricultural production at this farm is on the production of castor oil plant (*Ricinus comunis* L.), cowpea, fava beans (*Phaseolus lunatus* L.) and corn. Animal husbandry serves to fulfill the family's consumption of milk and meat from small ruminants such as sheep and goats. The mechanization level of the farm is very low. The farmer uses exclusively hand tools and family labor for all farm operations.

The income of the farm located in the „Serra“ for the drought year 1998 was negative. There is no income contribution from plant production activities and from animal husbandry. Due to the limited forage base and the exclusive utilization of animal husbandry products for family consumption, these farm activities cause only costs for the farmer. Considering the fixed costs the farmer had a negative income in the drought year. And, as the farmer has no possibilities to earn an additional salary from non-agricultural activities, the situation of the farm in drought years is very precarious putting in risk the existence of the family, because the household income was negative, too.

Table 3. Annual gross farm income of a farm at the „Serra“ in the years 1998 and 1995

	Drought year 1998	Normal year 1995
Plant production activities	–536.00 R$	+2671.00 R$
Animal husbandry activities	–85.00 R$	–85.00 R$
Fixed costs	–88.00 R$	–248.00 R$
Total gross farm income	**–709.00 R$**	**+2338.00 R$**
Income, non-agricultural activities	0.00 R$	0.00 R$
Total household income	**–709.00 R$**	**+2338.00 R$**

Source: own calculations 1995 and 1998

In the normal year 1995, the situation of the farm was different. From plant production activities the farmer can earn about 2671.00 R$. The income contribution from animal husbandry is negative, because the production is used for the consumption of the family. Considering the fixed farm costs the farmer can achieve a farm and household income of 2338.00 R$, that means a monthly income of 195.00 R$, which is one and a half times higher than the Brazilian minimum salary (see Table 3).

Income Analysis for a Farm Located at the Site Condition „Depressão Sertaneja“

The farm located at the ecological site condition „Depressão Sertaneja“ has a total area of 192 ha. The focus of agricultural production at the farm is on the production of corn, cowpea and fava beans. Animal husbandry serves on the one hand for the fulfillment of the family‘s consumption of milk, cheese, butter and meat, and on the other for selling small ruminants like sheep and goats. The mechanization level of the farm is very low. The farmer uses exclusively hand tools and family labor for all farm operations. Although the farm size is quite large the farmer does not use animal traction systems for soil tillage or weeding. In labor-intensive periods the farmer employs sporadically day laborers, especially for planting and harvesting cowpea and corn.

There is no contribution to the farm income from plant production activities and from animal husbandry in the drought year 1998. Subtracting the fixed farm costs of about 72.00 R$, the total farm income in the drought year is negative.

In this specific case, the farmer's family has a further income from the non-agricultural sector: two old-age pensions for rural workers (one Brazilian minimum salary per month and person). This means that the family is financing the farm activities by non-agricultural income sources in the drought year and the additional income from governmental rural pensions avoids, too, that the farmer has a negative household income endangering the existence of the family.

Table 4. Annual gross farm income of a farm at the „Depressão Sertaneja“ in the years 1998 and 1995

	Drought year 1998	Normal year 1995
Plant production activities	–333.00 R$	+98.00 R$
Animal husbandry activities	–1119.00 R$	+725.00 R$
Fixed costs	–72.00 R$	–151.00 R$
Total gross farm income	**–1524.00 R$**	**+672.00 R$**
Income, non-agricultural activities	+3040.00 R$	+2400.00 R$
Total household income	**+1516.00 R$**	**+3072.00 R$**

Source: own calculations 1995 and 1998

In the normal year 1995, the situation of the farm is a bit more positive. From plant production activities the farmer can earn about 98.00 R$. The income contribution from animal husbandry is about 725.00 R$. Subtracting fixed costs, the farm income in a normal year is positive, but at the specific site condition of the „Depressão Sertaneja“, with its scarce water resources, the total farm income is very low.

This shows how important non-agricultural income possibilities can be for agricultural enterprises in the semiarid „Sertão“ in the northeast of Brazil, especially for those enterprises coping with unfavorable site conditions as, e.g., water resources (see Table 4).

Discussion

The analysis of gross farm income of three investigated farms in the municipality of Tauá can serve to show how gross farm income can vary in a region which is characterized by extreme uncertainty in the precipitation regime with long-lasting and recurrent drought periods. Each of these three farms characterizes a certain farm type which is common for the municipality of Tauá. The main differences exist in the size of farm area and the suitability of the ecological site conditions for the realization of plant production and animal husbandry activities.

A detailed description and analysis of plant production and animal husbandry activities realized on each farm was carried out for all investigated farms; however, due to the large amount of data, it was not realistic to present these (still not aggregated) data sets in this contribution. The differences in the contribution of plant production and animal husbandry activities to the gross farm income are based primarily on the different physical productivity of cropping and animal husbandry systems. Large farms with better site conditions in terms of water resources and soil fertility normally can resist better a decline in yield than smaller farms with unfavorable site conditions. Even if there is a complete failure in marketable crop yield by a drought period, the possibility exists to compensate the income loss of plant production activities by animal husbandry due to the better forage base on large farms with favorable site conditions.

The partial profitability of each plant production or animal husbandry activity expressed by their gross margins per production unit can be very different due to the specific input–output relations on each farm in dependency of different strategies by the farmers to reduce variable production costs and thus improve the partial profitability. Partial profitability is really important from the prospective of introducing new production methods or technologies on farms which have to cope with uncertain natural site conditions influencing crop yield and animal husbandry production so strongly that in certain years, e.g., drought causes not only a decline, but also a total failure in crop yield. From this aspect the existing data sets need further analysis and aggregation for each farming system in order to select successful farm activities for introducing on a large scale to farmers in the investigated regions after they have been tested in linear programming models for suitability of integration in the existing farm organization.

As in Brazil, and especially in the northeast, credit policy for the agricultural sector is an important tool for the development of rural areas applying different financial service programs to facilitate the introduction of innovative crop production and animal husbandry systems to improve the farm income in the long run, it is important to analyze not only the partial profitability of farm activities, but also the financial situation in general. For example, a small farm can have an excellent profitability from the realized farm activities, but, due the small size of the farm, this income is only sufficient to cover the personal expenses needed by the farm family during the year. A very low or negative income does not allow any of the own capital to be saved to finance the variable expenses of agricultural activities or to realize long-term investments on the farm to stabilize future farm income. Also, without a reasonable financial basis, it is also impossible to pay back borro wed money received from credit programs, especially when the financed innovative farm activities do not reach the expected profitability because they are not well adapted to the organization of the farm or to drought periods. Many credit programs have failed in recent years because the farmers were not able to pay back the borrowed money due to credit programs not being well adapted to the financial possibilities of the farmers.

Acknowledgements

The research project was carried out under the auspices of the agreement on scientific-technological cooperation signed by the governments of Germany and Brazil. The German partner was sponsored by the Federal Ministry of Education and Research (BMBF – Förderkennzeichen 01 LK 9710/6). The Brazilian partner was sponsored by the Council for Scientific and Technological Development (CNPq).

References

Bazin F (1993) Evolução histórica dos sistemas de produção agropecuária da região de Tauá-CE. SUDENE/UPPA, Recife, 36 pp

ESPLARr (Centro de Pesquisa e Assesoria) (ed) (1993) Plano de desenvolvimento agroecológico e participativo dos pequenos produtores do município de Tauá (PDA – Tauá). Tauá, 56 pp

FUNCEME (Fundação Cearense de Meteorologia e Recursos Hídricos) (1995) Relatório de Pluviometria por faixa de anos - Estado do Ceará, Municípios de Parambu e Tauá, 1980–1995, Fortaleza

SUDENE (Superintendência do Desenvolvimento do Nordeste) (1990a) Dados pluviométricos mensais do Nordeste, Estado do Piauí, Série Pluviometria 2, Recife

SUDENE (Superintendência do Desenvolvimento do Nordeste) (1990b) Dados pluviométricos mensais do Nordeste, Estado do Ceará, Série Pluviometria 2, Recife

Structure and Possible Development of the Agricultural Sector of Ceará and Piauí

Sabine Höynck

Institute for Technology in the Tropics, University of Applied Sciences Cologne, Betzdorfer Str. 2, 50679 Cologne, Germany, sabine.hoeynck@gmx.de

Introduction

The agricultural sector of Ceará and Piauí represents a large share of the population in Ceará and Piauí. In 1998, 35.5% of the economically active population were primarily engaged in the agricultural sector of Ceará, 32.5% (Ceará in 1998) and 41.8% (Piauí in 1996) of the population were living in rural areas In the period 1985–1998, the share of the agricultural sector of the regional GDP has decreased from 15.3 to 5.3% in Ceará and from 27.0 to 10.5% in Piauí. The GDP per person employed in the agricultural sector has also decreased in relation to the other sectors.

While the industrial and service sectors have experienced substantial economic growth, the agricultural sector has remained rather stagnant, meaning that income decreases in relation to the income generated in the other sectors (Fig. 1).

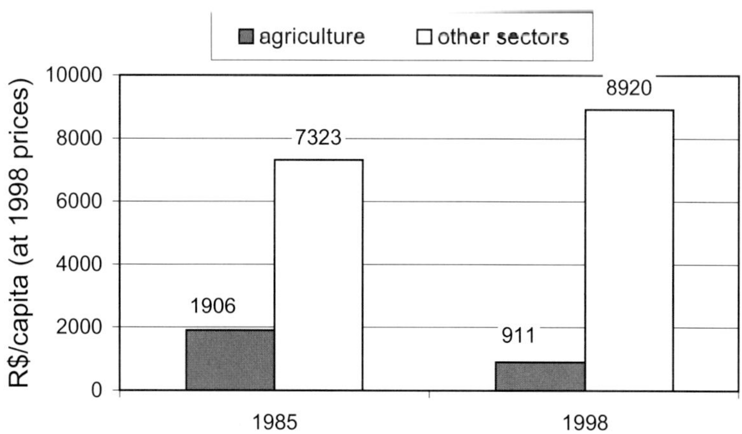

Fig. 1. Agricultural GDP per capita of the economically active population of Ceará in comparison to other sectors (Source: Governo do Estado do Ceará 1999)

More than any other sector, the agricultural sector depends on climatic conditions. The climatic conditions, specifically with respect to rainfall distribution and intensities, are highly insecure, with the prospect of an increasing variability for the future.

Problems faced by the agricultural sector vary depending on the farm type concerned. A general distinction should be drawn:

1. Capital-intensive crop production systems. These production systems can only be established when water resources are sufficiently and reliably available. Apart from those located at a few naturally favored locations (e.g., the „serras" and „brejos"), these farm types require irrigation. The profitability of this farm type depends basically on the product prices and the input costs, among which the costs for water and irrigation will play a more dominant role in the future.
2. Large-scale extensive animal production systems. These systems cover the largest areas in regions where water scarcity does not allow crops to be produced in a way that would be economically competitive. Beef production is the major product of these systems but sheep and goat production are becoming increasingly important.
3. Large-scale rainfed highly mechanized grain crop production dedicated for export out of the region. This system only occurs in the south of Piauí, favored by two factors: rainfall and soil conditions are suitable for this production system and population density, specifically land-use pressure by the subsistence sector, is low.
4. Small-scale subsistence farming in areas lacking irrigation opportunities. By number this system holds the largest share. Crop production is primarily dedicated to cover the farm household demand for food. Integrated farming systems with a share of subsistence crop production and a share of animal production are very common in this sector. Sheep and goats are an increasingly important cash income source for subsistence farms. Cattle are kept to serve the two purposes of milk and meat production.

Further distinctions might consider the possibilities to specialize under specific local conditions, e.g., for dairy farms combined with irrigated forage production located close to dairy processing units, or horticulture in the mountainous areas („serras"). Such a distinction would lead too far when the development of all the agricultural sector models is taken into account, although the local importance of such specialization should be kept in mind.

The Agricultural Sector Model RASMO

The agricultural sector model RASMO (Regional Agricultural Sector Model for Ceará and Piauí) has been developed to analyze the income situation of the farming sector and the land-use systems in their spatial distribution. Distinguishing among farm size classes, production intensity classes, and soil classes, agricultural production parameters have been included for all municipalities (for 1996) of the two states. The major data source is the agricultural census of 1995/1996 (IBGE 1998). Additional data on production technology applied in correspondence to the land-use systems and yields have been used to define production systems for 7 annual crops (corn, beans, rice, cassava, cotton, melons and tomatoes) and 5 perennial crops (cashew, banana, sugarcane, coconut and mangos). The production value of these 12 crops amounts to 93% of the overall crop production value of the region.

Husbandry systems are related to the natural pasture-land and fodder production, basically residuals from food crop production in the small-scale farming sector or forage plantations. For the latter, the production system „buffalo grass" has been introduced in the model.

RASMO simulates the adaptation of the agricultural sector to specific changing environmental conditions based on restrictions of productive factors: soil, capital, labor (also considering the seasonal variations of demand) and water, as well as the interaction of production systems within the land-use systems. As the objective of the decision function of farmers, income maximization is assumed.

Farmers, especially subsistence farmers, can only adopt income maximization strategies if subsistence requirements of food crops are fulfilled. On the subsistence production level, markets allowing market-oriented choices of farmers as consumers and producers do not exist. This problem is considered in the „subsistence restriction". Production potentials exceeding the minimum food production requirements are assumed to have been exploited in an income-maximization driven adaptation process to environmental changes.

This model has been used firstly to display the actual situation of production systems depending on farm size classes and secondly to simulate the adaptation of the sector to major environmental changes: water availability and corresponding yield levels and prices. For simulation of long-term adaptations of the sector, the production system module has been introduced into the integrated model SIM.

Present Situation of Farming Systems

Considering the agro-ecological conditions, the state of Ceará is dominated by the „Sertão" conditions that are characterized by the highest occurrence of drought events and the lowest capacity to compensate periods of water shortage within the regions of Ceará and Piauí. Much more favored are the regions characterized as „Litoral" (coastal regions) which dispose of groundwater resources and have a higher incidence and reliability of rainfalls. Cashew, as a cash crop that can be produced both on a large- and small scale, is a major activity in the litoral region of Ceará. The litoral regions are also the regions where irrigation development is most advanced. Piauí is dominated by the area „Meio-Norte", where rainfall events are rather reliable and abundant, but which suffers from low productivity potentials related to soil conditions. Piauí also disposes of a considerable share of „Cerrado" land, which is basically suitable for pasture systems, allowing pasture management with a share of artificial pastures. In the southern cerrados of Piauí, rainfall conditions and soil conditions favor the production of grain crops, predominantly soybean. These large-scale highly mechanized grain production systems have been established rather recently, predominantly by migrants from the south of Brazil.

The regional concentration of selected crops is shown in Fig. 2. Beans and corn are the prevailing crops of subsistence farms, concentrating on the Sertão of Ceará. In the respective municipalities, 20–60% of the cropping area is dedicated to one of these subsistence products; the situation in similar in the interior-east of Piauí. The west of Piauí, though, is a much more humid region, where the major subsistence crop is rice. In the south of Piauí, land use is characterized by grain production on a large scale, with a nucleus of rice production in the Parnaíba River region and, as mentioned before but not shown here, a large share of soybean production.

Animal production systems on a large scale are typical for the „Meio-Norte" and „Cerrado" landscape. In these specialized systems, potential cropping area is used for forage production, referred to in Fig. 2 as seeded pasture.

Concerning the intensity of production systems, water resources for irrigation are the most important restriction. Thus, a regional differentiation of farm production systems according to the irrigation water availability can be displayed to demonstrate the structure of farming systems in response to regional potentials (Fig. 3). As can be seen at first glance, the share of farm land preserved in a naturalistic status (farm land not used and natural forest) is much more prominent in Piauí than in Ceará. Land-use systems in the litoral area are characterized by perennial crops, predominantly non-irrigated plantations of cashew and irrigated plantations of fruit trees (mango, coconut).

The structure of farming systems is not only dependent on the natural environmental conditions and irrigation resources conditions, but also to a large extent on the farm size structure and population density. In the short to medium term, the optimum allocation of production resources within the agricultural sector is very much restricted. In most of the areas, food crop production, covered by the annual crop fraction in Fig. 4, concentrates on the small-scale farms to comply with the subsistence demand. The potential for capital generation is extremely low since marketable surplus is hardly generated in this fraction of farms. The available farm income generated in the production systems is shown in Fig. 5, calculated based on the agronomic results of 1995/1996, a year with „average" climatic and market conditions.

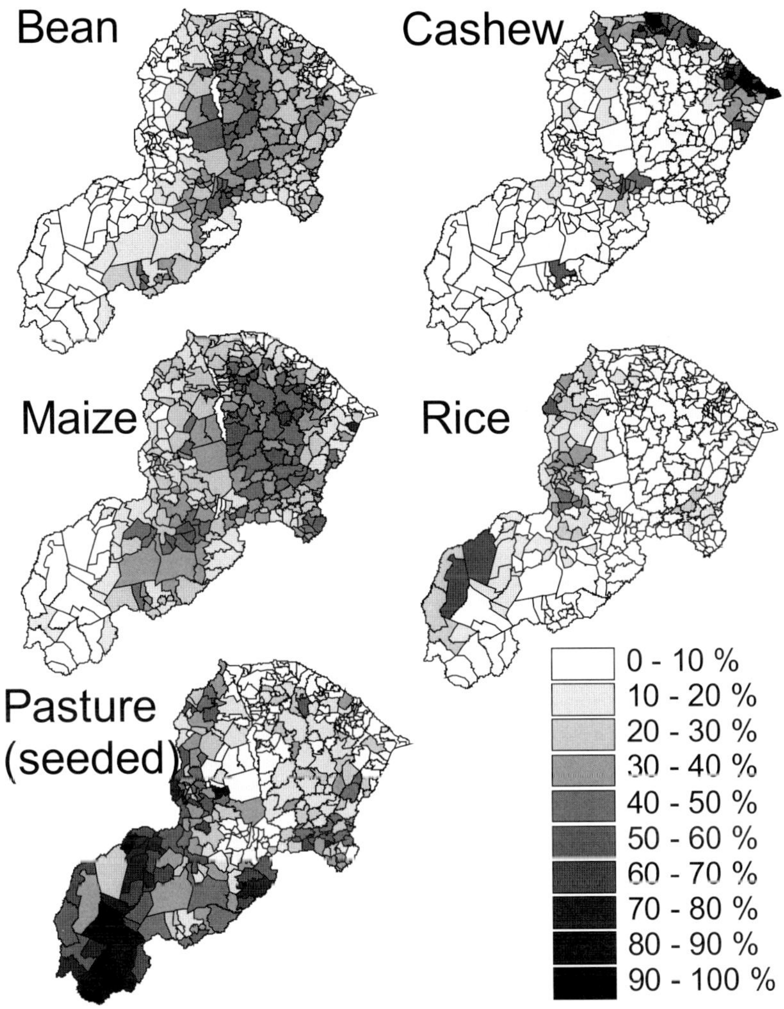

Fig. 2. Regional concentration of selected agricultural activities Data source: Agricultural Census 1995/96 (IBGE 1998)

From the viewpoint of decreasing farm sizes in the subsistence-dominated small-scale farming sector, an overexploitation of the available land resources is common. The actual system of continuously exploiting farm land for crop production and residual use as fodder without applying soil-enriching measures (biomass accumulation, fertilization) could be fatal for the land resources.

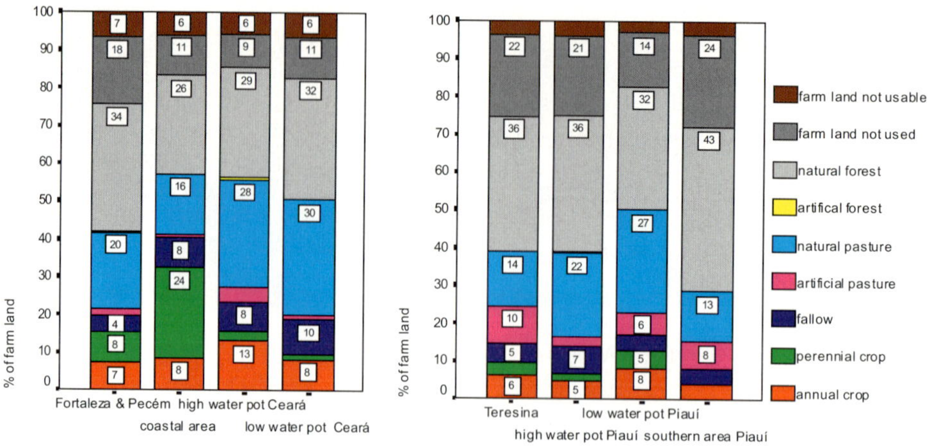

Fig. 3. Production structures depending on the regional aspect of irrigation water availability in Ceará and Piauí (Source: IBGE 1998)

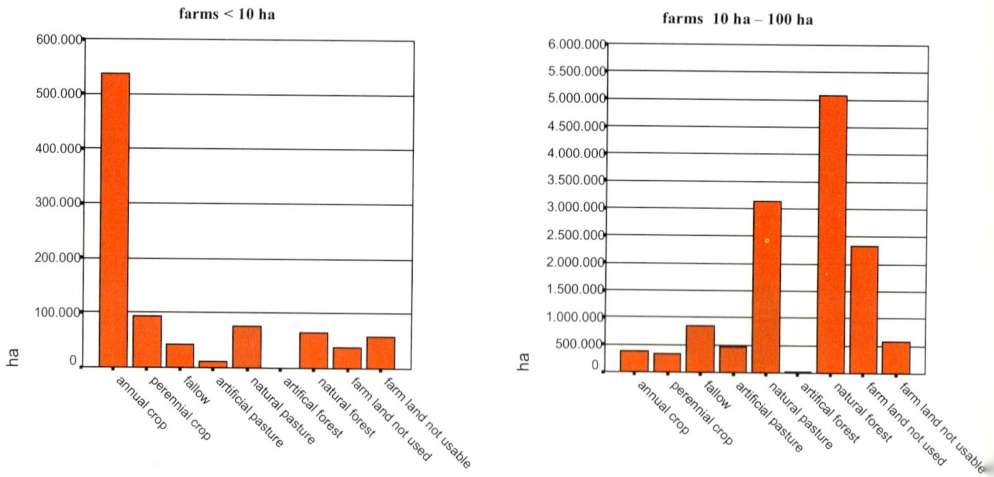

Fig. 4. Land use in farm areas of small-scale farms in Ceará and Piauí

Another time bomb is found also among the large-scale farms. Large-scale husbandry systems are typically beef cattle systems. Recent developments show an increase in the number of sheep and goats within the husbandry sector. The latest observations of deterioration processes of husbandry production areas are alarming. The implications of the increasing number of goats and sheep have not yet been assessed in relation to their impact on the soils. Pasture management is an urgent requirement for sustainable development of the region.

The carrying capacity of natural pasture systems in the Sertão region lies at more or less 1 large animal unit/10 ha (500 kg life weight). Complementary forage crop production and the mixed farming systems of annual crops might substantially increase this capacity. Still, the stock density found in some of the municipalities appears to be not realistic, specifically for the small farming sector. There the stock density in relation to the farms' pasture land exceeds the natural capacity up to 1000 times (Fig. 5). The explanation for this lies in the fact that normally farm land of small-scale farms does not include specific pasture areas. During periods of natural food crop availability, off-farm land is used (natural „forest", pasture area along the roads, farm land of large-scale farms).

Prospects for future development

Historically, the development of farm land has been different in Ceará and Piauí. Piauí is a state with extensive land resources for agricultural production systems in comparison with the population inhabiting the area. This has led to a constant increase in farm land in the last decades. In contrast, Ceará has suffered seriously from the shock of being berated of the extensively planted cash crop cotton that suffered simultaneously from exhaustion of the soil and pest infestation („bicudo"). Thus, beginning in the mid-1970s and cumulating in the mid-1980s, extensive farm land had to be given up. Deducting the cotton areas from the land development farm area development shows a similar constant growth of farmland extension as Piauí. It can be assumed that cotton areas will be reconquered for agricultural production in the future and that additional land will be made available since population pressure, especially in the small-scale farming sector, will remain a very strong driving force to extend agricultural activities.

A pessimistic preview, though, could assume new shock waves to come within the next two decades, since agricultural production systems are increasingly encroaching marginal land. Production systems are intensified at the expense of fallow areas. Experts warn that 20% of the farm area in Ceará and also extensive stretches in Piauí are endangered by irreversible desertification processes.

Agricultural research and extension services must extend their programs to promote sustainable agricultural production systems, specifically for the low income group of farmers, to avoid irrecoverable loss of production resources in the next 10–20 years. Promising approaches exist, e.g., integrated farming systems combining forage management strategies (e.g., Leucena as protein bank) for the critical dry season periods complementary with farm household alimentation strategies in an agro-forest-pasture system (EMBRAPA Caprinos 2001).

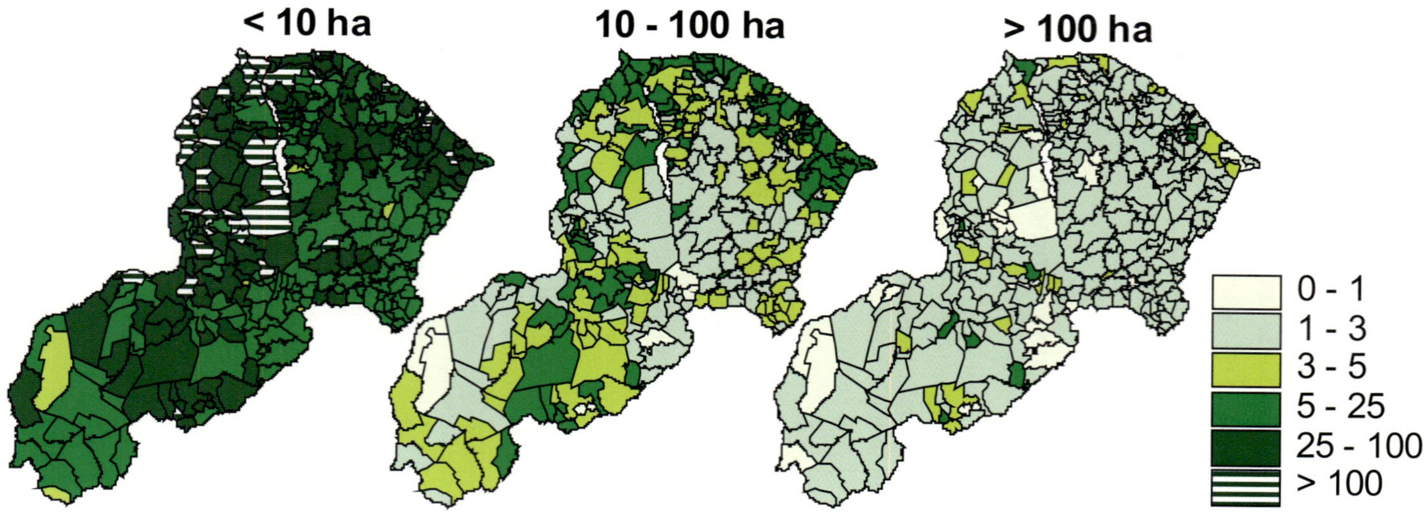

Source: H. Hinterthür (2000) Calculation of RASMO (unpubl.)

Fig. 5. Stock density (Animal Units/ 500 kg) per ha pasture land (as labeled in the agricultural census 1995/1996, IBGE 1998)

The planned extension of irrigated areas, concentrating on the regions identified as high potential areas, will increase the share of cropping area only to a small degree, since irrigation is most likely to be introduced in those areas already more suitable for agricultural production under rainfed conditions (alluvium).

Economic Situation of the Farm Households

Based on the agricultural census information on land use in the respective municipality and farm-size group and the corresponding RASMO assumptions on the gross margins of the production systems, RASMO has calculated the potential annual cash income (in subsistence farms primarily used for home consumption) for the base year 1995/1996. As can be seen in Fig. 6, most of the municipalities would attribute a farm income of less than 1000 R$/year for farms up to 10 ha [about 1000 US$ (12/95); this is less than a minimum salary – the minimum pension for retired persons is half a minimum salary]. This income is normally not sufficient to sustain a family. The real farm income of the low intensity production system farms (covered mostly by the group <10 ha) might be slightly higher than calculated here, since productivity levels are estimated for yield potential of specialized production systems. Subsistence systems are dominated by mixed cropping practices, with average yields of each crop being slightly higher than its share of the area. Up to 10% – in specific areas even more – might be added to the income levels below calculated based on the mono-cropping yields. Unfortunately, extensive primary farm data collection would be necessary to exactly quantify the effect of mixed farming system on the farm income, which could not be considered yet.

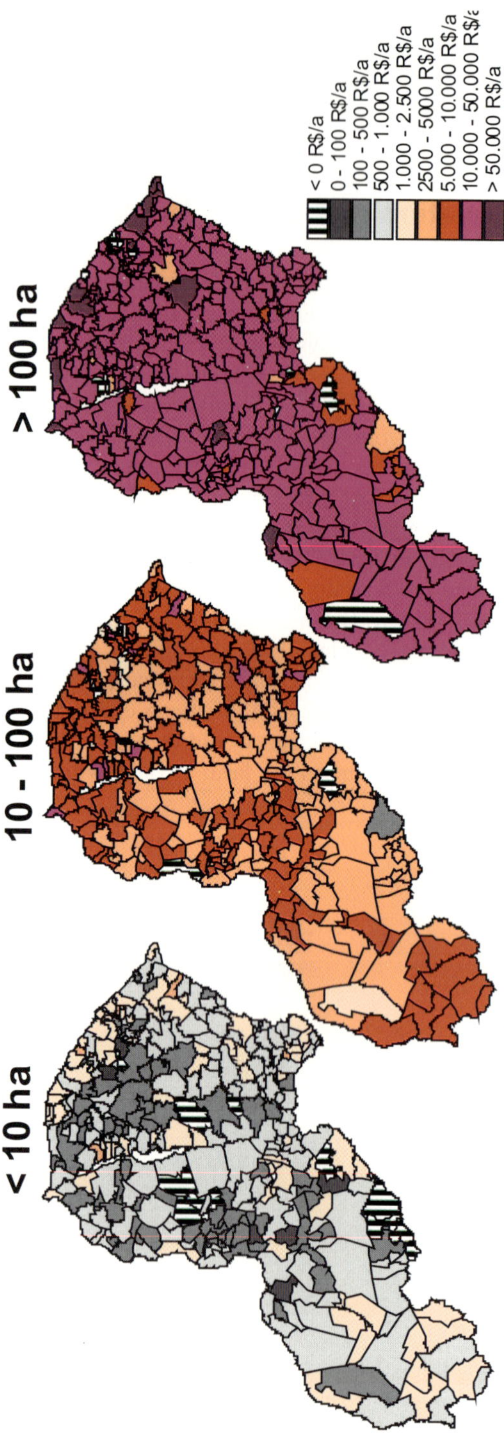

Source: H. Hinterthür (2000) Calculation of RASMO (unpubl.)

Fig. 6. Annual farm income 1995/1996

The income situation is comparatively better in those areas classified as high potential areas in terms of water resources.

A look at this comparative calculation of income depending on farm size classes shows that these classes can only approximately serve as potential income classes. Income depends rather on the farm type described above. An interesting example for this is the case of the municipality „Baixa Grande do Ribeiro“ in the southwest of Piauí which, in the RASMO calculation for 1996, shows the paradoxical situation of losses in large farms, while small farms earn on average up to 1000 R$ per year. In this area agriculture practices are extensive, meaning that fallow or farm forest areas are large. Extensive farming systems in such remote areas in Piauí are often characterized by subutilization of resources. Farms of less than 10 ha hold only a small share of about 15% of all farms. This rather small number of farms might well represent the share of agricultural activities taking place in the higher potential areas of this municipality, e.g., river valleys.

Economic Development

With the help of the RASMO model it has been estimated how agricultural production systems would adapt to conditions of decreased water availability (according to the rainfall pattern of the year 1942), and how such scarcity affects the average farm income for each municipality and farm size group (Fig. 7). These relative changes are expressed in percent of the farm income calculated for the base year 1995/1996. Farming systems adapt in a way that subsistence requirements on the municipality level are fulfilled while remaining resources are used in an income-maximizing way. Farm income would decrease most considerably for small farms: 100% or more loss of income in the drought-prone areas of the Sertão and in some cerrado areas.

Large-scale farms would suffer least from reduced water availability, since the prevailing husbandry production systems are less sensitive to water scarcity than the annual crops dominating the small-scale subsistence sector. White or green patches (Fig. 7) indicate that these municipalities and the prevailing production system in the respective farm size group are not affected by reduced water availability, either because water is available in abundance normally, or because the adaptation of production processes would be possible. Green patches indicate that, according to knowledge on production systems collected in the database and algorithms of RASMO, available resources might be used more efficiently, which might increase income even if production conditions in terms of water availability are less favorable than in the reference year 1995/1996.

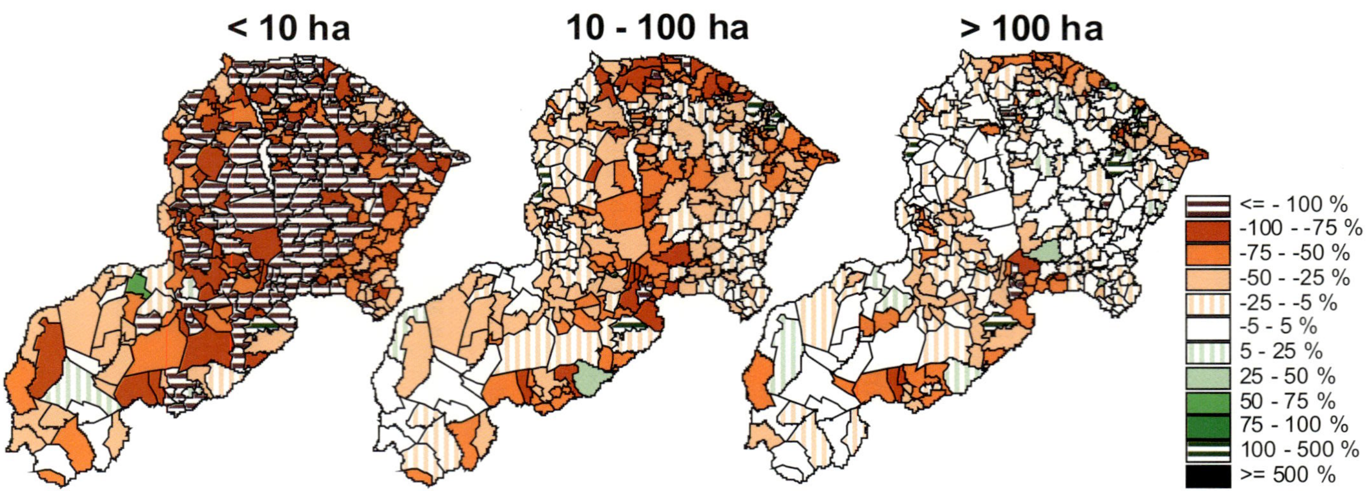

Source: H. Hinterthür (2000) Calculation of RASMO (unpubl.)

Fig. 7. Change of farm income in relation to farm income levels in reference year 1995/1996 in response to reduced water availability

Conclusions

The present structure and the potential adaptation to future developments of the agricultural sector of Ceará and Piauí assessed with the tool of the agricultural sector RASMO model leave the majority of farmers to live at a food production and income level which is, in a long-term view, not sufficient to sustain a family. The farm size structure prevailing presently and expected for the future and the related production structure inhibit an optimum allocation of productive resources. Most of all, development towards an income-generating agriculture is hampered by the requirement to produce food on land which is not appropriate for food crop production. RASMO was able to detect areas of overexploitation of land resources which would require further analysis and consequently political decision-making to preserve the production resources for future generations. Within the WAVES project, the regional model MOSDEL might be applied for more detailed analysis in the respective regions.

Economically, the subsistence-oriented production system is stagnant – or even recessive considering ecological constraints – since there is no range of decisions to optimize on-farm resource allocation and to invest in technical progress. Under the persisting population pressure, there are limited prospects for mitigating the poverty situation for the majority of people engaged in the agricultural sector.

Positive regional impulses originate from the growth of the irrigated agricultural sector. Being by nature regionally concentrated, irrigation areas facilitate a regional economic growth transferred by interwoven sector structures. In interaction with the surrounding rainfed agricultural area, the development of irrigation centers offers off-farm income opportunities and decentralized markets, reducing the rigidity of the subsistence constraint.

To support public efforts in regional planning oriented towards a growth in which all social groups might participate, future model-based research should further elaborate the constraint situation for the specific farm types (labor, cash, marketing) which should also consider the income situation of agricultural households from non-farming activities.

References

Beltrão V, Lamour C (1985) Uso atual e potencial dos solos do Nordeste. SUDENE, Projeto Nordeste 6, Recife

EMBRAPA Caprinos (2001) Sistema de produção sustentável para a região da Caatinga. Ongoing project no 09.1998.001 (João Ambrósio de Araújo Filho) (http://www.cnpc.embrapa.br/projetos.html)

Governo do Estado do Ceará (1999) Plano multianual 2000–2003 (http://www.seplan.ce.gov.br/)

Hinterthür H, Gaese H (1999) Die Berücksichtigung des Produktionsfaktors Arbeit in einem Agrarsektormodell für einen bevölkerungsreichen semiariden Raum in Entwicklung: das Beispiel Nordost-Brasilien. Gött Beitr Land Forstwirtsch Tropen Subtropen 133:99–105
IBGE (1998) Censo agropecuário 1995/96. IBGE, Rio de Janeiro

Groundwater Supply Conditions and Willingness To Pay for Desalinated Water in Tauá (Brazil)

José César Vieira Pinheiro, Luiz Artur Clemente da Silva, Marcelo Theophilo Folhes and Manuel-Osório de Lima Viana

Departamento de Economia Agrícola, Universidade Federal do Ceará, CEP-60150-100 Fortaleza-Ceará, Brazil, jcvpinhe@ufc.br

Introduction

This study deals with components of local water supply and demand. On the supply side, it presents an appraisal of the spatial and social distribution, together with water quality conditions of deep wells in the *municipal district* of Tauá, located in the southwestern region of the state of Ceará, where, as a typical aspect of the semiarid tropics in northeast Brazil, the climatic and spatial variability of rainfall is very high.

Surface water resources are the main water source for the urban area of Tauá. Groundwater plays a relevant role in the countryside in small communities, mainly during periodic droughts. Rivers and reservoirs are almost always intermittent, generally offering low water discharge with high contents of dissolved solids, i.e., their water is very saline. Since the communities are widespread throughout the countryside, there is a wide range of water quality.

The study also describes the use of water by families, considering both the concentration of salts in deep-well waters and inequality of access to drinking water; recent studies on the subject have been accomplished by Pinheiro and Silva (2001).

According to IRDB (1993), nearly all the strategies for tackling the subject of water have been supply-driven. The importance of water demand in the selection of appropriate policies has been virtually ignored. For that reason, on the demand side, the typical community of *Poço da Onça* („Jaguar Well") in Tauá was chosen as a case study for capturing the residents' perception towards water shortages. Thus, the study also seeks to estimate households' willingness to pay (WTP) for an improved water supply system in that semiarid region. Such a parameter would help to adjust the drinking water supply to its demand. An effort in the direction of the equilibrium price would also warrant, totally or partially, the financing of needed investments for improving the quality of water services.

Thus, even though this study is not a complete, integrated model of simultaneous equations for supply and demand (in reality, it is often not a question of supply-demand water balance, but rather of willingness to pay), it is of help for a water policy design by taking into account water ownership and tenure, purchase or catchment costs, solidarity schemes or reasons for conflict among residents, and also willingness to pay for water.

Methodology

Survey and data

In order to examine some components of supply, the study worked initially on data from the *Programa de Recenseamento de Fontes de Abastecimento de Água Subterrânea no Estado do Ceará* (CPRM – Companhia de Pesquisa de Recursos Minerais, 1997; census on groundwater supplies in the state of Ceará). This census provided information on the use of deep wells, on their discharge, concentration of dissolved solids in water, and the number of families living around the water sources.

On the other hand, in order to estimate demand components, a field survey was carried out in the community of Poço da Onça in March, 1999; 501 residents were found there. Most of the year, the population of Poço da Onça consumes water from a deep well. The content of dissolved solids in the ground water is measured to be 6.200 mg/L, six times the recommended limit. The survey collected questionnaires filled out by all (126) the local families. The data obtained covered a variety of topics, including water consumption, current water source, modalities of use, socioeconomic characteristics of the families and the price they would be willing to pay for a desalinated (through the process of reverse osmosis) water supply system.

Methods of analysis

Supply conditions: the usual measure of equality or inequality is the Gini coefficient, the difference between actual distribution and equal distribution. This index takes a value from zero to one, where 0 means equal distribution (Gastwirth 1972). Then, in the present case study, the water distribution inequality among the families in Tauá was measured using the GINI index and other measures of inequality.

According to Hoffmann (1984, 1991), considering a distribution of values X_i (i=1, 2, 3,..., N) in growing order, the Gini index can be obtained through the following expression (where μ is the distribution mean):

$$G = \frac{2}{N^2 \mu} \sum_i i X_i - \frac{1}{N} - 1 \,. \tag{1}$$

The Gini index relates classes of the accumulated number of families and accumulated average availability of drinking water. Ten classes of average availability of drinking water (well discharge in liters of water/hour) were designed, with the following minimum values: 0, 400, 800, 1200, 1600, 2000, 2400, 2800, 3200, and 3600.

The percentiles used were „50^-“, „20^+“, „10^+“, „5^+“, and „1^+“ (Spiegel 1994). The 50^- proportion means the percentage of families with a lower degree of access to drinking water from deep wells. The 20^+, 10^+, 5^+, and 1^+ represent the percentages of families with higher degrees of availability of drinking water from deep wells.

Theoretical principles were taken from Weber and Bailly (1997) saying that, in all societies, anthropologists found scarcity to be socially constructed by selection of needs and use of specific goods as a means for assuring or strengthening social status. Therefore, the attributes focused upon in this study are: (1) behavior and attitudes which assure social status; (2) water tenure and water sharing, including forms of solidarity and reasons for conflicts; (3) water source sites per type of use; and (4) costs of purchase or extraction, as a measure of the effort for collecting water.

Demand estimation contingent valuation studies have been used to value public goods and other non-market commodities. The essence of contingent valuation methods is the specification of a hypothetical market for a non-market commodity and a query for the value placed on increments of that good by individuals. In other words, a contingent valuation survey simply asks people an open-ended question about how much they would be willing to pay for that commodity (Randall 1987). Econometric studies of water demand take into consideration a linear utility function for each person:

$$U_{ij}=\alpha_0+\alpha_1 p_j+\alpha_2 y_i+\alpha_3 z_i+e_{ij} \, . \tag{2}$$

The observed utility (U) for a particular individual (i) of a particular water type or source (j) depends upon price (p; R$ per month), income (y; R$ per month per family), a set of socioeconomic variables (z), and a stochastic error (e_{ij}).

It is assumed that an individual chooses between two water sources based on maximizing two conditional indirect utility functions. The first describes the utility gained from using the new or proposed source (j=1) and, the second one, the utility derived from using the current source (j=0). The theoretical formulation by Hanemann (1984), presented as follows, allows the estimation of monetary values for non-market good/service benefits, in real markets, using binary response data. The case of an individual willing to pay a price p for a service, after inserting the binary variable (j) in the utility function and also the socioeconomic ones (z), is represented by:

$$u_1(1, y-p; z) \geq u_0(0, y; z) \tag{3}$$

where: u_1 and u_0 are random variables with specific probability distributions, and whose means are: $v(1, y–p; z)$ and $v(0, y; z)$, respectively.

Equivalently, they can be written as:

$$v(1, y-p; z) - v(0, y; z) \geq e_0 - e_1 \, . \tag{4}$$

The result is a random variable whose probability distribution is given by:

$$P_1 = Pr[v(1, y-p; z) + e_1 \geq v(0, y; z) + e_0] \tag{5}$$

or:

$$P_1 = Pr[\Delta v \geq \varphi] \tag{6}$$

and:

$$\Delta v = v(1, y-p; z) - v(0, y; z) \tag{7}$$

where $\varphi = e_0 - e_1$ and, finally, $P_0 = 1 - P_1$.

Since P_φ (.) represents the accumulated distribution of φ, the probability for an interviewee to be WTP the price mentioned for the commodity is:

$$P_1 = P_\varphi(\Delta v) \, . \tag{8}$$

Thus, P_φ (.) is a difference in utility under the functional form:

$$v(j, y; z) = a_j(z) + by \tag{9}$$

where $j=0$, 1 and $b > 0$ and

$$\Delta v = a_1(z) + b(y-p) - a_0(z) - by = a^* - bp \tag{10}$$

where Δv represents the difference in utility. The variables p, y, and z were described earlier, and Eq. (10) shows the trade-off between utility and income.

Considering that φ has a standard logistic distribution, the mean equals 0 at the point where $P_\varphi(0)=0.5$, i.e., where the individual is indifferent between accepting or refusing the offer proposed. Thus, p satisfies ($a^* - bp = 0$), resulting in:

$$p = \frac{a^*}{b} \tag{11}$$

where p is the benefit or the price that, in the present case, families are willing to pay (R\$/month) for an improved water supply system.

Results and Discussion

Groundwater Supply Conditions in Tauá: Well Discharge, Salinity Degrees and Social Inequality Measures

Figure 1 provides an overview of the spatial distribution of the deep wells in Tauá. There are a large number of wells intensively explored for domestic and agricultural purposes; however, without any rational management. The density of private deep wells with excessive pumping is very high in the urban area of Tauá. A total of 36% of the families are supplied by groundwater for drinking. Groundwater with high concentrations of salt is the main reason for the abandonment of 44% of the deep wells in Tauá. The water from 51% of the examined wells in operation is saline (>1000 mg/L of solids dissolved). For more than 3000 families depending on this resource, the water is unsuitable for drinking.

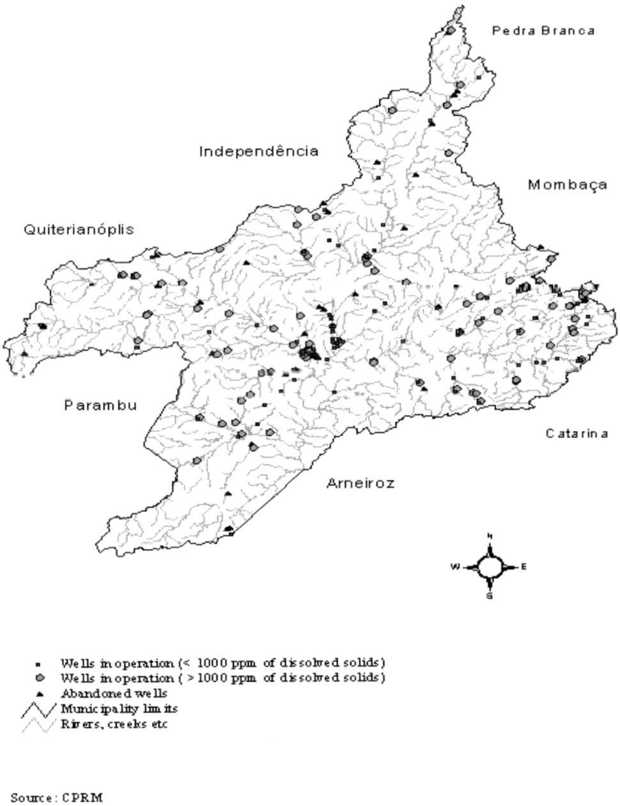

Fig. 1. Distribution of deep wells in Tauá

Table 1 shows that 49% of the deep wells have up to 1.000 mg/L dissolved solids, the maximum recommended for human consumption. This represents 55% of the available well discharge.

The results presented in Table 2 show that 36.5% of the deep wells have an average discharge of up to 800 liters per hour (L/h). Most families (75.8%) use this class of deep wells. On the other hand, discharges higher than 3.200 L/h, representing 24.6% of the total deep wells, supply only 2.9% of the families. This means a high family concentration around wells with low production.

Table 3 presents inequality measures of drinking water supply among families. It shows a high degree of inequality, since 50% of the families have at their disposal only 1.62% of the total drinking water in Tauá. In contrast, only 10% of the families have access to 78.41% of the available water; and only 1% of them hold 32.02% of the total supply of available good quality water. This means that a small proportion of the families own 19.8 times more water than 50% of the families, without considering those who have access to saline water only.

This high degree of concentration of the supply of drinking water for human consumption in Tauá is confirmed by the value of the Gini index, 0.854 (with minimum and maximum values of 0.843 and 0.861, respectively). While the large majority of the population suffers from low water availability, a privileged minority (1%) has most of that resource.

Table 1. Distribution of deep wells, families and water discharge by classes of total dissolved solids (TDS) in water in Tauá

Classes of total dissolved solids (mg/L)	Deep wells		Families		Well discharge (L/h)	
	Number	%	Number	%	Number.	%
Up to 500	22	12.6	406	7.5	80.0	11.4
500–1000	63	36.2	1560	28.6	309.4	44.3
1000–1500	43	24.9	782	14.3	135.1	19.2
1500–2000	29	16.7	1766	32.4	111.1	15.8
2000–2500	9	5.2	240	4.4.	40.9	5.8
2500–3000	1	0.5	250	4.5	10.5	1.5
3000–3500	4	2.2	441	8.1	11.4	1.6
>3500	3	1.7	3	0.1	2.8	0.4
Total	174	100.0	5448	100.0	701.2	100.0

Source: CPRM

Table 2. Distribution of good quality deep wells and families by classes of well discharge in Tauá

Well discharge classes (L/h)	Deep wells		Families	
	Number	%	Number	%
Up to 400	17	20.0	1144	58.2
400–800	14	16.5	346	17.6
800–1200	9	10.6	184	9.4
1200–1600	9	10.6	116	5.9
1600–2000	5	5.9	48	2.4
2000–2400	4	4.7	32	1.6
2400–2800	2	2.4	14	0.7
2800–3200	4	4.7	24	1.2
3200–3600	3	3.5	16	0.8
>3600	18	21.1	42	2.1
Total	85	100.0	1966	100.0

Source: CPRM (1997)

Table 3. Relative distribution of families and availability of ground water in Tauá

Percentage of families	Percentage of available drinking water from deep wells
50^{-}	1.62
20^{+}	88.99
10^{+}	78.41
5^{+}	64.77
1^{+}	32.02
Gini index	0.854
Minimum value	0.843
Maximum value	0.861

Source: Authors' calculation based on data from CPRM.

These indicators justify short-term palliative investments in order to improve water distribution by increasing the number of tank trucks. Thus, the needs of families that depend on low discharge and salty deep wells can be met by providing water from higher yielding wells. In addition, water distribution systems should be specially designed for drinking water supply. Improvement of water supply systems can be achieved by desalination, building of water tanks or provision of a piped distribution system. As an example, in Canindé (Ceará state), Pessoa (2000) estimated the following drinking water purchase costs per m³: R$ 0.68 for water from CAGECE (public supply), R$ 1.26 for desalinated water, and R$ 150.00 for bottled mineral water.

Water Supply Conditions in *Poço da Onça*: Uses, Tenure and Social Behavior

In *Poço da Onça* (the community where the demand case study was conducted), the supply of water presents the following elements: the families carry and store 83 L of water per day; 32% of this work is done by women. People use water for drinking or cooking, with no other alternative, from the salty deep well located in the community, but only in times of extreme water shortages. Other uses of noble water justify the sacrifice of collecting it from a dam at a distance of 2 km from the community. Users almost always use the salty water from their own deep well for cleaning and bathing. For washing clothes and dishes, they would rather go to the distant water sources (dam, creek, or saline wells), than do it at home.

Local power is attributed to the person who controls the water storage: the greater the owned water-tank capacity, the higher the status of the household. This study showed that the degree of kinship is the main factor of solidarity among residents. In addition, the gradual increase in water scarcity during the year is a source of conflict.

In order to make the analysis easier, the year was divided into three periods. Period I is composed of the months from February to April, with two-thirds of the annual rainfall. 10.4% of the families have water tanks for storing water, with capacities above 10 m^3. People, then, consume water collected from rainfall or sold by tank trucks. In general, families consume good quality water during the rainy season. Then, there is a high level of solidarity among them, due to the use of households' private water tanks regardless of their degree of kinship. This, obviously, is a function of easy water replacement by rain water during the rainy season.

Period II covers the next 4 months after the rainy season, from May to August. During this period, a medium level of solidarity may be found. Only families with some degree of kinship are able to consume good quality water from household water tanks.

Period III starts in September and finishes in January. A low level of solidarity among the families characterizes this period. Those who do not have water tanks consume water of doubtful quality or are forced to drink brackish water. During these months, the regular water sources are completely dry or with little water supply, eventually consumed by animals.

The effort to obtain water was measured by the cost of collecting it from existing sources, when the water volume accumulated will depend upon the user's storage capacity. A total of 107 owners of small water tanks, with up to 1 m^3 capacity each, are daily forced to carry water from distant sources, resulting in annual costs for water of R$ 13,482.00. There are only seven families with medium-sized water tanks, >1–10 m^3 capacity; they spend annually R$ 1295.00. Residents (12 families) who own water tanks larger than 10 m^3 capacity can make reasonable stocks at an annual cost of R$ 3600.00. The annual expense on water to provide for the basic needs of the residents of *Poço da Onça* is R$ 18,377.00. This amount means, on average, a monthly cost of R$ 12.15/family (about US$ 5.62 at the time of the study).

A Water Demand Model: Willingness To Pay (WTP) for an Improved Water System in *Poço da Onça*

According to the methodology described above, the parameter estimates and statistics of the logit model for good water all over the year, in the present study, were:

$$L=1.707-0.094\,p+0.032\,y-0.252\,s \qquad R2=0.96 \qquad F=47.01 \tag{12}$$

The coefficient of the price variable (p) is negative and statistically significant at 1%. Consistent with *a priori* expectations, it suggests that one additional Real (R$ 1.00) in the price, that households would pay monthly for a proposed desalination system, would reduce the logarithm of the probabilities division (Eq. 11) by 0.094, under the hypothesis of acceptance of that price. The results also confirm that individuals with higher income levels (y) are more likely to be willing to pay (WTP) the proposed monthly water tariff, at a significance level of 9%. As to family size(s), according to the Student's t-statistic, the result is significant (11%) in a one-tailed test, having a negative sign, which is in accordance with *a priori* expectations. It suggests that larger families are much more resistant to pay for a given water supply tariff than smaller families.

These parameters may be used for the estimation of benefits, considering the distribution previously described in Eq. (11). Thus, according to this demand model, the families would be WTP R$ 20.82 monthly (around US$ 9.63) for the water system proposed, providing good quality water all over the year, even in drought periods. However, in years of normal climate conditions, there is plenty of rainwater available (zero WTP) during the months of February to April, so that the monthly WTP would be equal to R$ 20.82 × ¾=R$ 15.61.

Conclusions

On the supply side, the results presented in this study demonstrate that, in Tauá, most of the rural population (64%) that depends on deep wells has no access to good quality drinking water (Tables 1 and 2). In addition, the analytical approach used here proves that the degree of the water supply concentration is very high. In fact, there is a privileged minority of families (1%) who own most of the deep wells.

There is evidence that water storage capacity is perhaps the main basis of social status for a family in the community. Factors such as the degree of kinship among users and the growing tendency for water to be scarce during the year have an effect on solidarity arrangements and, in certain circumstances, they represent potential sources of conflict.

On the demand side, the field survey at *Poço da Onça* shows that where only scarce and/or brackish water sources are accessible to people (at least partially, as an effect of an imperfect market supply), a harsh water market emerges involving strong efforts for collecting water, which means an increasing WTP for a desalination system, that supplies potable water. Thus, the benefits for the community should be obvious from the investment in a new water system whose monthly average costs per family would equal the WTP. It is estimated that families would be WTP on average R$ 20.82/month for an improved water system, during drought years, and R$ 15.61 (US$ 7.22) per month in years of normal weather conditions.

References

CPRM – Companhia de Pesquisa de Recursos Minerais (1997) Cadastro de poços profundos do Ceará (CD-ROM)

Gastwirth JL (1972) The estimation of the Lorenz curves and Gini index. Rev Econ Stat 54:306–316

Hanemann W (1984) Welfare evaluations in contingent valuation experiments with discrete responses. Am J Agricult Econ 66:322–341

Hoffmann R (1984) Estimation of inequality and concentration measures from grouped observations. Rev Econ 4:5–21

Hoffmann R (1991) Estatística para economistas. Editora Pioneira, São Paulo, 426 pp

IRDB (World Bank Water Demand Research Team) (1993) The demand for water in rural areas. Determinants and policy implications. World Bank Res Observ 8:47–70

Pessoa LCC (2000) Análise de desempenho e do impacto ambiental de dessalinizadores por osmose reversa em Canindé-CE. Dissertação de Mestrado em Engenharia Civil, Centro de Tecnologia, UFC, Fortaleza, CE

Pinheiro JCV, Silva LAC (2001) Apropriação da água subterrânea segundo sua qualidade para o uso Doméstico no Ceará: uma aplicação de medidas de desigualdade. Departamento de Economia Agrícola, CCA/UFC, 10 pp

Randall A (1987) Resource economics. An economic approach to natural resource and environmental policy. Ohio State Univ, Columbus, 434 pp

Spiegel MR (1994) Estatística. Makron Books do Brazil Publisher, São Paulo, 643 pp

Weber J, Bailly D (1997) Gestão dos recursos naturais não renováveis e desenvolvimento: novos desafios para a pesquisa ambiental. Editora Cortez, São Paulo, 275 pp

Part 8: Supporting Mathematical Tools

Visualization and 3-D Analysis of Sedimentation in Water Reservoirs

C. A. Vidal, N. J. B. Cavalcante, J. C. Araújo, A. S. Silva and E. A. Silva

Departamento de Computação, Universidade Federal do Ceará,Caixa Postal 12 166, CEP 60.455-760 Fortaleza-CE, Brazil, cvidal@lia.ufc.br

Introduction

The irregular distribution of rains in semiarid areas such as those of northeastern Brazil, where the rainy season consists of just a few months in the first semester of the year, makes the storage of water a problem of paramount importance. In several states of the northeast, the government has been building large dams in order to maintain the water supply of cities in the region and of irrigation projects. The precise control of the storage capacity of these reservoirs is of highest importance, especially during periods of lingering drought, since water is a vital element for both survival of the local populations and maintenance of their economic activities.

Sedimentation is a complex process (Hassanzadeh 1995) that, over several years, modifies the original topography of a reservoir's bottom, and can be considered one of the main factors responsible for the loss of a reservoir's storage capacity (Fan and Morris 1992). Therefore, sedimentation needs to be monitored carefully. In fact, as pointed out by Rao et al. (1997), correct estimation of sediment yield at any given point in space and time is of vital importance for water resources development and management.

Validated sedimentation models can be used in software systems for the simulation and management of water reservoirs. In 1993, Wu and Haith performed a Monte Carlo simulation experiment to explore the relationships between land use, climate, and water supply yields. Lo (1994) implemented a computer program for modeling the erosion process and simulating the water erosion and transport of sediment, nutrients and flow for watersheds ranging from 1 to 50,000 ha. He compared the simulated results with measured sediment yields for the Shihmen Reservoir watershed in Taiwan. More recently, Soncini-Sessa et al. (1999) have developed a software tool that implements a „decision support system“ for management of water reservoirs. At the planning level, it is used to generate management policies and release plans over various time horizons and, at the management level, it is used by decision-makers to take daily release decisions.

In addition, scientific visualization and scatter data modeling techniques have been successfully applied to analyze the sparse sampling data of sediments. In most cases, sediment core sampling locations are scattered. Therefore, it is very difficult to obtain a general view of distribution by just looking at the data itself, no matter if it is in tables or in scatter plots. Scientific data visualization and modeling are essential in order to provide management and engineers with detailed information on the distributions of the sediments. The most simple and straightforward way of displaying scattered data is to list them in a table, or place them on a scatter plot. The drawback of this kind of straightforward discrete presentations is that they fail to convey important information about the spatial characteristics represented by the data. Data modeling is necessary in order to produce quality presentations that impart this information.

Among the traditional data modeling methods, the inverse distanced weighted approximation is a relatively simple and computationally inexpensive approach to interpolate scattered data. The major drawback which significantly affects the quality of the output of a data interpolator rooted in this method is that it ignores any local shape properties implied by the data (Nielson 1993). One of the most used techniques for scattered data modeling is based on a spectral domain-decomposition numerical method. In recent years, the spectral domain decomposition method has been successfully applied in high-resolution geophysical fluid dynamics models (Korczac and Patera 1986; Ma 1993; H. Ma, 1998, A spectral domain-decomposition method for scattered data modeling, unpubl. manuscript). The superior computational properties of this method, such as geometrical flexibility, high convergence rate, and suitability for parallel computing, make it a good choice for data modeling as well. After the spectral domain-decomposition data model has poured the scattered sediment core sampling data onto a virtual grid, the rest of the data visualization task is carried out using visualization tools, such as the IBM Data Explorer (DX) visual programming tool (IBM 1995), which has been widely used in many works (Ma et al. 1998). Images with detailed information on the spatial distribution of sediments are thus created. This process of creating images must be used together with other visualization techniques in order to improve the sediment analysis to be performed.

In this paper, we discuss how digital terrain models (DTMs) coupled with visualization techniques can be used to monitor the process of sedimentation in reservoirs. All these techniques were implemented in a visualization tool, called AÇUDES (Vidal et al. 1999), developed by the authors. We then demonstrate the use of these techniques with a real reservoir located in the state of Ceará, northeastern Brazil.

Aspects of Sedimentation of Water Reservoirs

Over 90% of the water demand in the semiarid state of Ceará, Brazil, is supplied by surface water, since groundwater is scarce and usually salty. Indeed, water supply cannot rely on the rivers, for all the rivers in the state are intermittent, and run dry at least 2 months a year.

For a water supply with a minimum confidence level, it is necessary to build dams, which will form reservoirs and time climatic variability can be worked out. In the state of Ceará, there are more than 8000 dams, and maximum reservoir capacity is about 12 billion m^3. Maintenance of this infrastructure is a strategic and major task for water managers.

Sedimentation is one of the most important risks for water management systems in the Brazilian semiarid region, as it can affect the system both in qualitative and quantitative aspects. Sedimentation usually affects water quality negatively, by bringing suspended solids, reducing water transparency, and bringing pathogenic microorganisms. In terms of quantity, sedimentation reduces storage capacity of the reservoirs and makes them shallower, rendering them more susceptible to the high evaporation rates of the region. It is a fact that about 30% of stored water is lost to the atmosphere by evaporation. It is also important to note that higher evaporation means higher risk of salination, the main water quality problem in the semiarid.

According to Araújo and Oliveira (2000), after measuring sedimentation processes in seven dams in Ceará, it was possible to conclude that an average of 1.7% of storage capacity is lost every decade in the state. This means about 200 million m^3 is lost every 10 years just by sedimentation. For that reason, it is urgent for water research and management institutions to understand and manage the sedimentation process in the most important watersheds.

Visualization Techniques for the Study of Sedimentation

In this section, some visualization techniques that are well suited for studying sedimentation of water reservoirs are described. They are used to identify the characteristics of the sediment deposits based on two independent bathymetric surveys.

Color Contrast Contouring

Color contours are usually used in digital elevation models for helping the viewer's visual perception to identify the terrain's height variations (Fig. 1). Here, a color saturation scale is used in which two distinct hues are used to shade, respectively, the digital terrain model based on the old bathymetric survey and the one based on the more recent bathymetric survey.

The choice of which hues to use is a very important one, since a good choice of these colors strongly helps the visualization of the sediment deposits. Some preselected pairs of hues have been implemented from which the user can make a selection; however, it is also possible for the user to define new pairs of hues. In the examples presented in this work, a light-blue scale was used for the original model and a light-red scale for the new one, which gave a very good perception of the sediment deposits.

After the color selection, some fine adjustments are performed on the height interpolations at the regular grid. Then, the two shaded DTMs are superimposed to allow for the visualization of sediment deposits on the bottom of the reservoir. The places where the more recent model emerges from the old model characterize sediment deposits. The two superimposed models are initially viewed aerially (Fig. 2).

The user can take advantage of the zoom, pan, rotation and translation functions to manipulate the combined model. These functions considerably improve the capacity of identifying the sediment deposits in the reservoir, especially in situations where the aerial view is not clear enough (Fig. 3).

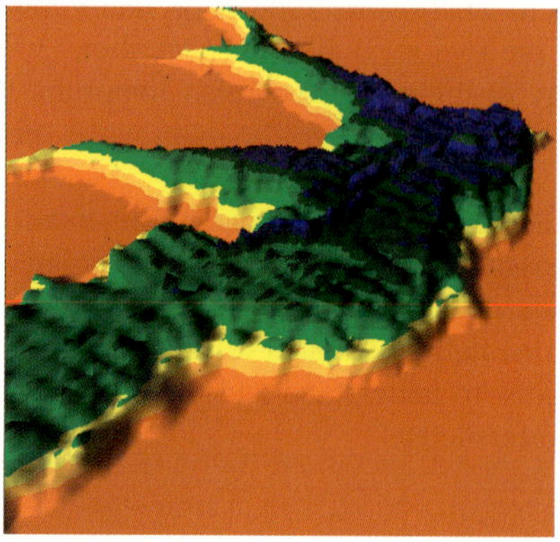

Fig. 1. Enhancement of visual perception by color contours

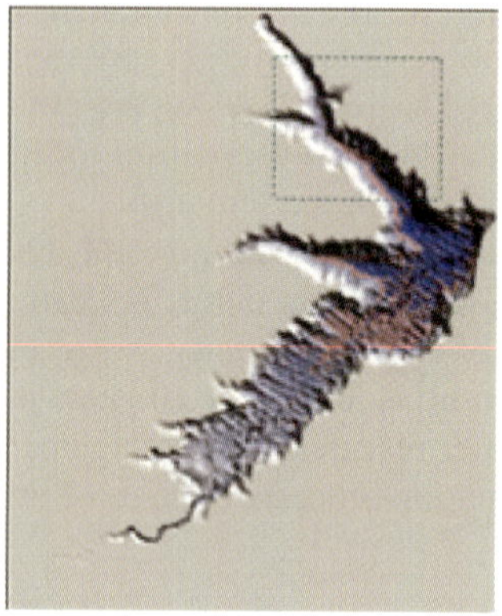

Fig. 2. Aerial view of the sediment deposits

Color Contrasted Contour Lines

This technique is just a less expensive alternative to the color contrast contouring presented in the previous section. Here, the contour lines of the old model are shown in one color and the emerging parts of the new model are displayed in another (Fig. 4). Zoom, pan, rotation and translation can also be used in combination with this technique to improve the visualization of the sediments.

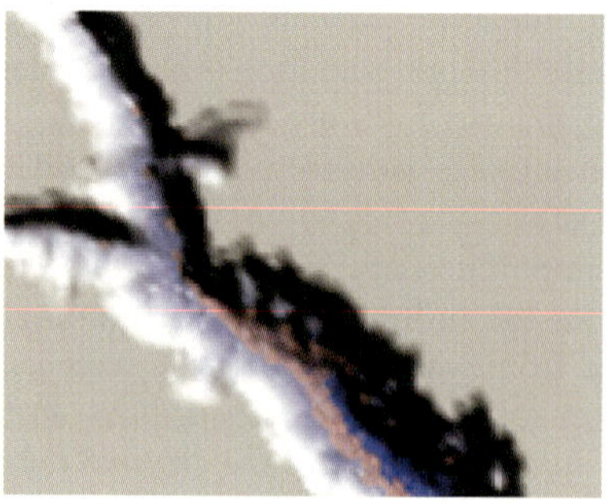

Fig. 3. Zoomed view of a sediment deposit

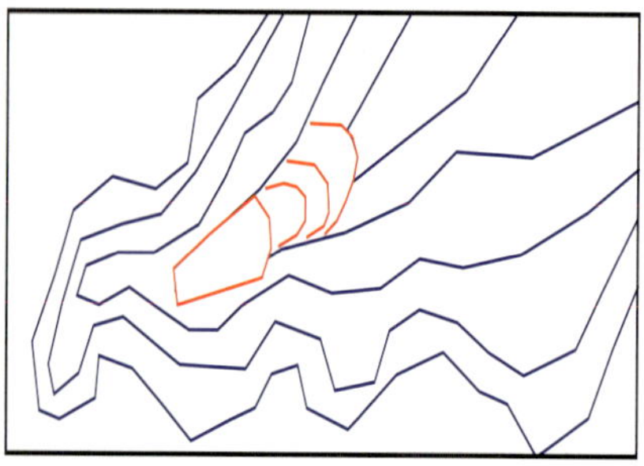

Fig. 4. Color contrasted contour lines of a sediment deposit

Cross Sectioning

In this section, two types of cross-sectioning techniques are presented that allow the user to study, in detail, a particular sediment deposit. The first technique is called direct cross sectioning and the second one is called path-generated cross sectioning. The ways by which cross sections are generated with these two techniques are conceptually different. With the first technique, the user directly generates the cross section to be visualized; while with the second technique, the user defines a path along which a series of cross sections are generated, based on a pre-defined spacing. In the case of direct cross sectioning, the user sees, in a display window, an aerial view of the reservoir's terrain with a colored line representing the sectioning plane. With path-generated cross sectioning, the user sees an aerial view of the DTM with a colored line representing the user-defined path and lines, normal to the path, representing all the generated cross-sectioning planes. In two separate windows, the user also sees two different representations of the cross section displayed. In one of them, the user sees a plot showing the intersection of the sectioning plane with the two DTMs. Shaded regions on the plot identify, with a given color, the sediment deposits and, with a different color, the erosions. In the other display window, the user sees a perspective view equivalent to that of a person standing on the bottom of the reservoir and looking at the center of the cross section. The sectioning plane is viewed as a semi-transparent wall behind which one can see the rest of the reservoir's floor in that direction.

Direct cross sectioning. In this case, the user directly defines the cross section where the sedimentation study needs to be performed. The position of the cross section is either loosely defined by clicking with the mouse in two different locations of the model, or precisely defined by specifying the coordinates of the cross-section's extremities. Direct cross sectioning is very useful in cases where the user wants to study a specific cross section; this is the most common situation found in practice. In this case, the user has total control over the cross-sectioning process.

Random cross sectioning. This is a particular case of the direct cross-sectioning technique where the user loosely defines a cross-section's position through the use of mouse-clicks. The user is allowed to define the cross section anywhere in the plan view of the reservoir, without any restrictions. Two successive mouse-clicks (points 1 and 2 in Fig. 5) create an edge, which represents the position of the cross section to be studied. Then, a third mouse-click (eye position in Fig. 5) defines the position from where the user looks at the cross section. This is very important, since the location of the user respective to the cross section can determine a completely different visualization, especially in the window where the perspective view is displayed. Figure 5 illustrates this process for a hypothetical model.

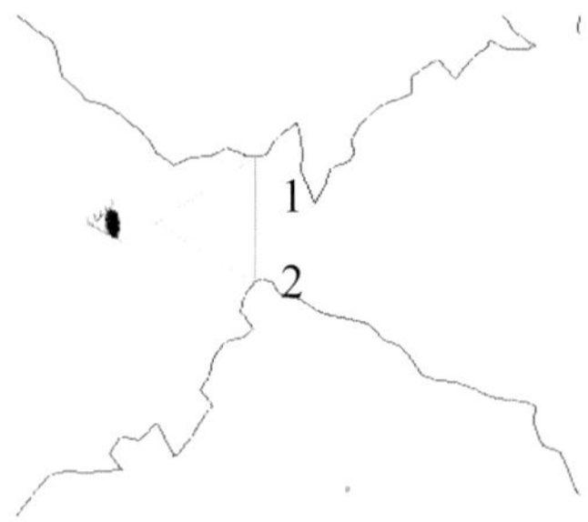

Fig. 5. Random cross sectioning

Precise cross sectioning. This is the case of direct cross sectioning where the user precisely defines a cross-section's position by specifying the two coordinates of its extremities. After the cross-section definition, a mouse-click is used to define the position from where the user looks at the cross section. Here, some restrictions apply: if any of the specified coordinates is located outside the model, the user is informed that the given coordinate cannot be used in this situation. Considering that it is very difficult to specify exact coordinates with mouse-clicks over the plan view of a reservoir, the use of the precise cross-sectioning technique is very useful in situations where the exact location of a cross section needs to be identified for study.

Path-generated cross sectioning. In this case, instead of directly defining the cross section to be studied, the user defines a path from where a number of cross sections will be created. The path can vary from a straight line (unidirectional cross sectioning) to a spline curve (directionally varying cross sectioning). This technique is used in cases where the user wants to study a variety of cross sections along a specified path.

Unidirectional cross sectioning. In this case of path-generated cross sectioning, the user can specify the path either as the direction of one of the coordinate axes of the plan view, or as the direction of the line segment defined by two successive mouse-clicks (points 1 and 2 of Fig. 6). Then, a series of cross sections perpendicular to the path will be automatically marked at a pre-defined spacing (Fig. 6) and the user can step back and forth to study the reservoir. This predefined spacing can be initialized with a default value or can be input by the user. Here, different from the direct cross-sectioning case, it is not necessary to determine the location from where the user is looking at the section. It is assumed that the user's view orientation is the same as that of a vector oriented from point 1 to point 2 along the path direction.

Directionally varying cross sectioning. In this case, the user can draw a spline curve through a series of successive mouse-clicks. Then, a set of cross sections perpendicular to the spline curve will be automatically marked at a pre-defined spacing (Fig. 7). Again, the user will be able to step back and forth to study the reservoir. Two aspects are considered here: first, the spacing can assume a default value or can be input by the user, and, second, the user can adjust the spline by changing the positions of its control points. The use of a spline curve allows the user to study areas that would be difficult or even impossible to study using only one unidirectional cross-sectioning investigation, as we can observe from Fig. 7. The directionally varying cross-sectioning technique gives a high flexibility to study the sedimentation along the whole model and, in some sense, most of the afore-mentioned techniques can be replaced by it. Indeed, the unidirectional cross-sectioning technique, for example, is a particular case of directionally varying cross sectioning, where the spline assumes the form of a straight line.

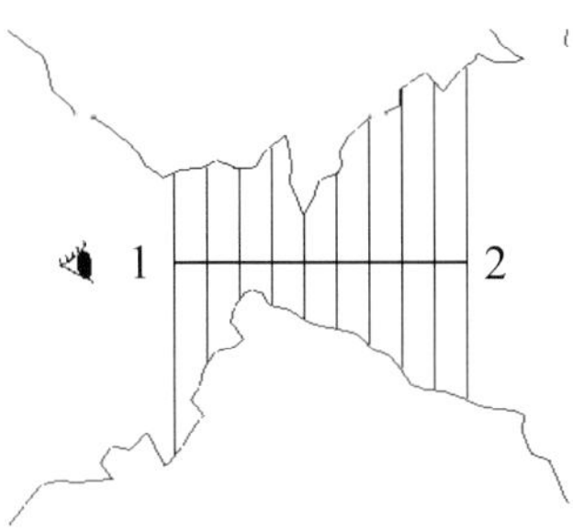

Fig. 6. Unidirectional cross sectioning

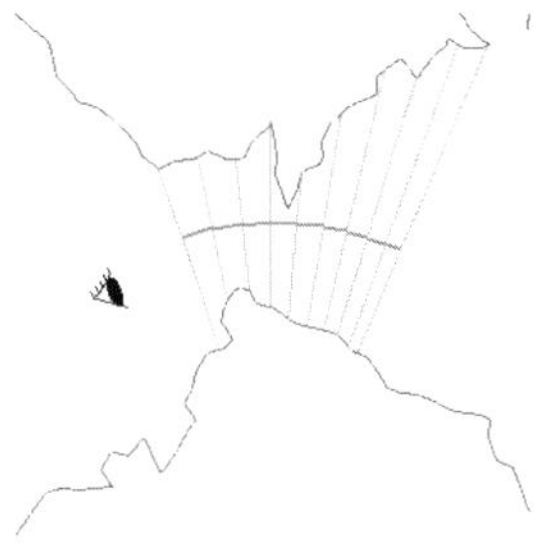

Fig. 7. Directionally varying cross sectioning

Cross-Sectioning Visualization

The cross sections generated with the techniques described above are visualized in two different display windows. In one of these windows, the cross section is displayed as superimposed plots with shaded areas; and, in the other window, it is displayed as the user's three-dimensional view of the terrain through the cross section.

Superimposed plots with shaded areas. When a cross-sectioning plane intersects the superimposed DTM, it defines two curves. The first curve represents the intersection of the plane with the older terrain model, and the second curve represents the intersection of the plane with the new terrain model. In one of the display windows, the user sees these two curves plotted and the areas between the curves are shaded (Fig. 8). Whenever the second curve is above the first curve, the region between them is shaded in red, indicating sedimentation. When the situation is reversed, the region between the curves is shaded in blue, indicating erosion.

Automatic viewer positioning and transparency. When a cross section is defined, it is generated as a view which is obtained as follows. A view angle is fixed and the distance from the observer to the cross-sectional plane is computed so that the viewer can see the whole intersection of this plane with the DTM. The viewer is centered with respect to the two extremities of the cross section and is positioned as if a person 1.80 m tall were standing on the bottom of the reservoir at that particular location. The view direction is always normal to the intersection plane. In the display window, the view of the observer just described is shown, considering the sectioning plane as a transparent panel (Fig. 8). Thus, the user can see not only the intersection of the terrain with the sectioning plane, but also the terrain model of the bottom of the reservoir that is behind the transparent panel.

Sedimentation studies

Pacajús Reservoir is a water reservoir located in Pacajús County, in the state of Ceará. It was constructed by the damming of the Choró River in 1992 and is part of the water supply system of Fortaleza, the capital of the state. In 1998, a bathymetric survey was performed and a total number of 36,000 data points were recorded. Using those data points, a DTM was constructed. Due to the lack of digitized data for the original terrain before the flooding of the reservoir or a useful bathymetric survey between 1992 and 1998, the data for a fictitious bathymetric survey was generated in order to simulate sedimentation.

Based on these two models, the visualization techniques described above were tested and very interesting results were obtained. These techniques were implemented in the software AÇUDES, in order to analyze the results. This software was developed by the Computer Graphics Group, from the Department of Computing of the Federal University of Ceará, and has been used to implement all visualization and DTM techniques derived from the studies carried out by this group. The AÇUDES software has many visualization options, such as the representation of the model only by its level curves (in intervals defined by the user), by its boundary together with the trace of the boat used in the bathymetric process, by its isosurfaces representing the different depths, and many other options. In the following, several screen shots taken from the studies based on the visualization techniques presented in this work illustrate each of these options.

Figure 8 shows three display windows together. The top window shows the automatic viewer positioning with transparency and the left bottom window displays a unidirectional cross sectioning over a branch of the reservoir where a green cross section is selected. In the right bottom window, a superimposed plot of the green cross section is shown with the sediment deposit colored in red. Figure 9 shows a directionally varying cross sectioning over the main valley of the Choró River in the reservoir.

Conclusions

The precise control of the storage capacity of reservoirs is of the highest importance in northeastern Brazil, especially during periods of lingering drought, since water is a vital element for both survival of the local populations and maintenance of their economic activities.

Sedimentation is a complex process that, over several years, modifies the original topography of a reservoir's bottom, and can be considered one of the main factors responsible for the loss of a reservoir's storage capacity. Therefore, sedimentation needs to be monitored carefully. In fact, as has been pointed out by some authors, correct estimation of sediment yield at any given point in space and time is of vital importance for water resources development and management.

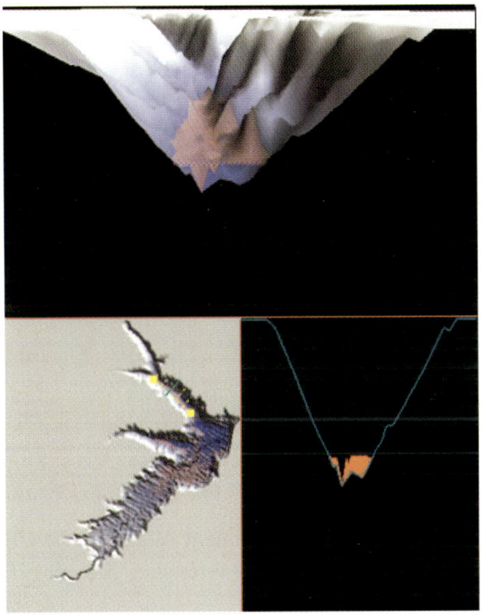

Fig. 8. Display windows

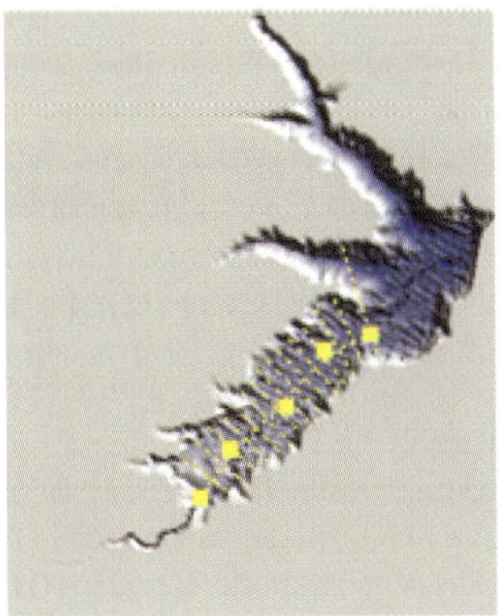

Fig. 9. Directionally varying cross sectioning of the reservoir

In this paper, we have discussed how digital terrain modeling coupled with visualization techniques can be used to monitor the process of sedimentation in reservoirs. Several visualization techniques were discussed: color contrast contouring, color contrasted contour lines, and cross sectioning. Four types of cross-sectioning techniques were presented. The use of these techniques was demonstrated with a real reservoir located in the state of Ceará, northeastern Brazil.

Acknowledgements

The authors would like to acknowledge the financial support of the following Brazilian institutions: Fundação Cearense de Amparo à Pesquisa (FUNCAP) and Conselho Nacional de Desenvolvimento Científico e Tecnológico (CNPq).

References

Araújo JC, Oliveira MRL (2000) Assessment of erosion and its impact on water availability in Brazilian semi-arid. Abstracts of German-Brazilian workshop on neotropical ecosystems vol 1, p 209

Fan JH, Morris GL (1992) Reservoir sedimentation. 1. Delta and density-current deposits. J Hydraul Eng ASCE 118:354–369

Hassanzadeh Y (1995) The removal of reservoir sediment. Water Int 20:151–154

IBM (1995) IBM visualization data explorer user's reference, 3rd edn. IBM Corporation, New York

Korczak KZ, Patera AT (1986) An isoparametric spectral element method for solution of the Navier-Stokes equations in complex geometry. J Comput Phys 62:361–382

Lo KFA (1994) Quantifying soil-erosion for the Shihmen reservoir watershed. Taiwan Agricult Syst 45:105–116

Ma HA (1993) Spectral element basin model for the shallow water equations. J Comput Phys 109:133–149

Ma H, Jones KW, Stern EA (1998) Scientific visualization and data modeling of scattered sediment contaminants. In: Proceedings of the conference on visualization 1998, New York/New Jersey Estuaries, pp 467–470

Nielson GM (1993) Scattered data modeling. IEEE Comput Graph Appl 13:60–70

Rao SVN, Rao MV, Ramasastri KS, Singh RNP (1997) A study of sedimentation in Chenab basin in western Himalayas. Nordic Hydrol 28:201–216

Soncini-Sessa R, Villa L, Weber E, Rizzoli AE (1999) TwoLe: a software tool for planning and management of water reservoir networks. Hydrol Sci J 44:619–613

Vidal CA, Sales ISA, Silva ERA, Silva ASQ, Medeiros Filho ES (1999) Modelagem Digital de Terrenos Aplicada ao Gerenciamento de Recursos Hídricos-Um Estudo do Açude de Pacajus. Anais GIS BRASIL 99:100–109

Wu RS, Haith DA (1993) Land-use, climate, and water-supply. J Wat Res Plann Manage ASCE 119:685–704

Effectiveness of Parallelization of Solution and Optimization Algorithms Applied to Large-Scale Linear Problems for the WAVES Project

Mauro C. Pequeno, Clécio A. T. Fontelles, Valdísio Viana, Andréa Linhares, Kíssia Carvalho, Alinson Donato, Auler Sousa, Rodrigo Ribeiro and Vitor Santos

Departamento de Computação, Universidade Federal do Ceará,Caixa Postal 12 166, CEP 60.455-760 Fortaleza-CE, Brazil, mauro@lia.ufc.br

Introduction

The aim of the WAVES program is to analyze the interactions between availability of water, natural resources and human activities with the purpose of predicting the impact of the variable availability of water in natural and social systems. The final results of these studies can lead to the development of integrated regional models that allow us to understand the dynamics of the ecosystem and serve as a tool to evaluate and propose strategies of sustainable rural development.

Within the WAVES project, we deal with problems such as the exploitation of the natural resources using a model that promotes the sustainable development of the population under study. The modeling of such problems suggests the need for optimization where the number of equations and/or variables can reach thousands. In problems of this magnitude, depending on the chosen method of solution, some problems can arise, mainly related to processing time. To overcome these limitations, techniques of parallel operations have been developed in order to facilitate the processing capacity of the program so that the time of problem resolution is significantly diminished.

The subgroup of integrated modeling studied algorithms and techniques to solve large-scale linear equations systems using linear programming, optimization techniques and parallel processing. The linear programming will be focused on here as a tool for modeling problems of exploiting available resources, with the purpose of introducing computational concepts, methods, algorithms and procedures for numerical solution of real problems, emphasizing diverse applications, through the use of optimization techniques. The optimization goes further to serve as a tool for predicting the future from data and information collected in the present, adding information obtained through bibliographical and documentary research.

Our work deals with large-scale problems using techniques to resolve and optimize them. In the first instance, we present four methods of linear systems resolution. Next, we discuss optimization techniques applied to large-scale problems.

The Dantzig-Wolfe decomposition was selected. We present its central idea and an approach to the algorithm. Then, we give brief comments about parallel computation. In the next section we describe the implementation of the methods discussed before. Algorithms were first implemented in a serial approach. As they can be perfectly parallelized, they were developed using parallel concepts. We also describe the procedure for test executions and will exhibit the results and its analyses. Finally, we document our conclusions.

Large-Scale Linear Problems

In a general sense, large-scale linear problems consist of problems whose models can be described by systems whose number of equations and/or variables can be in the thousands. Large-scale linear problems can be found in many research areas of mathematics. Here, we will focus on two applications: the resolution of large-scale linear equations systems and optimization techniques to solve large-scale linear programming problems.

Solution Methods for Large-Scale Linear Equations Systems

The resolution of linear systems is one of the central problems of computational mathematics and computer science, being mainly responsible for the construction of high-performance computers.

Suppose we have a linear system with n equations and n variables $Ax= b$, where A is the matrix (a_{ij}) of coefficients, $x=(x_1, \ldots, x_n)$ and $b=(b_1,\ldots, b_n)$ are, respectively, the vectors of unknowns and independent terms of the system. The matrix A is non-singular with inverse denoted by A^{-1}. We wish to find a value of x that satisfies the system $Ax=b$, in the smallest time and obtaining the best possible accuracy. In fact, no best method exists for solving systems of linear equations because each method depends on particularities presented by the linear system. Hestenes and Stiefel (1952) supply properties to verify whether a numerical method is a „*good*" method to solve linear systems:

- The method should be simple, composed of repetitions of elementary routines requiring the minimum of storage space.
- The method should converge rapidly, if the number of steps is infinite.
- The procedure should be stable with regard to rounding errors.
- At each step, the current solution should be shown to be better than the previous solution.
- e) Most of the original data should be used in each step of the procedure, to preserve the original properties of the system.

Based on these features, four methods of linear systems resolution were studied: the direct methods elimination of Gauss and LU factorization, and the iteratives methods of Gauss-Jacobi and conjugated gradient, presented in the following sections.

Direct Methods

Direct methods refer to techniques that perform a factorization of the matrix and the system using substitution algorithms.

Gauss Elimination

Gauss elimination stands between full elimination schemes like the Gauss-Jordan and triangular decomposition. Gauss elimination reduces the matrix A not all the way to the identity matrix but only halfway, to a matrix whose components on the diagonal and above remain nontrivial. This method reduces the matrix A into an upper triangular matrix U, and the numerical solution for x is calculated by backward substitution in the equation $Ux=b$.

LU Factorization

Briefly, this method consists of factorizing the matrix A into a lower triangular matrix L and an upper triangular matrix U, where all values on the diagonal of either L or U must equal 1. We use forward reduction to find $Ly=b$ and the numerical solution for x is calculated by backward substitution in the equation $Ux=y$.

Iterative Methods

The term *iterative method* refers to a wide range of techniques that use successive approximations to obtain more accurate solutions to a linear system at each step. This method requires arguments related to the tolerance for the residual, and number of iterations, but does not change the original matrix. The rate at which an iterative method converges depends greatly on the spectrum of the coefficient matrix. Hence, iterative methods usually involve a second matrix that transforms the coefficient matrix into one with a more favorable spectrum. The transformation matrix is called a *preconditioner*. A good preconditioner improves the convergence of the iterative method.

Gauss-Jacobi Algorithm

The Gauss-Jacobi method is derived by examining each of the n equations in the linear system $Ax = b$. Consider the i-th equation:

$$\sum_{j=1}^{n} a_{ij} x_j = b_i \tag{1}$$

Solving it for the value of x_i while assuming the other entries of x remain fixed, we obtain

$$x_i = \left(b_i - \sum_{j \neq i} a_{ij} x_j \right) \Big/ a_{ij} \tag{2}$$

So, the Gauss-Jacobi method can be defined by:

$$x_i^{(k)} = \left(b_i - \sum_{j \neq i} a_{ij} x_j^{(k-1)} \right) \Big/ a_{ij} \tag{3}$$

Note that the order in which the equations are examined is irrelevant, since the Jacobi method treats them independently.

In matrix terms, the definition of the Jacobi method can be expressed as:

$$x^{(k)} = D^{-1}(L+U)x^{(k-1)} + D^{-1}b \tag{4}$$

where the matrices D, $-L$ and $-U$ represent the diagonal, the strictly lower triangular, and the strictly upper triangular parts of A, respectively.

Conjugate Gradient Method

The conjugate gradient method is an effective method for symmetric positive definite systems. The method relies on generating vector sequences of iterates (i.e., successive approximations to the solution), residuals corresponding to iterates, and search directions used in updating the iterates and residuals. Although the length of these sequences can become large, only a small number of vectors need to be kept in the memory.

At each iteration of the method, two inner products are performed in order to compute update scalars that are defined to make the sequences satisfy certain orthogonality conditions. In a symmetric positive definite linear system these conditions imply that the distance to the true solution is minimized in some norm.

Optimization Techniques

In linear programming problems we have to optimize the objective function subject to a linear equality and/or inequality constraints, and others constraints on the variables. Special techniques have been developed to override these problems.

One of this is the simplex method, developed by George B. Dantzig in 1947. It is a pivotal method (based on accomplishing a sequence of pivotal operations in the matrix of coefficients of constraints system), and not polynomial bounded in the worst case. The simplex algorithm uses a canonical table, and many computations are performed in every pivotal step. Every time a pivot is chosen, it is accomplished on every column of the table. This can be very time-consuming. Subsequent extensions of the simplex method, such as the revised simplex and the decomposition principle, allow systems to be solved that contain tens of thousands and, finally, hundreds of thousands of variables.

In the revised simplex method, all the necessary computations are carried: one special structure of the pivot matrix makes it possible to store it compactly using little storage space, and restricting the pivotal operations to the inverse matrix. The revised method is purely an efficient computational scheme for applying the main ideas of the simplex method. At present, the revised simplex method is still the only approach that has proved to be computationally efficient and robust in practice.

The publication of the Dantzig-Wolfe decomposition principle (DW) in 1960 initiated the extensive work on large-scale mathematical programming that has followed. This decomposition approach can be used to solve LPs with a large number of constraints, by solving a finite sequence of smaller size problems. For these reasons, in this work we studied the Dantzig-Wolfe decomposition principle to solve large-scale linear programming problems.

Dantzig-Wolfe Decomposition Principle

The development of the Dantzig-Wolfe decomposition principle depends principally on two notions; the first is the column generation, and the second is a theorem stating that a point is in a closed bounded convex polyhedron if, and only if, it can be written as a convex combination of extreme points of the polyhedron. Consider the LP:

$$\text{Minimize } z(x)=cx \tag{5}$$

$$\text{Subject to } Bx=b \tag{6}$$

$$x \geq 0. \tag{7}$$

where B is a matrix of order $m \times n$. To solve this problem by the simplex method, we have to deal with bases of order m. If m is very large, the problem becomes unwieldy and cannot be handled even by large computers. The decomposition principle can be used to break down such a large problem into smaller problems, each of which contains fewer numbers of constraints.

Minimize $c_1x_1 + c_1x_1 + K + c_kx_k$

$A_1x_1 + A_2x_2 + \ldots A_kx_k$

Subject to

$B_1x_1 \quad > \quad b_1$

$= $

$B_2x_2 \quad b_2$

$<$

M

$B_kx_k \quad b_k$

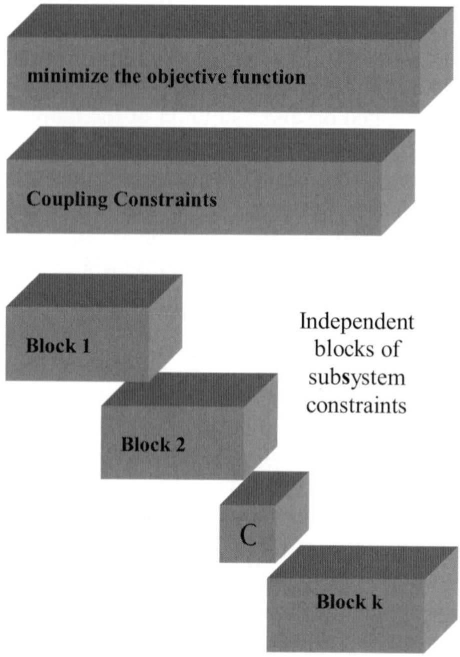

Fig. 1. The block angular structure model

where,

C_i – Vector of costs

B_i – Matrix of coefficients of block i

b_i – Vector of constraints of block i

x_i – Vector of decision variables

A_i – Coupling matrix

While the decomposition principle can be applied in general linear programming, its application can be expected to be beneficial only when the coefficient matrix is a matrix with a special structure. It has been observed that the most satisfactory results are obtained when the matrix B can be blocked, such that all the entries in it outside of those blocks are zero. This *block angular structure* arises in the optimization of a system consisting of coupled subsystems. The *block angular structure* model can be seen in Fig. 1, where the objective function is separable into blocks of variables for blocks, each one giving restrictions with local characteristics.

To apply the decomposition principle the constraints should be divided into two groups. One group leads to a problem, which contains general constraints, called the *master problem*. The other group is constituted by constraints with a special structure, called *subproblems*. The division of the constraints into two groups has to be done carefully, considering the structure of the matrix. So, the available computer can handle the number of constraints in both the subprogram and the master program.

The *master problem's* number of constraints is slightly greater than the number of coupling equations, but the number of columns can be very large. This program is solved without tabulating all these columns by generating them when the simplex method needs them; the resultant algorithm involves iteration between a set of independent *subproblems* whose objective functions contain variable parameters (simplex multipliers or prices) from the master program. They send their solutions to the master program, which combines these with previous solutions in an optimal way by computing new prices[i]. These are again sent to the *subproblems*, and the iteration proceeds until an optimality test is passed.

The Numerical Solution of Model Block Angular

The algorithm works with two linear problems: the *master problem* and the *subproblems*. The information about optimal solutions is changed between the two groups, until the solution of the original problem is found. The master program gives to the subproblems the cost's coefficients and receives the variable that should be entered in the base. This technique is called „*column generation*".

The Master Problem

Consider the linear problem (5)–(7) above. Let $X = \{x / Bx = b, x \geq 0\}$ be nonempty, and let x^i for $i > 0$ be its extreme points. Then a point, x, is in X if, and only if, it can be written as a convex combination of extreme points of X, i.e.,

$$x = \sum \lambda_j x_j, j = 1, \ldots, t \tag{8}$$

$$\sum \lambda_j = 1, j = 1, \ldots, t \tag{9}$$

$$\lambda_j \geq 0, j = 1, \ldots, t \tag{10}$$

Substituting Eqs. (8)–(10) into (5)–(7), we have:

$$\text{Minimize } \sum (cx_j) \lambda_j \tag{11}$$

$$\text{Subject to: } \sum (Bx_j)\lambda_j = b \tag{12}$$

$$\sum \lambda_j = 1, j = 1,...,t \tag{13}$$

$$\lambda_j \geq 0, j = 1,...,t \tag{14}$$

Equations (11)–(14) define the *master problem.*

The Subproblem

Let us suppose that a viable basic solution exists ($\lambda = (\lambda_B, \lambda_N)$) for the problem (11)–(14), in other words, for the simplex method, an inverse matrix exists of the base (B^{-1}). So, the dual variables w and α, corresponding to the restrictions of the problem (11)–(14), can be defined by:

$$(w, \alpha) = c_B B^{-1} \tag{15}$$

where c_B is the cost of the basic variables with $c_j = cx_j$ for each basic variable λ_j.

Using the revised simplex, we find that the criterion of entrance of the base to determine the optimal solution is given by:

$$\text{Maximize } z_j - c_j = (w, \alpha)\left[Bx_j, I\right] - cx_j \text{, i.e.,} \tag{16}$$

$$\text{Maximize } z_j - c_j = \text{Maximize } wBx_j + \alpha - cx_j; j = 1,..t \tag{17}$$

If $z_k - c_k > 0$ for any column k, then the variable k can be entered into the base. However, to determine the column k directly is not feasible, since t is large, so the extreme points corresponding to the non-basic variables are not known. To avoid this problem, the condition is used that X belongs to a limited set, which guarantees that the maximum of the linear function is found at one of the extreme points, in other words, Maximize $z_j - c_j =$ Maximize $wBx_j + \alpha - cx_j; j = 1,..t$. This is equivalent to:

$$\text{Maximize } (wB - c)x + \alpha \tag{18}$$

$$\text{Subject to } x \in X \tag{19}$$

Equations (18) and (19) are called the *subproblem.*

The Algorithm

Begin

The initial basic feasible solution for the master program (11-14) must be found.

Calculate the dual variables w and α that satisfy $(w,\alpha)=c_B B^{-1}$, where B^{-1} is the inverse matrix of the base and $c_j = cx_j$.

Principal Step

While (the solution is not optimal) **do**

Obtain the solution, $z_j - c_j$, of the problem (18,19)

If ($z_j - c_j > 0$) **then**

The solution found is optimal

Else

$y_k := B^{-1}\left[Bx_j, I\right]^{-1}$

Insert the column $(z_k - c_k, y_k)$ in the master problem (11-14)

Do the pivotal operation in y_{rk} where r is determined by

$$\frac{b_r}{y_{rk}} = \min\left\{\frac{b_r}{y_{rk}}; y_{rk} > 0\right\}$$

Make the pivotal operation

The column y_k is removed

End if

End while

End

Parallelism

Many tasks can be accomplished faster by dividing them into sets of subtasks assigned to multiple workers. Parallel processing works on precisely the same principles. Dividing a large task among workers by assigning a set of the information (data and tasks) to them is called task partitioning, and passing information to each other between subtasks is called communication (Fig. 2). Jája (1992) defines a parallel computer as a collection of processors, typically of the same type, interconnected in a certain fashion to allow the coordination of their activities and the exchange of data.

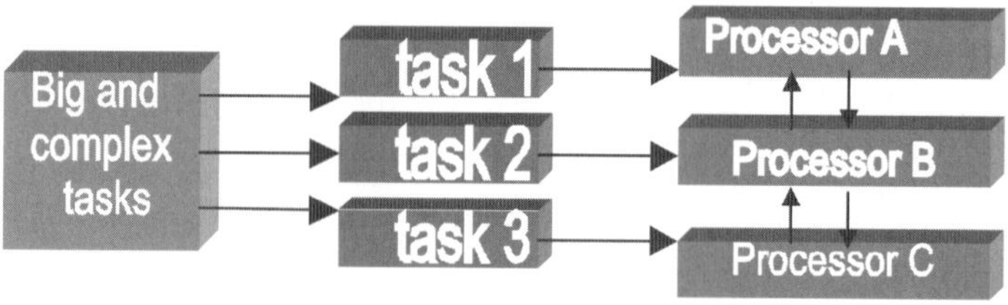

Fig. 2. Parallel processing

Three main factors have contributed to the current strong trend in favor of parallel processing. First, the cost of hardware has been falling steadily. Second, there have been significant advances in very large-scale iteration (VLSI) technology. Third, the fastest cycle time of a Von Neumann-type processor seems to be approaching fundamental physical limitations beyond which no improvement is possible; in addition, as higher performance is squeezed out of a sequential processor, the associated cost increases dramatically. All these factors have pushed researchers into exploring parallelism and its potential use in important applications. For example, the Dantzig-Wolfe decomposition principle has an algorithm naturally parallel, the idea of creating subproblems allows an algorithm to become parallel, because a processor can do one subproblem solution and column generation (see Fig. 3).

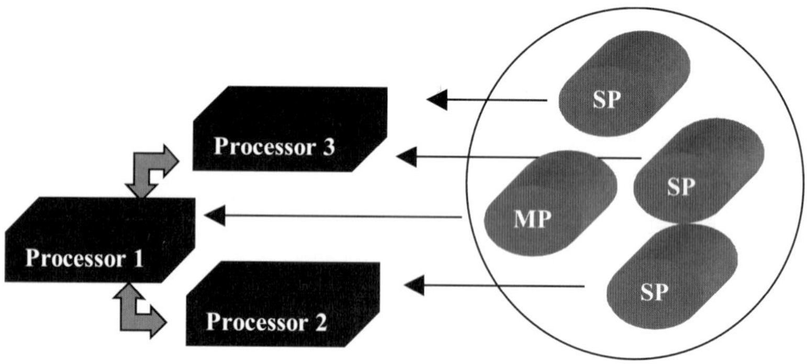

Fig. 3. Dantzig-Wolfe distribution in parallel processors

A parallel algorithm is implemented on parallel computers using a programming language, for example: PVM (parallel virtual machine) and MPI (message passing interface). This language must be flexible enough to allow efficient implementation and must be easy to program in. Parallel processing is making a tremendous impact on many areas of computer applications. New languages and programming paradigms are being developed and they try to achieve these goals. With the high computing power of parallel computers, it is possible to address many applications that were until recently beyond the capacity of conventional computing techniques.

Considerations and Results

Our tests were divided in two phases. Initially, the algorithms were implemented sequentially, using the languages FORTRAN and/or C. After that, some were implemented in a parallel approach, combining the libraries MPI and/or PVM to languages C and FORTRAN.

The MPI is a method of communication among some processors that have their own memory and is based on the transmission of data through a network interconnection. As well as the PVM, the MPI was carefully projected to execute efficiently on different machines. However, two features distinguish it from other tools of parallelism: security and scalability. So, we decided to use routines MPI and language C because of results obtained in previous studies, where we concluded that the junction of the library of parallelism MPI and language C resulted in better performance.

Solution Methods for Large-Scale Linear Equations Systems

Our tests applied to large-scale linear equations systems used six different implementations for each method. The algorithms were submitted to several tests in many different situations to accomplish a subsequent comparison between the processing time and the accuracy obtained in each case. The processing times were obtained by using routines in their own libraries of parallelism, while the accuracy was measured through infinite norm (see Fig. 4).

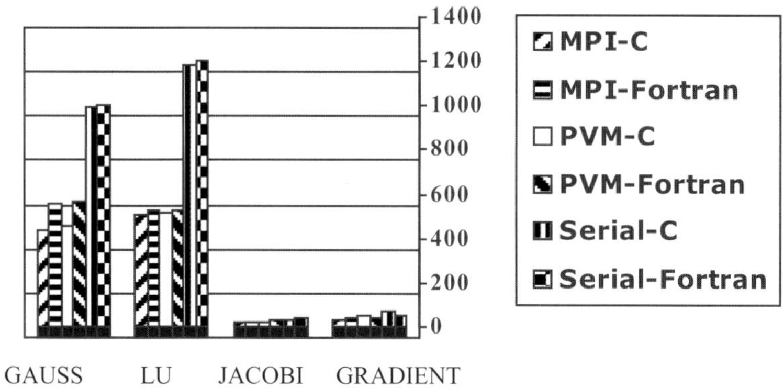

Fig. 4. Methods (Matrix -2000) X Time (sec)

The mathematical and technical restrictions of the methods were also considered and, when possible, we have tried to outline those problems. For instance: the Gauss-Jacobi method, although it presented the best performance, cannot be applied because of its convergence restrictions. The gradient method presents the largest gain, but its application is restricted to positive-definite matrices. To solve this problem, a literature search was made of the conjugated gradient method, to study its restrictions and to see whether the possibility existed of extending the method for any matrix A.

Dantzig-Wolfe Technique

In the first instance, we generated random data to test the algorithms. The implementation of a generator of problems for the phase of tests was necessary. Such program demands were: input data, the number of subproblems, the number of constraints and the number of variables in each subproblem, the number of restrictions in the master problem and the intervals of values of the coefficients. After the input of such values the program generates as output the archive with the block angular linear system.

In an initial approach, we intended just to develop a parallel algorithm able to solve a linear programming problem using the Dantzig-Wolfe decomposition. So, this approach is a little limited with regard to cases that can be solved by algorithm. Each subproblem is solved by just one process and, because of that, the simplex method was not implemented in parallel. As discussed above, the algorithms were subjected to several tests in many different situations to accomplish a subsequent comparison between the processing time and the accuracy obtained in each case. We only used combinations MPI-C and MPI-FORTRAN in these tests, obtaining the results presented in Fig. 5.

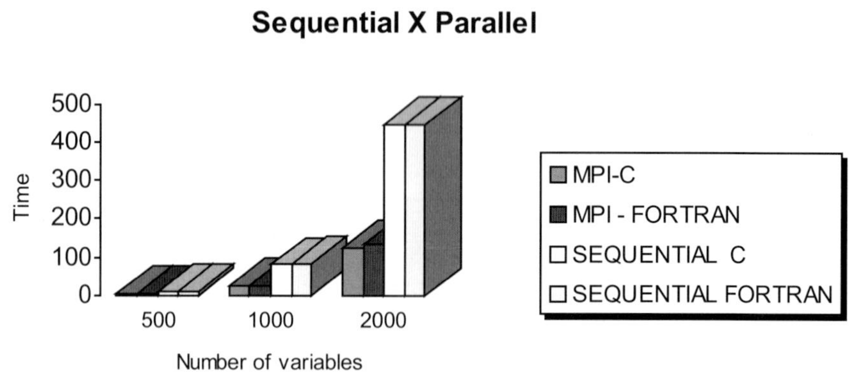

Fig. 5. Comparison between serial and parallel, with four processors

In a second instance, we dealt with real data provided by the referential model developed in collaboration between Brazilian and German researchers. This model can be described by a large-scale system with approximately 2,000,000 coefficients, configuring a 497×4089 matrix (497 equations and 4089 variables). The original structure of the matrix suggests the existence of coupling constraints and convexity constraints. So, we could use the Dantzig-Wolfe approach again. To solve the subproblems, we adjusted the revised simplex method to treat bounded variables. The new tests associated with this model are being conducted in a group with some workstations (RISC/IBM) in the laboratory of the CENAPAD/NE. A special structure to read/write files was developed based on the data organization. We also used just static memory allocation because we had an extra memory space to make these tests. The final analysis of the results has as feedback the resolution of this same model using the optimization software GAMS. Our main points of focus are: use of memory space; time of response; quality of the solution; convergence and limits of precision; and propagation errors.

Conclusions

Parallel programming does not involve just the development itself; it is also necessary to perform several considerations in the sense of optimizing parallel programs. It was observed that the combination of the language C, in a parallel approach, and the tool MPI was the most efficient for the direct methods, according to tests.

In the sense of solution algorithms to solve large-scale systems problems, the iterative methods converge much more quickly than the direct methods, by preserving the features of the coefficient matrix A. It was observed that, for the iterative method conjugated gradients, the use of parallel programming is only feasible when the system has a high dimension, at least 500×500, mainly because the convergence of these methods is already quite fast, even using a sequential approach. In the case of optimization algorithms, the first instance of tests simulated to the Dantzig-Wolfe algorithm showed that the parallel approach is very advantageous, and could have relative savings greater than 200%. In the worst case, when we used only two processors, the relative savings amounted to 70%.

Our goal is to prepare a tool to find a solution to optimization problems linked to the WAVES project. Some of these problems are being solved in Germany, using the software GAMS; however, these problems are consuming a great time of CPU. After ending the implementation, we will compare our results with those obtained in Germany. The analyses will be based on tests that will involve technique analyses, analysis of sensitivity and other tests necessary for the validation of the model.

References

CENAPADNE, http://www.cenapadne.br

Chaudhuri P (1992) Parallel algorithms – design and analysis. Prentice Hall, Englewood Cliffs

Hestenes RM, Stiefel E (1952) Methods of conjugate gradients for solving linear systems. J Res Natl Bureau Stand 49(6)

Hilier FS, Lieberman GJ (1991) Introduction to mathematical programming. McGraw-Hill, New York

JaJá J (1992) An introduction to a parallel algorithms. Addison-Wesley Publ, Reading, MA

Lasdon LS (1970) Optimization theory for large systems. Macmillan Publ, New York

Maculan Filho N, Pereira MVF (1980) Programação linear. Atlas, São Paulo

Murty KG (1983) Linear programming. John Wiley, New York

Nakamura S (1993) Applied numerical methods in C. Prentice Hall, Engelwood Cliffs

Pequeno MC, Carvalho K, Donato A, Sousa A, Ribeiro R, Santos V (2000) Using the parallel processing in solution of large scale linear equations systems for the project WAVES. In: Printz A (ed) IIIrd international WAVES workshop, TU Munich, Freising-Weihenstephan, March 2000

Richard B (1994) Templations solution for the solution of linear systems: building blocks for iterative methods. SIAM, Philadelphia, PA

Viping K (1994) Introduction to parallel computing – design and analysis of algorithms. Benjamin/Cummings Publ, Redwood, CA

WAVES, http://sw.npd.ufc.br/cai/waves.htm and http://www.usf.uni-kassel.de/waves

Yong DM (1971) Iterative solution of large linear systems. Academic, New York

Authors and Affiliations

Name	Adress
Abreu, C.B.R.	Universidade Federal do Ceará, Departamento de Hidráulica e Ambiental, 60451-970, Fortaleza, Ceará
Albino, R.S.	Núcleo de Referência em Ciências Ambientais do Trópico Ecotonal do Nordeste (TROPEN), Pró-Reitoria de Pesquisa e Pós-Graduação. Departamento de Biologia, Universidade Federal do Piauí (UFPI). Av. Universitária,1310. Campus da Ininga. 64049 550, Teresina, Piauí, Brasil
Albuquerque, B.L.	Faculty of Education of the Federal University of Ceará, Rua Tibúrcio Cavalcante, 2324, ap.1102 Fortaleza, Ceará
Araujo, J.C. de	Universidade Federal do Ceará, Departamento de Hidráulica e Ambiental, 60451-970, Fortaleza, Ceará, jcaraujo@ufc.br
Araújo Mendes, M.R. de	Núcleo de Referência em Ciências Ambientais do Trópico Ecotonal do Nordeste (TROPEN), Pró-Reitoria de Pesquisa e Pós-Graduação. Departamento de Biologia, Universidade Federal do Piauí (UFPI). Av. Universitária,1310. Campus da Ininga. 64049 550, Teresina, Piauí, Brasil
Bach, B.	Institute of Agricultural Economy, University of Hohenheim.(410 B), D-70593 Stuttgart, Germany, bachb@uni-hohenheim.de
Barbosa, C.P.	Universidade Federal do Ceará, Departamento de Hidráulica e Ambiental, 60451-970, Fortaleza, Ceará
Barros, I. de	Institute of Plant Nutrition, University of Hohenheim, D-70593 Stuttgart, Germany, inbarros@uni-hohenheim.de
Bezerra, C.L.F.	Universidade Federal do Ceará, Departamento de Biologia, Bloco 906, Centro de Ciências, 60451-970, Fortaleza, Ceará, lineu@ufc.br; eliedir@bol.com.br
Boehm, U.	University of Potsdam, Department Physics, P.O. Box 601553, 14415 Potsdam, Germany; boehm@pik-potsdam.de
Bronstert, A.	University of Potsdam, Department of Geoecology, Chair for Hydrology and Climatology, P.O. Box 601553, 14415 Potsdam, Germany, axelbron@rz.uni-potsdam.de
Cavalcante; N.J.B.	Departamento de Computação, Universidade Federal do Ceará, Campus do PICI, Bloco 910, Caixa Postal 12.166, CEP 60.455-760,.Fortaleza, CE, Brazil,
Campos, J.N.B.	Department of Water Resources and Environmental Engineering, Universidade Federal do Ceará, P.O. Box 6018, 60451-970 Fortaleza, Ceará, Brazil, , nilson@ufc.br
Carvalho, R.	Departamento de Computação - Universidade Federal do Ceará, Campus do PICI, bloco 910, Caixa Postal 12.166, CEP 60.455-760, Fortaleza, CE, Brazil
Carvalho Studart, T.M. de	Department of Water Resources and Environmental Engineering, Universidade Federal doCeará, P.O. Box 6018, 60451-970 Fortaleza, Ceará, Brazil, ticiana@ufc.br
Castro, A.A.J.F.	Núcleo de Referência em Ciências Ambientais do Trópico Ecotonal do Nordeste (TROPEN), Pró-Reitoria de Pesquisa e Pós-Graduação. Departamento de Biologia, Universidade Federal do Piauí (UFPI). Av. Universitária,1310. Campus da Ininga. 64049 550, Teresina, Piauí, Brasil. aajfcastro@uol.com.br.
Castro, A.S.F.	Departamento de Biologia, Bloco 906, Centro de Ciências, Universidade Federal do Ceará, 60455-970, lineu@ufc.br; eliedir@bol.com.br
Chagas, P.F.	Department of Water Resources and Environmental Engineering, Universidade Federal doCeará, P.O. Box 6018, 60451-970 Fortaleza, Ceará, Brazil, pfchagas@yahoo.com
Döll, P.	Center for Environmental Systems Research, University of Kassel, 34109 Kassel, Germany, doell@usf.uni-kassel.de
Döring, A.	Center for Environmental Systems Research, University of Kassel, Kurt-Wolters-Strasse 3, 34109 Kassel, Germany, doering@usf.uni-kassel.de

Name	Adress
Donato, A.	Departamento de Computação - Universidade Federal do Ceará, Campus do PICI, bloco 910, Caixa Postal 12.166, CEP 60.455-760, Fortaleza, CE, Brazil
Farias, R.R.S. de	Núcleo de Referência em Ciências Ambientais do Trópico Ecotonal do Nordeste (TROPEN), Pró-Reitoria de Pesquisa e Pós-Graduação. Departamento de Biologia, Universidade Federal do Piauí (UFPI). Av. Universitária,1310. Campus da Ininga. 64049 550, Teresina, Piauí, Brasil
Fernandes, L.	Universidade Federal do Ceará, Departamento de Hidráulica e Ambiental, 60451-970, Fortaleza-Ceará
Ferreira, L.G.R.	Coordenadoria de Assuntos Internacionais, Universidade Federal do Ceará, Fortaleza Ceará, Brazil, inter@ufc.br
Fontelles, C.A.T.	Departamento de Computação - Universidade Federal do Ceará, Campus do PICI, bloco 910, Caixa Postal 12.166, CEP 60.455-760, Fortaleza, CE, Brazil
Folhes, M.T.	Federal University of Ceará (UFC), Department of Agricultural Economics, 60150-100 Fortaleza, CE, Brazil
Frischkorn, H.	Universidade Federal do Ceará, Departamento de Hidráulica e Ambiental, 60451-970, Fortaleza-Ceará, cariri@ufc.br
Frota, Décio	DHME – SEAAB-PI – R. João Cabral S/N Bairro Pirajá CEP 64002-150, seab@seaab.pi.gov.br
Fuhr, D.	Center for Environmental Systems Research, University of Kassel, 34109 Kassel, Germany, fuhr@usf.uni-kassel.de
Gaese, H.	University of Applied Sciences at Cologne, Institute for Technologies in the Tropics, Betzdorfer Straße 2, 50679 Köln, Hartmut Gaese [bg075@dvz.fh-koeln.de]
Gaiser, T.	Soil Science and Land Evaluation (310), University of Hohenheim D-70593 Stuttgart, Germany, tgaiser@uni-hohenheim.de
Gerstengarbe. F.-W.	Potsdam Institute for Climate Impact Research, P.O. Box 60 12 03, 14412 Potsdam, Germany, gerstengarbe@pik-potsdam.de
Graef, F.	Zentrum für Agrarlandschafts- und Landnutzungsforschung (ZALF) e.V., Eberswalder Str. 84, 15374 Müncheberg, Germany, graef@zalf.de
Grebe, M.	Center for Environmental Systems Research, University of Kassel, Kurt-Wolters-Strasse 3, 34109 Kassel, Germany
Güntner, A.	Potsdam Institute for Climate Impact Research (PIK), P.O. Box 601203, 14412 Potsdam, Germany, guentner@pik-potsdam.de, and: University of Potsdam, Department of Geoecology, Chair for Hydrology and Climatology, P.O. Box 601553, 14415 Potsdam, Germany
Halm, D.	Eberhard-Karls-University of Tübingen, Centre for Applied Geoscience, Sigwartstr. 10, 72076 Tübingen, Germany, dietrich.halm@uni-tuebingen.de
Hauffe; D.	Potsdam-Institut für Klimafolgenforschung, Postfach 60 12 03, 14412 Potsdam, Germany
Hauschild, M.	Center for Environmental Systems Research, University of Kassel, 34109 Kassel, Germany
Heinrichs, G.	Internationales Büro des BMBF (Bundesministerium für Bildung und Forschung), Koenigswinterer Str. 522-524, 53227 Bonn, heinrichs@dlr.de
Herfort, J.	Institute of Plant Production and Agroecology in the Tropics and Subtropics (380), University of Hohenheim, 70593 Stuttgart, Germany, jherfort@uni-hohenheim.de
Hilger, T.	Institute of Plant Production and Agroecology in the Tropics and Subtropics (380), University of Hohenheim, 70593 Stuttgart, Germany, t-hilger@uni-hohenheim.de
Hoeynck, S.	Institute for Technology in the Tropics, University of Applied Sciences Cologne, Betzdorfer Str. 2, 50679 Cologne/Germany, hoeynck@netcologne.de
Jaeger, A.-K.	Potsdam Institute for Climate Impact Research (PIK), P.O. Box 601203, 14412 Potsdam, Germany, jaeger@pik-potsdam.de

Name	Adress
Joca, E.L.L.	Universidade Federal do Ceará, Departamento de Hidráulica e Ambiental, 60451-970, Fortaleza, Ceará
Krol, M.	Potsdam Institute for Climate Impact Research (PIK), P.O. Box 601203, 14412 Potsdam, Germany, Maarten.Krol@pik-potsdam.de
Kücken; M.	Potsdam-Institut für Klimafolgenforschung, Postfach 60 12 03, 14412 Potsdam, Germany
Külls, C.	Hydroisotop GmbH, Woelkestr. 9, 85301 Schweitenkirchen, Germany, hydroisotop@t-online.de
Lang, R.	uismedia, Biernerstraße 32, D – 85354 Freising, Germany, R.Lang@uisgruppe.de
Lange, F.-M.	University of Hohenheim, Institute of Soil Science and Land Evaluation, 70593 Suttgart, Germany, fmilange@gmx.de
Lantermann, E.-D.	Center for Environmental Systems Research, University of Kassel, Kurt-Wolters-Strasse 3, 34109 Kassel, Germany
Lima Viana, M.-O. de	Regional Graduate Program on Development and Environment (UFC).
Linhares, A.	Departamento de Computação - Universidade Federal do Ceará, Campus do PICI, bloco 910, Caixa Postal 12.166, CEP 60.455-760, Fortaleza, CE, Brazil
Lopez Gil, M.M.	Departamento de Hídrometeorologia, SEEAB, CEP 64002-150 Teresina, Piauí-Brazil, lopez_margot@hotmail.com
Machado Júnior, J.C.	Universidade Federal do Ceará, Departamento de Hidráulica e Ambiental, 60451-970, Fortaleza, Ceará
Mayorga, M.I. de O.	Universidade Federal do Ceará, Departamento de Economia Agrícola (CCA), 60150-100 Fortaleza, CE, Brazil, irles@ufc.br
Mendiondo, E.M.	Center for Environmental Systems Research, University of Kassel, 34109 Kassel, Germany, mendiond@usf.uni-kassel.de
Österle, H.	Potsdam-Institut für Klimafolgenforschung, Postfach 60 12 03, 14412 Potsdam, Germany
Oliveira, J.G.B. de	Núcleo de Referência em Ciências Ambientais do Trópico Ecotonal do Nordeste (TROPEN), Pró-Reitoria de Pesquisa e Pós-Graduação. Departamento de Biologia, Universidade Federal do Piauí (UFPI). Av. Universitária,1310. Campus da Ininga. 64049 550, Teresina, Piauí, Brasil
Oliveira, J.O.S.	Núcleo de Referência em Ciências Ambientais do Trópico Ecotonal do Nordeste (TROPEN), Pró-Reitoria de Pesquisa e Pós-Graduação. Departamento de Biologia, Universidade Federal do Piauí (UFPI). Av. Universitária,1310. Campus da Ininga. 64049 550, Teresina, Piauí, Brasil
Oliveira, M.R.L.	Former WAVES Program scholar, CNPq
Oliveira, V.P.V. de	Federal University of Ceará (UFC), Department of Geography, Fortaleza, CE, Brazil, vlpinto@ufc.br
Pequeno, M.C.	Departamento de Computação, Universidade Federal do Ceará, Campus do PICI, Bloco 910, Caixa Postal 12.166, CEP 60.455-760,.Fortaleza, CE, Brazil, mauro@vdl.ufc.br
Pinheiro, J.C.V.	Department of Agricultural Economics, Federal University of Ceará (UFC), 60150-100 Fortaleza, CE, Brazil, jcvpinhe@ufc.br
Printz, A.	Munich Polytechnic University (TUM), Insitute for Landscape Ecology, Am Hochanger 6, 85350 Freising, Germany, pepe@dec.loek.agrar.tu-muenchen.de
Ramos, V.M.	Núcleo de Referência em Ciências Ambientais do Trópico Ecotonal do Nordeste (TROPEN), Pró-Reitoria de Pesquisa e Pós-Graduação. Departamento de Biologia, Universidade Federal do Piauí (UFPI). Av. Universitária,1310. Campus da Ininga. 64049 550, Teresina, Piauí, Brasil
Ribeiro, R.	Departamento de Computação - Universidade Federal do Ceará, Campus do PICI, bloco 910, Caixa Postal 12.166, CEP 60.455-760, Fortaleza, CE, Brazil
Rocha, F.M. da	Center for Environmental Systems Research, University of Kassel, Kurt-Wolters-Strasse 3, 34109 Kassel, Germany

Name	Adress
Sales, M.C.L.	Núcleo de Referência em Ciências Ambientais do Trópico Ecotonal do Nordeste (TROPEN), Pró-Reitoria de Pesquisa e Pós-Graduação. Departamento de Biologia, Universidade Federal do Piauí (UFPI). Av. Universitária,1310. Campus da Ininga. 64049 550, Teresina, Piauí, Brasil, mclsales@uol.com.br
Santiago, M.M.F.	Universidade Federal do Ceará, Departamento de Hidráulica e Ambiental, 60451-970, Fortaleza-Ceará, marlucia@fisica.ufc.br
Santos, V.	Departamento de Computação - Universidade Federal do Ceará, Campus do PICI, bloco 910, Caixa Postal 12.166, CEP 60.455-760, Fortaleza, CE, Brazil
Silva, L.A.C. de	Department of Agricultural Economics, Federal University of Ceará (UFC), 60150-100 Fortaleza, CE, Brazil
Soares, F.A.R.	Núcleo de Referência em Ciências Ambientais do Trópico Ecotonal do Nordeste (TROPEN), Pró-Reitoria de Pesquisa e Pós-Graduação. Departamento de Biologia, Universidade Federal do Piauí (UFPI). Av. Universitária,1310. Campus da Ininga. 64049 550, Teresina, Piauí, Brasil
Sousa, A.	Departamento de Computação - Universidade Federal do Ceará, Campus do PICI, bloco 910, Caixa Postal 12.166, CEP 60.455-760, Fortaleza, CE, Brazil
Sousa , T.C.	Former WAVES Program scholar, CNPq
Souza, M.J.N.de	State University of Ceará (UECE), Department of Geosciences, 60150-100 Fortaleza, CE, Brazil
Schneider, L	Institute of Plant Production and Agroecology in the Tropics and Subtropics (380), University of Hohenheim, 70593 Stuttgart, Germany
Semrau, L.A.de S.	Universidade Federal do Ceara, Departamento de Hidráulica e Ambiental, 60451-970, Fortaleza-Ceará, ligia@secrel.com.br
Stahr, K.	Soil Science and Land Evaluation (310), University of Hohenheim, 70593 Stuttgart, Germany, kstahr@uni-hohenheim.de
Trigueir, E.R.C.	Departamento de Biologia, Bloco 906, Centro de Ciências - Universidade Federal do Ceará
Viana, V.	Departamento de Computação - Universidade Federal do Ceará, Campus do PICI, bloco 910, Caixa Postal 12.166, CEP 60.455-760, Fortaleza, CE, Brazil
Vidal, C.A.	Universidade Federal do Ceará, Departamento de Computação, Campus do PICI, Bloco 910, Caixa Postal 12.166, CEP 60.455-760,.Fortaleza CE, Brazil, vidal@vdl.ufc.br
Voerkelius, S.	Hydroisotop GmbH, Woelkestr. 9, 85301 Schweitenkirchen, Germany, hydroisotop@t-online.de
Voerkelius, U.	ulli@dec.loek.agrar.tu-muenchen.de, Voerkelius@t-online.de
Werner, P.C.	Potsdam Institute for Climate Impact Research, P.O. Box 60 12 03, 14412 Potsdam, Germany
Zeddies, J.	Institute of Agricultural Economy, University of Hohenheim.(410 B), D-70593 Stuttgart, Germany, i410b@uni-hohenheim.de

Printing (Computer to Film): Saladruck Berlin
Binding: Stürtz AG, Würzburg